Forgotten People, Forgotten Diseases

Forgotten People, Forgotten Diseases

The Neglected Tropical Diseases and Their Impact on Global Health and Development

Third Edition

PETER J. HOTEZ, MD, PHD
Professor of Pediatrics and Molecular Virology & Microbiology
Texas Children's Hospital Chair of Tropical Pediatrics
National School of Tropical Medicine, Baylor College of Medicine
Houston, Texas

ASM
PRESS
WASHINGTON, DC

Editorial Correspondence:
ASM Press, 1752 N Street, NW, Washington, DC 20036-2904, USA

Registered Offices:
John Wiley & Sons, Inc., 111 River Street, Hoboken, NJ 07030, USA

For details of our global editorial offices, customer services, and more information about Wiley products, visit us at www.wiley.com.

Wiley also publishes its books in a variety of electronic formats and by print-on-demand. Some content that appears in standard print versions of this book may not be available in other formats.

Library of Congress Cataloging-in-Publication Data

Names: Hotez, Peter J., author.
Title: Forgotten people, forgotten diseases : the neglected tropical diseases and their impact on global health and development / Peter J. Hotez.
Description: Third edition. | Washington, DC : ASM Press, [2022] | Series: American society for microbiology
Identifiers: LCCN 2021028002 (print) | LCCN 2021028003 (ebook) | ISBN 9781683673873 (paperback) | ISBN 9781683673880 (adobe pdf) | ISBN 9781683673897 (epub)
Subjects: LCSH: Tropical medicine.
Classification: LCC RC961 .H68 2022 (print) | LCC RC961 (ebook) | DDC 362.196/9883–dc23
LC record available at https://lccn.loc.gov/2021028002
LC ebook record available at https://lccn.loc.gov/2021028003

Cover design by: Owen Design Company
Interior design by: Susan Brown Schmidler

Set in 11/13pt MinionPro by Straive, Chennai, India

SKY10071955_040524

Dedicated to my wife, Ann Hotez, and my four adult kids,
Matthew Hotez, Emily Hotez, Rachel Hotez, and Daniel Hotez

To my mother, Jean Hotez, and brother and sister, Lawrence Hotes, M.D., and
Elizabeth Kirshenbaum, J.D., and their families

To the memory of my brother and father,
Richard Hotes, M.D.
Edward Joseph Hotez

To Mark Wallace and Paul Klotman, M.D.,
and the leadership of Texas Children's Hospital and Baylor College of Medicine

And to the Bill & Melinda Gates Foundation,
the U.S. National Institutes of Health,
and the Kleberg Foundation
for the opportunity to devote my life
to the Neglected Tropical Diseases

Contents

Preface

Ever since junior high school, I have been fascinated by the application of scientific knowledge for solving tropical public health problems of global importance. Starting with an M.D.-Ph.D. dissertation begun in 1980, my adult life has been a quest to develop experimental vaccines for diseases of the poor, beginning with human hookworm infection. More than 20 years ago, thanks to the support of the Bill & Melinda Gates Foundation, I had the opportunity and good fortune to head a multidisciplinary team to develop and manufacture those vaccines and test them in areas of Brazil and Gabon and where hookworm was endemic. Since then our group, now co-headed by my science partner for the last 20 years, Dr. Maria Elena Bottazzi, has led the development of vaccines against schistosomiasis, Chagas disease, and other neglected tropical diseases (NTDs). Starting in 2011 we began using this same approach to develop coronavirus vaccines, including vaccines against SARS and MERS, and beginning in 2020 we turned our attention to COVID-19. As a result a new low-cost recombinant protein, COVID-19 vaccine is being scaled up for production in India, with the hope that it will fill a troubling gap in terms of COVID-19 vaccines for Africa and Latin America. While that work was intensely satisfying on both a professional and personal level, I realized that completing early-stage development of a new product for an NTD such as hookworm was in many ways the easy part! It was apparent that unless there was greater general awareness about the public health and economic importance of NTDs there would never be the political will and large-scale financial investment necessary to ensure global access to a hookworm vaccine, or indeed any other product for the diseases of poverty.

The first edition of *Forgotten People, Forgotten Diseases* focused on summarizing in mostly nontechnical language the major concepts about NTDs and how they cause human suffering, as well as their global importance and the unique and unusual opportunity we had to lift the world's poorest people out of poverty through low-cost and highly cost-effective control measures. Along with Professor David H. Molyneux of the Liverpool School of Tropical Medicine, Professor Alan Fenwick from Imperial College London, Dr. Lorenzo Savioli from the World Health Organization (as well as some of his close colleagues there, including Drs. Dirk Engels and Jacob Kumaresan), Professor Jeffrey Sachs and Dr. Sonia Ehrlich Sachs of Columbia's Earth Institute, and Dr. Eric Ottesen (then at Emory University), I formed an informal NTD working group, and in a series of policy papers published in *PLoS* and the *New England Journal of Medicine*, we were able to articulate the concept of the NTDs and how we could control or eliminate them through a global scale-up of access to essential medicines. We also established a Global Network for NTDs to coordinate global advocacy and resource mobilization efforts for these conditions.

By the time of the second edition, published in 2013, much had already begun to change. In the area of public health control in developing countries, and through support from the United States Agency for International Development (USAID), approximately 250 million people had been treated with all or part of an integrated "rapid-impact package" of essential medicines for seven of the most common NTDs—ascariasis, hookworm infection, trichuriasis, schistosomiasis, lymphatic filariasis, onchocerciasis, and trachoma. The World Health Organization estimated that more than 700 million people annually were receiving essential medicines against one or more NTDs—almost all of whom were living in the poorest parts of Africa, Asia, and the Americas—representing some of the largest public health control efforts ever undertaken. The successes of both mass drug administration and product development activities rely heavily on a substantial alliance of private-public partnerships, including product development partnerships and nongovernmental development organizations, as well as international advocacy efforts to raise awareness about the NTDs (including the Global Network for Neglected Tropical Diseases) and parallel resource mobilization initiatives. Another major development was the realization that NTDs also occur among the poor living in wealthy countries, especially the United States and, to some extent, Europe. In 2011, I committed my life and work to this problem by relocating a group of more than a dozen scientists to Texas in order to establish the Texas Children's Hospital Center for Vaccine Development and the new National School of Tropical Medicine at Baylor College of Medicine. Through the hard work of our faculty and scientists, we uncovered an extraordinary disease burden from NTDs in Texas and adjacent Gulf Coast states, including Chagas disease, dengue, Zika virus infection, murine typhus, toxocariasis, trichomoniasis, and hookworm infection. NTDs and poverty are inextricably linked.

This third edition of *Forgotten People, Forgotten Diseases* coincides with the third decade of the NTDs movement and ecosystem that began in the early 2000s. Now mass drug administration/preventive chemotherapy is taken for granted as a recognized international standard for advancing global health and for addressing the plight of people who live in profound poverty. But this approach truly represents hard-fought efforts from our small group of tropical and parasitic disease experts, who included my "three musketeer" colleagues Alan Fenwick and David Molyneux, as well as the leaders of the WHO's Department of NTDs Control—Lorenzo Savioli, Dirk Engels, Mwele Ntuli Malecela, and so many others.

Today, more than 1 billion people benefit from access to essential NTD medicines, and also the many collateral benefits in terms of therapeutic effects on diseases that we did not necessarily intend to target. This book tells the story of how the NTD space evolved and control was implemented on a global scale. It discusses some of the major non-governmental development organizations committed to NTDs and advocacy for the NTDs, and the important contributions of the U.S. and U.K. governments, as well as the Bill & Melinda Gates Foundation, as well as other organizations. It tells how we measured the health and economic impact of the NTDs through the Global Burden of Disease (GBD) Study of the Institute for Health Metrics and Evaluation at the University of Washington. It also highlights the role of innovation in the development of new treatments and vaccines for NTDs, and the role of important product development partnerships, including ours, and others such as DNDi, IDRI, IVI, and FIND, to name some. It explains how science and vaccine diplomacy ensures that a new generation of these biotechnologies reaches the world's poorest people. Most of all, it tells the story of the world's people who live in extreme poverty and what it means for them to live with NTDs.

PETER J. HOTEZ
Houston, Texas

Acknowledgments

This book and my career in tropical medicine owe so much, to so many people. I had the unique opportunity to thank many of them during my 2011 Presidential Address to the American Society of Tropical Medicine and Hygiene.[1] I again want to thank my bosses at Baylor College of Medicine and Texas Children's Hospital, Dr. Paul Klotman and Mark Wallace, respectively, and the boards of those two institutions. I also thank my long-standing colleagues and partners in battle against the neglected tropical diseases (NTDs), including Professors David Molyneux and Alan Fenwick; and Drs. Lorenzo Savioli, Dirk Engels, and Mwele Ntuli Malecela—past and current heads of the Department of Control of NTDs of the WHO. Also thanks to TDR, the Special Programme for Research and Training in Tropical Diseases, of the World Health Organization. I also thank my science partner for the last 20 years, Dr. Maria Elena Bottazzi, and our team of amazing scientists at the Texas Children's Center for Vaccine Development, and the heads and directors of the many organizations committed to NTDs, which include the areas of implementation, product development, and advocacy. Along those lines I want to thank the heads of the important non-governmental development organizations, public-private partnerships, and product development partnerships committed to the NTDs, and my good colleagues at *PLoS Neglected Tropical Diseases*. A special thank you to Drs. Patrick Soon-Shiong and Gary Michelson for their commitment and interest in NTDs. I also extend my appreciation to the work of the Institute for Health Metrics and Evaluation of the University of Washington for its Global Burden of Disease Study (GBD) 2019. This book presents the results and mapping from the GBD 2019 for each of the major NTDs. I also want to thank Alyssa Milano for her long-standing commitment to NTDs and both Alyssa and Soledad O'Brien for their willingness to contribute forewords for the previous two editions. Many thanks to Nathaniel Wolf for his editorial assistance and insights and to Ashish Damania for his help with new maps and other related materials. I also thank Douglas Soriano Osejo for his help. Finally, many thanks to the donors and partners that made it possible for me to pursue a career in NTDs, including the Bill & Melinda Gates Foundation, the U.S. National Institutes of Health (especially the National Institute of Allergy and Infectious Diseases and the Fogarty International Center), the Robert J. Kleberg, Jr. and Helen C. Kleberg Foundation, the Carlos Slim Foundation, JPB Foundation, Tito's Vodka, Southwest Electronic Energy Medical Research Institute, Blavatnik Charitable Trust, Rebecca Marvil and Brian Smyth, Mendell Family Fund, MD Anderson Foundation, Rawley Foundation, John S. Dunn Foundation, Jay H. Newman and Newman Family Foundation, Jerold B. Katz Foundation, Jesse W. Couch Charitable Foundation, and others. Finally, I want to thank my wife Ann Hotez and my family for bearing with me through another edition of *Forgotten People, Forgotten Diseases*, and Christine Charlip for "rolling the dice" once again with me at ASM Press.

Peter J. Hotez
Houston, Texas

Note

1. **Hotez PJ.** 2012. ASTMH Presidential Address. Four Horsemen of the Apocalypse. *Am J Trop Med Hyg* **87**:3–10.

Introduction to the Neglected Tropical Diseases: the Ancient Afflictions of Stigma and Poverty

The age of hypocrisy has been succeeded by that of indifference, which is worse, for indifference corrupts and appeases: it kills the spirit before it kills the body. It has been stated before, it bears repeating: the opposite of love is not hate, but indifference.

ELIE WIESEL, *A JEW TODAY*, P. 17

It is a trite saying that one half the world knows not how the other lives. Who can say what sores might be healed, what hurts solved, were the doings of each half of the world's inhabitants understood and appreciated by the other?

MAHATMA GANDHI

Since the beginning of the 21st century, we have seen unfold a new sense of urgency about the plight of the world's poorest people in developing countries. Today, the average well-educated layperson living in "the North" (North America, Europe, and Japan) is far more aware than ever before about the suffering of the people living in "the South" (the developing countries of sub-Saharan Africa, Asia, and the Americas). Almost certainly, the human catastrophe of HIV/AIDS in sub-Saharan Africa, known as the "plague of the 21st century," and epidemics or pandemics from Ebola virus and Zika virus infections, and most recently coronavirus disease 2019 (COVID-19), have helped to focus world attention on health threats from infectious diseases, especially problems in the world's low- and middle-income countries (LMICs).[1]

Simultaneously, an unprecedented and extraordinary advocacy effort led by some highly influential international leaders and celebrities has helped to fuel a 21st-century global health movement. Throughout the decade of the 2000s, Bono, Angelina Jolie, Brad Pitt, George Clooney, Oprah Winfrey,

Forgotten People, Forgotten Diseases: The Neglected Tropical Diseases and Their Impact on Global Health and Development, Third Edition. Peter J. Hotez.
© 2022 American Society for Microbiology. DOI: 10.1128/9781683673903.ch01

Annie Lennox, Bob Geldof, and other actors, celebrities, and musicians; Bill Gates, Melinda French Gates, Warren Buffett, Carlos Slim and his family, and other philanthropists; Jeffrey Sachs; Chelsea Clinton; Prime Ministers Tony Blair, Gordon Brown, David Cameron, Theresa May, and Boris Johnson of the United Kingdom; and Secretaries of State Hillary Clinton and John Kerry and Presidents Jimmy Carter, Bill Clinton, George W. Bush, and Barack Obama of the United States have donated their time and energy to advocate for the health of the world's poorest people. These efforts captivated world attention and have even infused an element of glamour into solving global health problems. Between 2005 and 2006 alone, Bono, Bill Gates, and Melinda Gates were named *Time* magazine Persons of the Year; the Time Global Health Summit in New York was branded the "Woodstock of global health"; Brad Pitt narrated a 6-hour-long documentary, *Rx for Survival, a Global Health Challenge*, for PBS; former President Clinton featured global health issues at his annual Clinton Global Initiative; and Bono and Bobby Shriver launched Product RED to support HIV/AIDS, malaria, and tuberculosis relief at the 2006 World Economic Forum in Davos, Switzerland.

As a university professor and now as a dean, I can attest that these activities stimulated an unprecedented level of interest in global health issues from both undergraduates and graduate public health and medical students. With the important exception of our 2020–2021 years of COVID-19, almost every week during the academic year I have been visited by one or more young persons who request advice on how they can help solve a health problem in an LMIC. I am not the only faculty member to have this experience—today, new university-wide global health institutes are springing up at Duke, Baylor, Brown, Yale, Vanderbilt, Harvard, Emory, Washington University in St. Louis, the University of California campuses, University of Washington, and elsewhere, as university deans and presidents scramble to keep up with student interest.

Like any movement, the one in global health that I benchmark as beginning in 2000 was stimulated by a *manifesto*, which is defined by Webster as "a public declaration of motives and intentions by a government or by a person or group regarded as having some public importance."[1] For the global health movement, we can point to at least three landmark 21st-century policy documents that have effectively served as manifestos.

The first had its origins in January 2000, when then-World Health Organization (WHO) Director-General Gro Harlem Brundtland launched the Commission on Macroeconomics and Health (CMH) and appointed the international macroeconomist Jeffrey Sachs to serve as its chair. Jeff and his colleagues were charged with analyzing the impact of health on development. Their *Report of the CMH*, illustrated with examples of how health investments translate into economic development, elegantly articulated a profound relationship between disease and chronic poverty. As a result, the world's most influential finance ministers and policymakers began to regard improvements in global health as an important tool for poverty reduction. A second initiative was also launched in 2000 when the General

Table 1.1 The MDGs

1.	Eradicate extreme poverty and hunger
2.	Achieve universal primary education
3.	Promote gender equality and empower women
4.	Reduce child mortality
5.	Improve maternal health
6.	Combat HIV/AIDS, malaria, and other diseases
7.	Ensure environmental sustainability
8.	Develop a global partnership for development

Assembly of the United Nations convened in New York City to adopt a resolution known as the UN Millennium Declaration. The Declaration was a renewed call for sustainable development and for the eradication of poverty, and its core was a set of eight specific Millennium Development Goals (MDGs) along with a set of specific targets for the year 2015. As shown in Table 1.1, three of the goals (MDGs 4, 5, and 6) specifically emphasize health. Finally, a third manifesto was *Our Common Interest: Report of the Commission for Africa*, commissioned by British Prime Minister Tony Blair to provide specific recommendations on how to accelerate development and reduce poverty in Africa. The report served as an important blueprint for commitments by the Group of Eight (G8) nations at their 2005 summit in Gleneagles, Scotland.

Unlike many UN and international declarations, which too often are forgotten by the global community almost as soon as they are written, the CMH report, the MDGs, and the *Report of the Commission for Africa* continue to exert a major influence on global policymakers. Although the MDGs ended in 2015, they have since continued under a new set of Sustainable Development Goals (SDGs), sometimes just referred to as the "Global Goals". Equally important, together with the new advocacy by leaders and celebrities, the global health manifestos have stimulated high-level efforts to invent innovative financial instruments for supporting disease control, including some very substantial funding initiatives from both the G7 nations and some prominent private philanthropic organizations such as the Bill & Melinda Gates Foundation.

MDG 6 (to "combat HIV/AIDS, malaria, and other diseases") has been a particular target of these new funds, with approximately US$90 billion committed so far by the U.S. Congress for HIV/AIDS through the U.S. President's Emergency Plan for AIDS Relief (PEPFAR),[2] together with more than US$6 billion for malaria through the U.S. President's Malaria Initiative (PMI). Internationally, the Global Fund to Fight AIDS, Tuberculosis, and Malaria now commits more than US$4 billion annually to support interventions against these infections (http://theglobalfund.org), while the Gates Foundation has also committed vast sums. Practically speaking, these extraordinary new financial commitments mean that unprecedented numbers of poor people in Africa and elsewhere are receiving lifesaving antiretroviral medications for the treatment of HIV/AIDS or drugs and bed nets for the treatment and prevention of malaria. Such interventions are producing significant positive changes to the global health landscape under the auspices of the SDGs.

Unfortunately, with the exception of some important support from the Gates Foundation, the flurry of global health advocacy and resource mobilization occurring over the past few years has, until recently, largely bypassed the third, "other diseases" component of the original MDG 6. This neglect is particularly true for a group of exotic-sounding tropical infections that represent a health and socioeconomic problem of extraordinary dimensions but one that world leaders and global health advocates are only now waking up to. Beginning in 2005, an original core group of the 13 major so-called neglected tropical diseases, or NTDs, was proposed,[3] which has since been expanded by the WHO to a list of 20 diseases and conditions (Table 1.2).

Table 1.2 The NTDs (core group of 20)[a]

Infection type	Disease or pathogen name (common name)
Helminth (worm) infections	
Soil-transmitted helminth infections[b]	Ascariasis (roundworm infection)
	Hookworm infection
	Trichuriasis (whipworm infection)
	Others: Strongyloidiasis (threadworm) and toxocariasis
Other helminth infections	Schistosomiasis (snail fever)
	Lymphatic filariasis (elephantiasis)
	Onchocerciasis (river blindness)
	Food-borne trematode infections (liver fluke, lung fluke, intestinal fluke)
	Cysticercosis
	Human echinococcosis (hydatid cyst)
	Dracunculiasis (guinea worm infection)
Protozoan infections	Leishmaniasis
	Chagas disease
	Human African trypanosomiasis (sleeping sickness)
Ectoparasitoses	Scabies and other ectoparasitoses
Bacterial and fungal infections	Trachoma
	Buruli ulcer
	Leprosy (Hansen's disease)
	Yaws and endemic treponematoses
	Mycetoma, chromoblastomycosis, and other deep mycoses
Snakebite envenoming	Snakebite envenoming
Viral infections	Dengue[c]
	Rabies

[a] Data from the WHO (https://www.who.int/teams/control-of-neglected-tropical-diseases).
[b] Five major soil-transmitted helminth infections are listed, although they are typically considered as a single entity by the WHO.
[c] Some modify "dengue" to include "dengue and severe dengue" or "dengue and other arbovirus infections," which might include Zika virus, yellow fever, and West Nile virus infection, among others.

They include the major parasitic worm infections of humans, such as the major soil-transmitted helminth infections, e.g., ascariasis (roundworm infection), hookworm infection, and trichuriasis (whipworm infection); and lymphatic filariasis (LF or elephantiasis), schistosomiasis (snail fever), onchocerciasis (river blindness), food-borne trematode infections (liver fluke, lung fluke, and intestinal fluke), cysticercosis, echinococcosis, and dracunculiasis (guinea worm infection). In addition, the NTDs include an important group of infections caused by single-celled protozoan parasites, such as Chagas disease, leishmaniasis, and human African trypanosomiasis (sleeping sickness). Several nonparasitic infections are also prominent, including some atypical bacterial infections, such as trachoma, yaws, and endemic treponematoses; the mycobacterial infections Buruli ulcer and leprosy; mycetoma and related fungal diseases; and selected viral infections, such as dengue and rabies. More recently, snake envenomation and ectoparasitic conditions, especially scabies, were added. Still other tropical infections can also be considered NTDs, and there is an expanded list of these conditions included in the appendix.

While many educated people have since learned something about HIV/AIDS and malaria, and their impact in Africa and elsewhere in the developing world, far fewer have heard about this core group of NTDs. Therefore, it may come as a surprise to learn that the NTDs represent some of the most common infections of the world's poorest people. Today, of the almost 8 billion people living on our planet, an estimated 750 million people (10%) live on less than US$1.90 per day, which is considered the World Bank poverty threshold. Paul Collier, the Oxford University economist, helped to popularize the term "the bottom billion" to describe this group of people living in extreme poverty at the beginning of this century, a number now declining due to a global assault on poverty. Over the last 2 decades, China has accounted for some of the greatest reductions in those living in extreme poverty.

As shown in Table 1.3, most of the world's population living below World Bank poverty levels suffers from at least one NTD. The most common include worm infections, also known as helminth infections, led by ascariasis, trichuriasis, or hookworm infection—parasitic worm infections that are transmitted through the contaminated warm and moist soil of tropical developing countries (and are known as the soil-transmitted helminth infections)—and schistosomiasis. Essentially all of the "bottom 750 million" (to borrow from Paul Collier) are affected by one or more of the seven most common NTDs currently listed by the WHO.[3,4]

Shown in Fig. 1.1 are the countries in which the NTDs occur.[3] The extensive geographic overlap of these conditions means that many of the NTDs are *coendemic* and that it is common for poor people to be simultaneously infected with multiple NTDs. NTDs occur globally wherever poverty is widespread. While sub-Saharan Africa, South Asia, and Latin America dominate in terms of endemic NTDs, these diseases also occur in areas of poverty in the United States, especially in Texas and the Gulf Coast states,

Table 1.3 The WHO 20 NTDs ranked by prevalence[a]

Disease	Estimated global prevalence	Regions of highest prevalence
Soil-transmitted helminth infections		
Ascariasis	446 million	LMICs globally
Trichuriasis	360 million	LMICs globally
Hookworm disease	173 million	LMICs globally
Scabies	187 million	LMICs globally
Schistosomiasis	140 million	Africa and Middle East
Lymphatic filariasis	72 million	Africa and LMICs in Asia
Dengue	57 million[b]	Global
Food-borne trematode infections	34 million	East Asia
Onchocerciasis	19 million	Africa
Animal envenomation[c]	17 million	LMICs globally
Chagas disease	6–7 million	The Americas, including southern United States
Leishmaniasis	5 million	LMICs in Middle East, South Asia, Africa, and the Americas
Cysticercosis	5 million (with epilepsy)	LMICs globally
Trachoma	2 million[d]	LMICs globally
Cystic echinococcosis	0.9 million	LMICs globally
Leprosy	0.5 million	LMICs globally
Rabies	<0.1 million[c]	LMICs globally
Human African trypanosomiasis	<0.01 million[c]	Africa
Mycetoma and other mycoses	Not determined	LMICs globally
Buruli ulcer	Not determined	Africa
Yaws and endemic treponematoses	Not determined	LMICs globally
Dracunculiasis	Near eradication	Africa

[a] Data from Global Burden of Disease Collaborative Network, 2020 (http://ghdx.healthdata.org/gbd-results-tool/result/c1466642d6e4379c7465508afd12d904).
[b] Incidence figures used instead of prevalence estimates because of the acute nature of the illness.
[c] GBD 2019 does not specifically list snake envenomation, but reports this information as "animal envenomation," which includes snake envenomation as well as other causes.
[d] Only includes those with blindness or visual impairments.

and in Australia among aboriginal populations. Africa stands out because it accounts for 100% of the world's few remaining cases of dracunculiasis, 99% of the cases of onchocerciasis, more than 90% of the world's cases of schistosomiasis, approximately 40% of the cases of LF and trachoma, and one-third of the world's hookworm infections.[5] The impoverished areas of Asia, especially Southeast Asia and the Indian subcontinent, account for more than one-half of the world's cases of hookworm, ascariasis, and LF. Hookworm, schistosomiasis, LF, and onchocerciasis also remain endemic in focal regions of American tropics and subtropics, especially in Venezuela and elsewhere in Central Latin America and Brazil, where it has been suggested that these NTDs represent a living legacy of the transatlantic slave trade.[5] Today, these NTDs still primarily afflict the poor and marginalized people living in the region.[5]

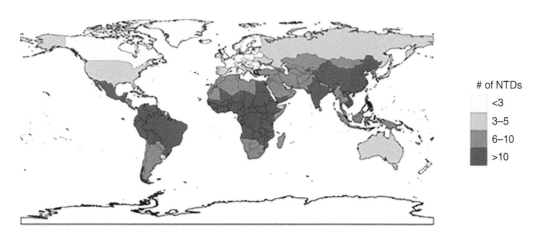

Figure 1.1 Map of the world's coendemic NTDs. (Drawn by Ashish Damania, National School of Tropical Medicine, Baylor College of Medicine. Data from Global Burden of Disease Collaborative Network, 2020.)

In addition to their geographic overlap and coendemicity, the major NTDs exhibit a remarkable set of common features, all of which adversely affect the health and socioeconomic status of the world's poorest people (Table 1.4).[6]

To summarize these common features:

1. *The NTDs have high prevalence.* As discussed above, today the NTDs rank among the most common infections of the poorest people in developing nations or LMICs.[3] A new trend is the observation that NTDs are not only found in poor nations, but are surprisingly common in the Group of 20 (G20) countries. In this case, the NTDs still mostly strike people who live in poverty. For example, NTDs such as hookworm infection, toxocariasis, trichomoniasis, and Chagas disease are common among the poor living in impoverished areas of the U.S. Gulf Coast states.[3]

2. *The NTDs are linked to rural poverty—but this situation may be shifting.* The high prevalence of the NTDs is frequently not widely appreciated by policymakers or sometimes even by many government officials from the countries where NTDs are endemic. An important reason for the lack of awareness about these conditions is that the NTDs are seldom noticed in capital cities, where the government officials work and live. Instead, the NTDs are widespread in poor rural and agricultural areas, particularly in regions where subsistence farming is practiced.[6] Therefore, unlike HIV/AIDS or other better-known infections, the NTDs are frequently both out of sight and out of mind. They truly are forgotten diseases afflicting forgotten people. There are important exceptions, such as dengue fever and leptospirosis, which are also found in urban slums. These conditions will be addressed separately (in chapter 8), but historically the NTDs occur in the setting of rural poverty. Along those lines, a new trend may be emerging. Increasingly, urban foci of

Table 1.4 Major attributes of the NTDs

Most prevalent among poor people
Endemic in rural areas (some poor urban areas) of low-income countries
Ancient ("the biblical diseases")
Chronic
Disabling (growth delays, blindness, or disfigurement)
Associated with high disease burden but low mortality
Stigmatizing
Poverty promoting

soil-transmitted helminth infections, schistosomiasis, leishmaniasis, and other NTDs are being reported.[6] This parallels observations that for the first time the majority of the world's population lives in urban areas. What remains uncertain is whether urbanization of NTDs reflects some type of reporting bias versus true adaptation of parasites or their insect vectors or snail hosts. However, the confluence of expanding slums and urban poverty and NTDs is a worrisome trend. Some demographers predict that the world's poor will eventually coalesce into massive "mega-cities" across Africa, Asia, and Latin America, and increasingly NTDs will be the predominant infections found in those areas.

3. *The NTDs are ancient conditions.* Another interesting feature of the NTDs is their nonemerging character. By this phrase, I mean the NTDs are just the opposite of better-known *emerging infections*, such as avian influenza, SARS (severe acute respiratory syndrome), COVID-19, Ebola, Lyme disease, and HIV/AIDS, which have either newly appeared in the population or have rapidly increased in incidence or geographic range. Instead, the NTDs have been around seemingly forever, as they have plagued humankind for centuries. This historical link is well documented through the accounts and descriptions of some of the dramatic clinical manifestations of the NTDs, particularly leprosy, dracunculiasis, schistosomiasis, hookworm infection, and trachoma, in ancient texts including the Bible, Talmud, Vedas, writings of Hippocrates, and Egyptian medical papyri.[7] One exception to this persistent state is selected NTDs that can sometimes reappear because of public health breakdowns resulting from civil or international conflicts. Later (in chapter 7), we will see how this situation has tragically unfolded in Angola, the Democratic Republic of the Congo, and Sudan and has resulted in the rise of human African trypanosomiasis and kala-azar.

4. *The NTDs are chronic conditions.* Another distinguishing feature of the NTDs is that unlike many infectious diseases with which we are familiar, they are mostly chronic infections lasting years and sometimes even decades. In some cases, poor people can suffer from NTDs for their entire lives.[6] The viral NTDs—dengue and related arbovirus infections and rabies—represent important exceptions to this feature.

5. *The NTDs cause disability and disfigurement.* Even though they are infectious diseases because they are caused by microbial or

multicellular pathogens, which are transmitted either from person to person or through contact with contaminated soil or water or through exposure to arthropod vectors (e.g., mosquitoes, sandflies, assassin bugs, and copepods), the NTDs frequently do not exhibit the classic features of most infections. That is to say, they do not typically cause acute febrile illnesses, which either resolve or kill. Instead, the NTDs mostly cause chronic conditions that lead to long-term disabilities and, in some cases, disfigurement.[6] I will highlight the specific disabling features of each of the NTDs when they are treated separately (in chapters 2 to 9), but to provide some specific examples here, the long-term effects of chronic hookworm infection and schistosomiasis in childhood produce a long-standing anemia, which is associated with physical growth retardation, impaired memory, and cognitive growth delays; in pregnant women, the anemia from hookworm infection and from schistosomiasis results in poor birth outcomes such as low neonatal birth weight and increased maternal morbidity and mortality. Onchocerciasis and trachoma cause impaired vision and blindness. Chagas disease causes a chronic and severely disabling heart condition. LF, onchocerciasis, guinea worm infection, leishmaniasis, Buruli ulcer, and leprosy cause either limb disuse or profound disfigurement (including genital deformities), which often prevent afflicted individuals from either obtaining or maintaining employment (Fig. 1.2).

6. *The NTDs have a high disease burden but low mortality.* An estimated 100,000 to 200,000 people die annually from the NTDs.[8] These numbers are derived from the Global Burden of Disease Study (GBD) 2019, an initiative led by the Institute for Health Metrics and Evaluation of the University of Washington, Seattle. Throughout this book, we will refer frequently to GBD 2019 estimates of disease prevalence and geographic distribution. While this number of people is significant and approximates the number estimated to have perished in the 2004 Christmas tsunami that hit the beaches of Thailand, Sri Lanka, and Indonesia, for example, the reality is that these numbers pale in comparison to the number of annual deaths from HIV/AIDS or malaria. Therefore, placing NTDs on the global health radar screen of world leaders and policymakers and motivating them to tackle these conditions in a substantive way require focusing advocacy efforts on something more than simply looking at deaths as an endpoint. While it is obvious that the individuals shown in Fig. 1.2 are having their lives ruined by the long-term consequences of their NTDs, these compelling images by themselves do not provide an obvious metric that we can use to justify to the global community investments either in this group of diseases or in the people who suffer from them. Instead, we need another mechanism to convince policymakers that the "other diseases" deserve the same international attention as HIV/AIDS and malaria.

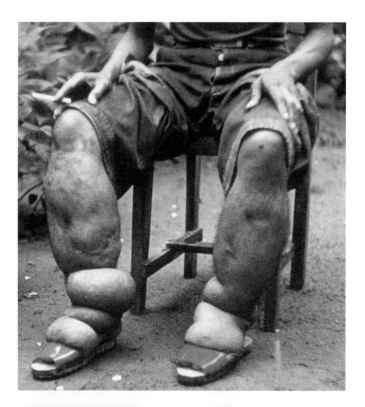

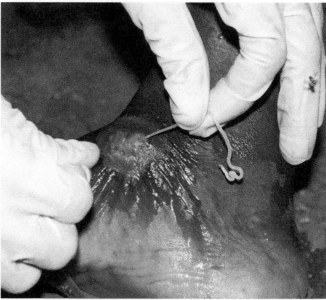

Figure 1.2 Disfiguring effects of the NTDs. (Top) Elephantiasis of the leg due to filariasis, Luzon, the Philippines. (Image from CDC-PHIL [ID#373]/CDC, 1962.) (Bottom) Guinea worm infection, with female worm emerging from the patient's foot. (© The Carter Center.)

One approach to measuring the full health impact of the NTDs is to use the disability-adjusted life years, or DALYs, which consider the number of healthy life years lost from either premature death or disability. Because of the chronic, disabling, and disfiguring components of the NTDs, the DALYs ascribed to them are substantial. Shown in Table 1.5 is a ranking of HIV/AIDS, malaria, tuberculosis, and the NTDs by deaths and DALYs, according to the GBD 2019. One of the greatest values of DALYs is that they facilitate the comparison of one condition with another. The data illustrate that the total disability resulting from the NTDs is almost as great as the disability from HIV/AIDS, malaria, or tuberculosis.[8]

The devastating comparison between the NTDs and the "big three" diseases—HIV/AIDS, malaria, and tuberculosis—has multiple implications for international efforts to control or eliminate infectious diseases. Today, much of the global enterprise targeting infections focuses primarily on HIV/AIDS, malaria, and tuberculosis. The DALY measurements suggest a strong rationale for considering the NTDs an important fourth leg of the chair. The rationale goes beyond merely comparing DALY estimates and pointing out the high disease burden resulting from the NTDs. Instead, an increasing body of evidence indicates not only that the NTDs exhibit geographic overlap and coendemicity with each other but also that the NTDs are coendemic with AIDS and malaria. The geographic overlap and coendemicity between the NTDs and malaria and AIDS will be further elucidated elsewhere (chapter 10). However, to briefly mention it here, there is new evidence that the morbidities resulting from the NTDs are additive with malaria and that some NTDs actually increase susceptibility to HIV/AIDS. The HIV/AIDS and NTD link is especially apparent for a condition known as female genital schistosomiasis, now considered one of the most important gynecologic conditions of adolescent girls and young women

Table 1.5 Ranking of the "gang of four" by deaths and DALYs

Condition	Initial 2006 estimates of DALYs annually[a]	Updated 2019 estimates of DALYs annually[b]
HIV/AIDS	84.5 million	47.6 million
NTDs	56.6 million	16.5[c]–25.4[d] million
Malaria	46.5 million	46.4 million
Tuberculosis	34.7 million	47.0 million

[a] Data from Hotez et al., 2006a.

[b] Data from Global Burden of Disease Collaborative Network, 2020.

[c] The lower number is obtained by subtracting the DALYs for malaria from the combined NTDs and malaria total (62.9 million DALYs) provided by the GBD 2019.

[d] The higher number is obtained by also including the DALYs from scabies (4.84 million DALYs) and venomous animal contact (4.09 million DALYs, including snakebite envenomation). The GBD 2019 does not currently include these two conditions in their NTDs category, but the WHO does. The WHO and IHME are working to harmonize their disease category assignments.

on the African continent.[8] Therefore, there is an important rationale for not simply tackling the big three conditions in isolation, as currently advocated by the Global Fund, PEPFAR, and PMI, but also for embracing the NTDs to take on what is really a "gang of four." This concept of integrating NTD control measures with those for malaria and HIV/AIDS will become clearer when we outline possible intervention strategies for NTD control (in chapter 10) and give the reason why we need to consider bundling treatment strategies for the NTDs together with those for HIV/AIDS and malaria (and even possibly why the Global Fund should incorporate NTD control into its programs).

7. *The NTDs are stigmatizing.* Not surprisingly, the blinding and disfiguring features of NTDs are stigmatizing and cause individuals to be ostracized by their families, their communities, and sometimes even health care professionals.[6] In some societies NTDs are considered a sign of a curse or an "evil eye." The social stigma of the NTDs strikes young women particularly hard, and as a result, these women are frequently abandoned by their husbands, prevented from holding or kissing their children, or unable to marry altogether. In some instances for example, community health workers have accused adolescent girls with female genital schistosomiasis of sexual promiscuity when in fact they acquired the condition by standing in water contaminated with schistosome larval stages.[8] Specific examples of these stigmatizing consequences of the NTDs will be illustrated in the chapters dealing with schistosomiasis; LF, onchocerciasis, and podoconiosis; Buruli ulcer; leishmaniasis; and scabies, yaws, and mycetoma (chapters 3, 4, 6, 7, and 12, respectively).

8. *The NTDs have poverty-promoting features and other socioeconomic consequences.* The health impact of the NTDs may also represent only the tip of the iceberg in terms of their adverse effects on international development. Because of their chronic and disabling features, the NTDs also produce important and serious socioeconomic consequences that keep affected populations mired in poverty. The NTDs not only occur in the setting of poverty; they also actually promote poverty. For example, the cognitive and intellectual impairments resulting from hookworm-associated iron deficiency and anemia severely affect childhood education in terms of school performance and school attendance. Reduced school attendance leads to reduced future wage-earning capacity, while chronic hookworm infection among agricultural workers reduces worker productivity in Africa, Asia, and the Americas. Similarly, LF has a huge impact on productive capacity and costs a significant percentage of India's gross national product, trachoma causes US$5.3 billion in worldwide losses annually, and leishmaniasis is responsible for 0.43% of French Guiana's social security budget.[9] We are only beginning to understand the full economic impact of the NTDs, but these nascent studies indicate that the effects are likely to be profound.

However, even a full consideration of the enormous disability, disfigurement, and economic impact does not adequately convey the total devastation wrought by the NTDs. In an interview with a Sri Lankan LF-affected patient suffering from a severe limb deformity, we can get a palpable sense of the enormous shame and stigma from the limb or genital deformities caused by her disease and how they in turn promote an inexorable slide into poverty.[10]

> I got this big leg when I was engaged to be married. When they heard it was filarial, they backed out of the marriage. I was earning Rs 2,500 (US$25) a month from sewing, but when the leg got worse, the hospital doctor told me I should not pedal the machine. So I lost my income as well. When my parents died and my sister got married, only my brother and I lived in the house. My brother married and left the house, but my sister become widowed so came to live with me and her child. She had no money to buy a bandage as instructed by the clinic. So I went to a house to cook. When they saw my leg, they asked me not to come there anymore and found fault with me for hiding such a dirty illness from them. When I get fever, I cannot walk to the hospital, so I take paracetamol for 2 days and walk to the hospital when I feel less pain.

According to the Sri Lankan health care team investigating such cases of LF, the woman in this vignette, who previously lived on earnings of approximately US$1 per day, lost even this meager income and became totally dependent on her brother-in-law.[10] An important theme in the succeeding chapters is how stigma actually contributes to the morbidity of the NTDs and creates not only a medical crisis for the affected individual but also a tragic cycle of social and economic devastation for both the individual and his family. According to Swiss Tropical and Public Health Institute's Mitchell Weiss, the stigma of the NTDs contributes to suffering, delays the seeking of help, promotes nonadherence to treatment, negatively affects families and communities, and ultimately lessens support for services, control, and research.[11] Later, we will even see how, with some of the NTDs such as leishmaniasis and schistosomiasis, the stigma is particularly acute for young women, often leading to their verbal and physical abuse (in chapters 3 and 7), or how the stigma associated with Buruli ulcer is linked to beliefs about witchcraft (in chapter 6).

In summary, the health impact of the NTDs reflects their chronic and disabling features. But there are also educational and socioeconomic consequences that may even be greater. Neglect occurs at many different levels: at the community level because the NTDs arouse fear and inflict stigmas, at the national level because the NTDs occur in remote and rural areas and are often a low priority for health ministers, and at the international level because they are not perceived as global health threats equivalent to the high-mortality big three conditions.[12] Paul Hunt, previously the UN Special Rapporteur on the right to the highest attainable standard of health, pointed out that relief from the suffering caused by the NTDs is a fundamental human right, which unfortunately has been largely ignored.[13] Despite their global importance, we so far have no Bono equivalent to champion the plight of the 1 billion of the world's poorest people who suffer from NTDs, and the total dollars thus far committed to NTD control are currently measured in the millions, not the billions.

Fortunately, this picture of neglect may one day turn an important corner, in part because of a new resolve by the WHO and national ministries of health, together with several key public-private partnerships dedicated to NTD control. Further, many of the organizations involved in NTD control have begun to partner through the WHO and new organizations including the END Fund and Uniting to Combat NTDs (discussed in chapter 10).[14] These new organizations work to mobilize resources for the NTDs and to promote high-level advocacy from global leaders and well-placed individuals and organizations. At the same time, student groups are beginning to voice their concerns about the urgency of addressing the NTDs. These important, nascent efforts are about to lead to a modest revolution in global health and to make a huge impact on the world's poorest people.

SUMMARY POINTS   **Introduction to the Neglected Tropical Diseases**

- The NTDs are among the most common infections of the world's poorest people, those living on less than US$1.90 per day.
- Nonemerging, ancient conditions
- Chronic and disabling features
- High morbidity, low mortality
- DALYs almost comparable to those for HIV/AIDS, malaria, and tuberculosis

- Coendemicity of the NTDs and with HIV/AIDS and malaria
- The "gang of four"
- Poverty-promoting features that keep populations destitute
- Associated with profound stigma
- Urgent need for stepped-up advocacy and resource mobilization

---

Notes

1. The designation of HIV/AIDS as the "plague of the 21st century" is found in Skolnik, 2007, p. 191. The definition of "manifesto" is from Agnes, 2000, p. 874.

2. Kaiser Family Foundation, The U.S. President's Emergency Plan for AIDS Relief (PEPFAR), May 27, 2020; https://www.kff.org/global-health-policy/fact-sheet/the-u-s-presidents-emergency-plan-for-aids-relief-pepfar/.

3. The original core list of 13 NTDs was shaped and refined by Molyneux et al., 2005; Hotez et al., 2006a; and Hotez et al., 2007. The WHO's list of 17 NTDs was first reported in World Health Organization, 2010, and later revised to the current list of 20 NTDs; see https://www.who.int/teams/control-of-neglected-tropical-diseases. Details of NTDs among the poor in the United States and G20 nations can be found in Hotez, 2016.

4. The numbers of people infected with NTDs are now updated regularly by the Global Burden of Disease Study (GBD) based at the Institute of Health Metrics and Evaluation at the University of Washington, Seattle (Global Burden of Disease Collaborative Network. 2020). The latest GBD 2019 and previous estimates can be found at http://www.healthdata.org, by going to the GBD Results Tool, or by searching using the terms "GBD 2019" and "disease" and "level 3."

5. Further details regarding Africa's disease burden are found in Molyneux et al., 2005; and Hotez and Kamath, 2009. Information about the link between NTDs and slavery from Africa is found in Lammie et al., 2007.

6. Many of these features are excellently summarized in World Health Organization, 2003. The urbanization of NTDs was reported in Hotez et al., 2017.

7. Specific citations of ancient references on NTDs can be found in Hotez et al., 2006b.

8. Details of the overall estimates for deaths and DALYs resulting from malaria and the NTDs can be found in the GBD 2019 at http://www.healthdata.org/results/gbd_summaries/2019/neglected-tropical-diseases-and-malaria-level-2-cause (Global

Burden of Disease Collaborative Network, 2020). The deaths and DALYs combine NTDs and malaria; however, it is possible to delink the NTDs values from malaria by searching for malaria as a level 3 illness, or by going directly to the GBD 2019 by using the GBD Results Tool. This provides a value of 115,000 deaths for NTDs. However, now that snakebite envenomation has been added as one of the newest NTDs, the death estimate is closer to 200,000. The paper by Herricks et al., 2017 reports on some of the differences in DALYs and death estimates between the GBD studies and the community of NTD investigators. Details of female genital schistosomiasis can be found in Hotez et al., 2019.

9. Specific references for these data can be found in Hotez et al., 2007; Hotez and Ferris, 2006; and Hotez et al., 2009.

10. Interviews and qualitative analysis of patients with LF are described in Perera et al., 2007.

11. Some of the details about the relationships between stigma and health are outlined in Weiss and Ramakrishna, 2006. In addition, there are excellent descriptions of the stigmatizing aspects of NTDs in World Health Organization, 2003. In May 2008, a special issue of *PLoS Neglected Tropical Diseases* was devoted to the links between stigma and NTDs, with a lead article by Mitchell Weiss (Weiss, 2008). The effects of NTDs on the general and specific aspects of mental health were recently summarized in Litt et al., 2012.

12. The three levels of neglect are described in greater detail in World Health Organization, 2006, p. 3.

13. Hunt, 2006.

14. Details on Uniting to Combat NTDs can be found at https://unitingtocombatntds.org/. Details on the END Fund can be found at https://end.org/.

References

Agler E, Crigler M. 2019. *Under the Big Tree: Extraordinary Stories from the Movement to End Neglected Tropical Diseases.* Johns Hopkins University Press, Baltimore, MD.

Agnes M (ed). 2000. *Webster's New World College Dictionary*, 4th ed. Wiley, New York, NY.

Bethony J, Brooker S, Albonico M, Geiger SM, Loukas A, Diemert D, Hotez PJ. 2006. Soil-transmitted helminth infections: ascariasis, trichuriasis, and hookworm. *Lancet* **367:**1521–1532.

Budke CM, Deplazes P, Torgerson PR. 2006. Global socioeconomic impact of cystic echinococcosis. *Emerg Infect Dis* **12:**296–303.

Fürst T, Keiser J, Utzinger J. 2012. Global burden of human food-borne trematodiasis: a systematic review and meta-analysis. *Lancet Infect Dis* **12:**210–221.

Global Burden of Disease Collaborative Network. 2020. Global Burden of Disease Study 2019 (GBD 2019). GBD Results Tool. Institute for Health Metrics and Evaluation, Seattle, WA.

Herricks JR, Hotez PJ, Wanga V, Coffeng LE, Haagsma JA, Basáñez MG, Buckle G, Budke CM, Carabin H, Fèvre EM, Fürst T, Halasa YA, King CH, Murdoch ME, Ramaiah KD, Shepard DS, Stolk WA, Undurraga EA, Stanaway JD, Naghavi M, Murray CJL. 2017. The global burden of disease study 2013: what does it mean for the NTDs? *PLoS Negl Trop Dis* **11:**e0005424.

Hotez PJ. 2011. New antipoverty drugs, vaccines, and diagnostics: a research agenda for the US President's Global Health Initiative (GHI). *PLoS Negl Trop Dis* **5:**e1133.

Hotez PJ. 2012. The Four Horsemen of the Apocalypse: tropical medicine in the fight against plague, death, famine and war. *Am J Trop Med Hyg* **87:**3–10.

Hotez PJ. 2016. *Blue Marble Health: an Innovative Plan to Fight Diseases of the Poor amid Wealth*. Johns Hopkins University Press, Baltimore, MD.

Hotez PJ. 2017. Global urbanization and the neglected tropical diseases. *PLoS Negl Trop Dis* **11:**e0005308.

Hotez PJ, Engels D, Gyapong M, Ducker C, Malecela MN. 2019. Female genital schistosomiasis. *N Engl J Med* **381:**2493–2495.

Hotez PJ, Fenwick A, Savioli L, Molyneux DH. 2009. Rescuing the bottom billion through control of neglected tropical diseases. *Lancet* **373:**1570–1575.

Hotez PJ, Ferris MT. 2006. The antipoverty vaccines. *Vaccine* **24:**5787–5799.

Hotez PJ, Kamath A. 2009. Neglected tropical diseases in sub-Saharan Africa: review of their prevalence, distribution, and disease burden. *PLoS Negl Trop Dis* **3:**e412.

Hotez PJ, Mistry N, Rubinstein J, Sachs JD. 2011. Integrating neglected tropical diseases into AIDS, tuberculosis, and malaria control. *N Engl J Med* **364:**2086–2089.

Hotez PJ, Molyneux DH, Fenwick A, Kumaresan J, Sachs SE, Sachs JD, Savioli L. 2007. Control of neglected tropical diseases. *N Engl J Med* **357:**1018–1027.

Hotez PJ, Molyneux DH, Fenwick A, Ottesen E, Ehrlich Sachs S, Sachs JD. 2006a. Incorporating a rapid-impact package for neglected tropical diseases with programs for HIV/AIDS, tuberculosis, and malaria. *PLoS Med* **3:**e102.

Hotez P, Ottesen E, Fenwick A, Molyneux D. 2006b. The neglected tropical diseases: the ancient afflictions of stigma and poverty and the prospects for their control and elimination. *Adv Exp Med Biol* **582:**23–33.

Hunt P. 2006. The human right to the highest attainable standard of health: new opportunities and challenges. *Trans R Soc Trop Med Hyg* **100:**603–607.

Lammie PJ, Lindo JF, Secor WE, Vasquez J, Ault SK, Eberhard ML. 2007. Eliminating lymphatic filariasis, onchocerciasis, and schistosomiasis from the Americas: breaking a historical legacy of slavery. *PLoS Negl Trop Dis* **1:**e71.

Litt E, Baker MC, Molyneux D. 2012. Neglected tropical diseases and mental health: a perspective on comorbidity. *Trends Parasitol* **28:**195–201.

Molyneux DH, Hotez PJ, Fenwick A. 2005. "Rapid-impact interventions": how a policy of integrated control for Africa's neglected tropical diseases could benefit the poor. *PLoS Med* **2:**e336.

Molyneux DH, Malecela MN. 2011. Neglected tropical diseases and the millennium development goals: why the "other diseases" matter: reality versus rhetoric. *Parasit Vectors* **4:**234.

Murray CJ, et al. 2012. Disability-adjusted life years (DALYs) for 291 diseases and injuries in 21 regions, 1990-2010: a systematic analysis for the Global Burden of Disease Study 2010. *Lancet* **380:**2197–2223.

Nash TE, Garcia HH. 2011. Diagnosis and treatment of neuro-cysticercosis. *Nat Rev Neurol* **7:**584–594.

Perera M, Whitehead M, Molyneux D, Weerasooriya M, Guna-tilleke G. 2007. Neglected patients with a neglected disease? A qualitative study of lymphatic filariasis. *PLoS Negl Trop Dis* **1:**e128.

Rajshekhar V, Joshi DD, Doanh NQ, van De N, Xiaonong Z. 2003. Taenia solium taeniosis/cysticercosis in Asia: epidemiology, impact and issues. *Acta Trop* **87:**53–60.

Skolnik R. 2007. *Essentials of Global Health.* Jones and Bartlett Publishers, Sudbury, MA.

Weiss MG. 2008. Stigma and the social burden of neglected tropical diseases. *PLoS Negl Trop Dis* **2:**e237.

Weiss MG, Ramakrishna J. 2006. Stigma interventions and research for international health. *Lancet* **367:**536–538.

Wiesel E. 1978. *A Jew Today.* Vintage Books, New York, NY.

World Health Organization. 2003. Neglected diseases that disable millions, p 104–153. *In* Kindhauser MK (ed), *Communicable Diseases 2002: Global Defence against the Infectious Disease Threat.* World Health Organization, Geneva, Switzerland.

World Health Organization. 2006. *Neglected Tropical Diseases: Hidden Successes, Emerging Opportunities.* World Health Organization, Geneva, Switzerland.

World Health Organization. 2010. Working to Overcome the Global Impact of Neglected Tropical Disease: First WHO Report on Neglected Tropical Diseases. World Health Organization, Geneva, Switzerland.

2

"The Unholy Trinity": the Soil-Transmitted Helminth Infections Ascariasis, Trichuriasis, and Hookworm Infection

As it was when I first saw it, so it is now, one of the most evil of infections. Not with dramatic pathology as are filariasis, or schistosomiasis, but with damage silent and insidious. Now that malaria is being pushed back, hookworm remains the great infection of mankind. In my view it outranks all other worm infections of man combined . . . in its production, frequently unrealized, of human misery, debility, and inefficiency in the tropics.

NORMAN STOLL, 1962

The neglected tropical diseases (NTDs) are the most common infections of the world's poorest people, and the soil-transmitted helminth (STH) infections are the most common NTDs. The word *helminth* comes from the Greek ἑλμίνς, meaning "worm,"[1] and the phrase *soil-transmitted* refers to the human acquisition of these worms through contact with soil contaminated with either parasite eggs or immature larval stages. STHs are also sometimes called intestinal helminths or intestinal worms because the adult stages of the parasite live in the human gastrointestinal tract. The STHs are also *nematodes*, a type of parasitic worm distinguished by their elongate and cylindroidal shape.

The three most important STH infections of humans, based on their prevalence and global disease burden, are:

- *Ascaris* infection (also known as roundworm infection or ascariasis)
- Hookworm infection (hookworm)
- *Trichuris* infection (whipworm infection or trichuriasis)

Together, these helminth infections afflict almost 1 billion people in low- and middle-income countries (LMICs).

Humans have been infected with STHs since ancient times. We know this from accurate descriptions found in Egyptian medical papyri and the writings

Forgotten People, Forgotten Diseases: The Neglected Tropical Diseases and Their Impact on Global Health and Development, Third Edition. Peter J. Hotez.
© 2022 American Society for Microbiology. DOI: 10.1128/9781683673903.ch02

of Hippocrates in the 5th century BCE, including reports of large *Ascaris* roundworms being expelled from infected people and of the characteristic pallor and sallow complexion of people with hookworm.[1] In addition, STH eggs have been recovered from coprolites, mummified feces thousands of years old, found in both the Old World and New World.[1] Today, an estimated 446 million, 173 million, and 360 million people are infected with ascariasis, hookworm, and trichuriasis, respectively (Table 2.1).[2] More often than not, a single individual living in a developing country, especially a school-age child, is infected with two and sometimes all three types of STH parasites simultaneously. Practically speaking, this observation means that the intestines of hundreds of millions of children living in Africa, Asia, and the Americas harbor a menagerie of worms. Harold Brown, the late former parasitology professor at Columbia University College of Physicians and Surgeons, frequently referred to *Ascaris*, *Trichuris*, and hookworms as "the unholy trinity" to indicate that it was extremely common for a child to be infected with all three parasites simultaneously. Typically, *Ascaris* roundworms and hookworms inhabit the small intestine, while *Trichuris* whipworms inhabit the large intestine.

How can we fathom the notion of almost 1 billion people infected with STHs? To understand this concept better, we need to travel to an LMIC where STH infections are *endemic*, meaning that the infections are constantly present in a particular region. Figure 2.1 shows children living in a rural village in Minas Gerais State, Brazil. The families of these children are mostly subsistence farmers involved with cultivation of manioc and beans. Looking at these children, one might not think that they appear terribly ill, unless one examines them more closely. The STH-infected children living in this Brazilian village are stunted in both weight and height because they are not growing normally. Moreover, they also do poorly on tests of cognition, memory, and intelligence. There is now strong evidence that such physical and mental disabilities result from the presence of intestinal worms.[3]

The reason we know that most of the children of Americaninhas, Minas Gerais State, Brazil, harbor intestinal worms is that we can diagnose their STH infections by examining their feces under a microscope. The adult male and female roundworms, whipworms, and hookworms mate in the intestines and produce eggs that exit the body in feces. Each type of STH produces characteristically shaped eggs that are easy to identify through microscopy. If we now do this test for all children in this particular rural Brazilian village, we get a result that is shown in Fig. 2.2, in which more than 70% of the children between the ages of 5 and 11 are infected with *Ascaris* worms and hookworms. It turns out that we can repeat this study in almost any rural Brazilian village or indeed almost any rural village in the tropical regions of the Americas, including Central America, and probably obtain a similar result or find that just as many children are also infected with *Trichuris* whipworms. Indeed, if we were to conduct fecal examinations on most of the rural villages in sub-Saharan Africa, on the Indian subcontinent, or in Southeast Asia, really wherever people live in poverty and depend on subsistence agriculture and where the soil and climate are suitable for survival of the parasite eggs and immature

Table 2.1 The "unholy trinity"

Major species	Common name	Length as adult (males and females)	Major location in the gastrointestinal tract	No. of cases worldwide[a]	Major disease other than impairment of child growth and development	Global distribution
Ascaris lumbricoides	Roundworm	15–35 cm (5–14 in.)	Small intestine	446 million	Intestinal obstruction	LMICs
Necator americanus and *Ancylostoma auodenale*	Hookworm	7–13 mm (0.3–0.5 in.)	Small intestine	173 million (approximately) 85% *N. americanus*)	Iron deficiency anemia	LMICs and the southern United States
Trichuris trichiura	Whipworm	3–5 cm (1–2 in.)	Large intestine (colon)	360 million	Colitis, dysentery	LMICs

[a] Worldwide case estimates from Global Burden of Disease Collaborative Network, 2020. The data can also be directly accessed by going to each information for each of the helminth infections. The "level 3" summary for intestinal nematode infections is available at http://www.healthdata.org/results/gbd_summaries/2019/intestinal-nematode-infections-level-3-cause, or for the individual STH infections ("level 4") by going to http://www.healthdata.org/results/gbd_summaries/2019/ascariasis-level-4-cause, http://www.healthdata.org/results/gbd_summaries/2019/hookworm-disease-level-4-cause, and http://www.healthdata.org/results/gbd_summaries/2019/trichuriasis-level-4-cause.

Figure 2.1 Children (left) living outside the Brazilian village of Americaninhas, Minas Gerais State (right). About 75% of people living in the area are infected with hookworm. The effects of the disease—malnutrition and anemia—are worse in children. (Brigid McCarthy/NPR.)

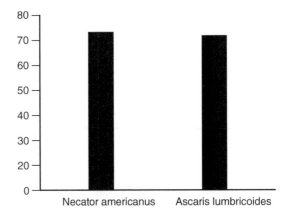

Figure 2.2 Prevalence of STH infections among school-age children in Americaninhas, Brazil. (Data courtesy of Jeff Bethony and David Diemert, Human Hookworm Vaccine Initiative; modified from graph prepared by Sophia Raff.)

larval stages (typically the warm and moist soil of the tropics), we would find a similar paradigm of extraordinarily high rates of STH infections. Such observations suggest how it can be that hundreds of millions of people harbor the unholy trinity in their bellies.

Beginning in the late 1980s, parasitologists of the Chinese Academy of Preventive Medicine, now known as the Chinese Center for Disease Control and Prevention, conducted a 4-year study of intestinal parasites on an almost unimaginable scale by performing fecal examinations on 1,477,742 individuals in every province of China. The results were impressive and demonstrated that approximately 531 million cases of ascariasis, 212 million cases of trichuriasis, and 194 million cases of hookworm infection had occurred in that country.[4] In collaboration with the Institute of Parasitic Diseases in Shanghai, I began working in China shortly after the completion of this nationwide survey of parasites. What particularly impressed me was the very tight link between high endemicity of STH infections in rural China and the level of economic under-development.[5] Wherever rural poverty was extreme and the villagers were engaged in subsistence agriculture, and provided there were suitable moisture and warmth, it was almost guaranteed that high levels of hookworm and other STH infections were present. Conversely, in areas of rapid economic gains, the STH infections disappeared. For instance, during the late 1990s when I visited a village in Jiangsu Province, not too far from Shanghai, there had been a steep decline in the prevalence of hookworm from just a decade previously.[6] The decline coincided with the building of new factories such that fewer villagers were engaged in agricultural activities. Moreover, there was even a new Kentucky Fried Chicken franchise, as well as a pirated version with the same red-and-white logos—called KCF instead of KFC! Hookworm occurs only in the setting of poverty, and in a sense, the factory, KFC, and KCF represent indicators of economic development.

As shown in Fig. 2.3, the relationship between STH prevalence and poverty is extremely tight,[7] and together with colleagues we even invented a "worm index" of economic development as a poverty indicator.[7] However, a clear understanding of the specific mechanisms underlying the link between a high prevalence of STH infections and poverty is still somewhat elusive. At least three possible factors linking poverty to STH infections have been identified so far, including (i) inadequate sanitation, because survival of the environmental stages of STH parasites depends on the deposition of human feces on soil; (ii) poor housing construction, because dirt floors allow propagation of STHs in households, whereas cement floors prevent parasite transmission; and (iii) inadequate access to essential medicines, because better-off families can afford deworming drugs.[8] Urbanization is also a potent factor in reducing the prevalence of STH infections. In eastern China, for example, rapid economic growth has brought with it a significant decline in prevalence, whereas in the poor and largely rural southern and southwestern provinces of China, such as Hainan, Sichuan, Yunnan, Guizhou, and Guangxi, hookworm and other STH infections remain highly endemic.[5,8]

In addition to their enormous global prevalence and their intimate link with rural poverty, another important feature of STH infections is their predilection for affecting children more than adults. For reasons that are not well understood, children between the ages of 4 and 15 on average harbor larger numbers of STHs than do any other group; i.e., children are wormier than adults. This propensity is particularly true for *Ascaris* roundworms and *Trichuris* whipworms, less so for hookworms. For example, shown in Fig. 2.4 is a little girl from Paraguay who simultaneously is emaciated and has a distended

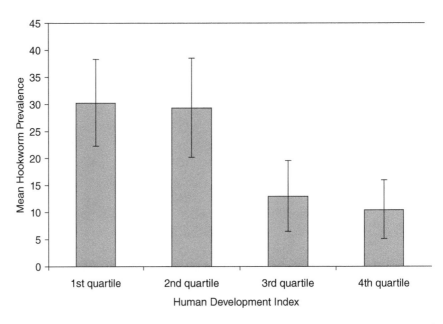

Figure 2.3 The relationship between prevalence of hookworm and poverty. The socioeconomic status of 94 countries was assessed according to a number of commonly used indicators, with poverty measures divided into quartiles including the poorest (first quartile), very poor (second quartile), poor (third quartile), and least poor (fourth quartile). (Modified from de Silva NR, et al. 2003. Soil-transmitted helminth infections: updating the global picture. *Trends Parasitol* 19:547–551, with permission from Elsevier.)

abdomen. It is sometimes possible to gently palpate the abdomens of children like her and feel the presence of worms in their intestines. Figure 2.4 also shows the *Ascaris* roundworms that she expelled after treatment with an anthelmintic drug (a process often referred to as *deworming*). It is easy to grasp how this girl could get into medical trouble if the roundworms were allowed to remain and obstruct the intestine or in some cases migrate from the intestine and into the liver or pancreas.

Although such clinical pictures are dramatic, they actually represent only a small portion of the global pediatric pathology caused by STHs. Far more important is the observation that in hundreds of millions of children the STHs stunt physical growth, physical fitness, and development. These processes probably operate at least partly through parasite-induced malnutrition, as all three major STHs can live in the intestines of children for years, where they can rob children of essential nutrients. For example, *Ascaris* roundworm infections most likely impair growth by impairing the digestion of protein, causing the malabsorption of fat, lactose, and vitamin A, as well as reducing appetite; *Trichuris* whipworm infections also result in reduced appetite, as well as in protein losses; and hookworms impair growth by causing blood loss that leads to profound protein and iron losses and ultimately to anemia.[9] Through these mechanisms, it is possible that STHs represent the world's leading cause of growth retardation and stunting!

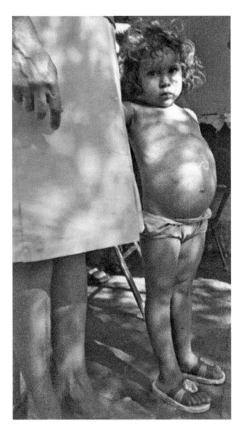

Figure 2.4 (Left) Little girl from Paraguay with severe *Ascaris* worm infection. (Right) Worms expelled after anthelmintic treatment. (Photos courtesy of Nora Vish.)

Moreover, the unholy trinity also adversely affects the neuropsychiatric activities of children, in turn damaging school performance and reducing school attendance.[10,11] The mechanisms by which school performance is impaired are not well established, but a number of clinical studies have shown that STHs can adversely affect cognition and memory and in some cases possibly lower intelligence.[11] Therefore, chronic infections with STHs destroy the lives of children not by shortening their lives but instead by impairing their physical growth, mental development, and ability to learn in school. Each of the NTDs not only occurs in the setting of poverty but also promotes poverty. In the case of STH infections, roundworms, whipworms, and hookworms promote poverty primarily through their impact on overall child development. Presumably, these processes account for the observation that chronic infection with hookworm during childhood is associated with a 43% reduction in future wage-earning capacity (similar studies for ascariasis and trichuriasis are not yet available).[12] Therefore, STH infections have a huge impact not only on health but also on education, and like other NTDs they are economic threats.

As suggested by the opening quotation from the late Norman Stoll, hookworm is probably the most significant STH. New Global Burden of Disease

(GBD) 2019 information confirms this observation, with preliminary indications that hookworm is responsible for almost one-half of the disability-adjusted life years (DALYs) lost from all of the STH infections.[13] Hookworms are 1-cm-long parasites that live in the small intestine, where they suck blood from the small blood vessels lining the gut mucosa and submucosa. Approximately 173 million people, almost one-quarter of the world's poorest people, are infected with hookworm. The greatest concentration of cases occurs in rural areas of sub-Saharan Africa, East Asia and the Pacific region, the Indian subcontinent, and tropical regions of the Americas, especially Brazil and Central America (Fig. 2.5).[2,14] Infection rates are often particularly high in coastal areas, an observation that most likely reflects the unique requirements of the soil-dwelling environmental stages of these parasites.

Nearly as striking as the high prevalence of hookworm in developing countries is the almost complete absence of hookworm in highly developed countries, including the United States. However, up until the 1930s, hookworm infection (as well as many other NTDs, such as malaria and typhoid fever) was endemic in the southern United States.[15] Shown in Fig. 2.6 is a map of the distribution of hookworm in the American South during the first decades of the 20th century, when high rates of hookworm infection occurred along the Gulf Coast and the Atlantic seaboard (the basis for the high rates of hookworm in coastal areas will become clearer when we discuss the hookworm life cycle). In some regions where more than 50% of the children were infected, it was shown that hookworm was a major reason why children were malnourished, why their growth was stunted, and why they did poorly in school and were prevented from reaching their full economic potential.[12,15]

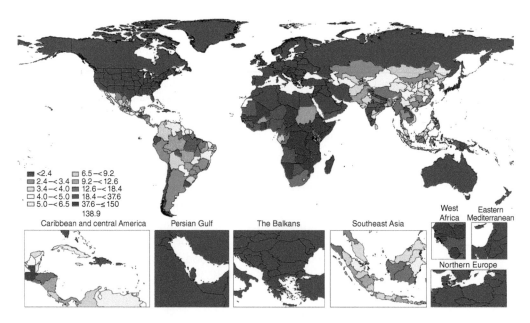

Figure 2.5 Global distribution of human hookworm infection. (From Global Burden of Disease Collaborative Network, 2020 [http://www.healthdata.org/results/gbd_summaries/2019/hookworm-disease-level-4-cause].)

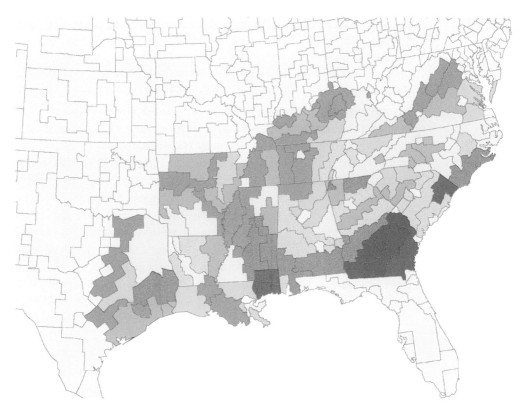

Figure 2.6 Distribution of human hookworm infection in the American South at the turn of the 20th century. The map displays the rates of hookworm infection among children by county groups. Red areas indicate the highest infection rates, followed by orange, yellow, and green. (Data from Bleakley, 2006.)

After Charles Wardell Stiles and Bailey K. Ashford identified *Necator americanus* as the predominant hookworm in the United States, it became known as the "germ of laziness" or the "vampire of the South."[1,15] It is believed that hookworm was introduced into the United States when *N. americanus* was imported by infected slaves from sub-Saharan Africa during the 17th, 18th, and 19th centuries.[1,15] Up until the 1950s, hookworm was also common in Japan and South Korea. In each of these now developed countries, reductions in the prevalence of tropical infections occurred primarily because of overall reductions in poverty and a shift to a more urbanized economy. In her book *Malaria, Poverty, Race, and Public Health in the United States*, the medical historian Margaret Humphreys argues that the Agricultural Adjustment Act and other New Deal legislation, which Congress passed in 1933, promoted rural depopulation by providing investment capital for the purchase of machinery.[16] This took agricultural workers out of cotton and tobacco production.[16] Such legislation caused landlords to tear down rural shacks and forced former dwellers to move either north or into southern cities.[16] There is a common misconception that during the first 2 decades of the 20th century, the Rockefeller Foundation and its forerunner, the Rockefeller Sanitary

Commission, eradicated hookworm in the American South and later in parts of Asia and South America through a combination of aggressive sanitation and the widespread distribution of shoes. For reasons that we will see below, it turns out that shoes may not be an effective hookworm prevention measure, while sanitation in the absence of parallel economic development frequently has little impact on the transmission of STH infections.[17] Instead, rural depopulation, urbanization, and economic development in the United States during the 1930s and in Japan and Korea in the years following World War II were probably the major elements leading to control of STH infections. In Asia, control was further hastened through widespread deworming by using first-generation anthelmintic drugs. Similar changes in human ecology probably account for the reductions observed in eastern China over the last 2 decades. Therefore, urbanization and economic development represent two of the most powerful forces responsible for the control of hookworm infection and other NTDs. Far more than the Sanitary Commission, the major health legacy of John Rockefeller was his foresight in establishing The Rockefeller University as a biomedical research powerhouse and in endowing the first generation of public health schools in the United States, beginning with the flagship school at Johns Hopkins University.

Recently, two of my Baylor College of Medicine colleagues, Drs. Rojelio Mejia and Megan McKenna, traveled to rural Alabama and found that hookworm infection remains endemic there.[15] Their visit was the product of a series of conversations and meetings I had with Catherine Coleman Flowers, a renowned African American environmental activist and 2020 MacArthur Fellow. Catherine is a pioneer in the concept of environmental justice who was working to address the problem of open raw sewage that runs through areas of the southeastern United States. When Catherine and I compared notes, I quickly came to the conclusion that it was plausible that hookworm infection was still present and perhaps was never really eliminated from the United States. Rojelio and Megan (now on the infectious diseases faculty at University of Texas Southwestern Medical School in Dallas) indeed confirmed this. Their studies published in the *American Journal of Tropical Medicine and Hygiene* are a poignant reminder that NTDs can be found wherever extreme poverty is widespread.

Humans become infected with hookworm through contact with infective larvae that live in the soil.[18] The major cause of human hookworm infection is the nematode parasite *N. americanus*, although a second but less common species, *Ancylostoma duodenale*, also causes hookworm infection. The life cycle of *N. americanus* is shown in Fig. 2.7. Soil-dwelling infective hookworm larvae exhibit the ability to directly penetrate human skin. The larvae are less than 1 mm long (Fig. 2.8) and are therefore largely invisible to people working in the fields or children playing on the ground. Larvae enter through any exposed skin, including the hands, the arms, the buttocks, the legs, and yes, sometimes even the feet. The ability of *N. americanus* larvae to penetrate all aspects of the skin explains why shoes have minimal if any impact on reducing the hookworm prevalence in affected communities. The higher

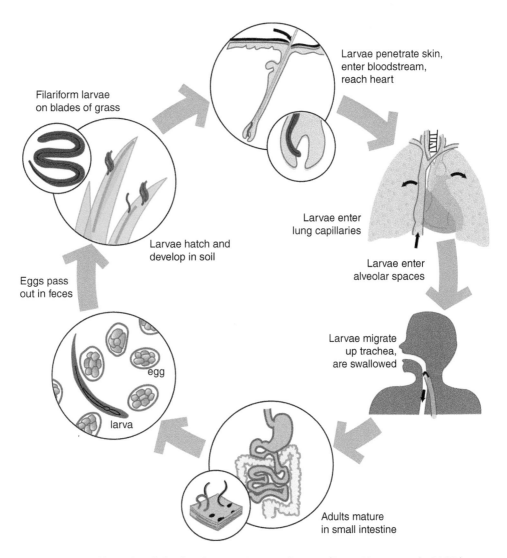

Filariform larvae
on blades of grass

Larvae hatch and
develop in soil

Larvae penetrate skin,
enter bloodstream,
reach heart

Larvae enter
lung capillaries

Larvae enter
alveolar spaces

Eggs pass
out in feces

Larvae migrate
up trachea,
are swallowed

egg

larva

Adults mature
in small intestine

Figure 2.7   Life cycle of the hookworm *N. americanus*. (From Hotez et al., 2005.)

rates of hookworm infection in coastal areas reflect the sandy soils present in these regions. Hookworm larvae can migrate through sandy soils better than through soils with a high clay content.[19]

It is common for people exposed repeatedly to hookworm larvae in the soil to acquire a pruritic (itchy) inflammatory condition of the skin known as ground itch or dew itch. The larvae then follow a 5- to 8-week migratory path through the body tissues, which includes an obligatory migration through the lungs that results in a cough (in contrast, when *Ascaris* larvae migrate through the lungs, they cause wheezing and other allergic symptoms that resemble asthma, a subject under investigation by Dr. Jill Weatherhead on our Baylor National School of Tropical Medicine faculty). Eventually, the infective hookworm larvae pass up the respiratory tree, crawl over the epiglottis, and are swallowed, before they enter the intestine and develop into adult hookworms.

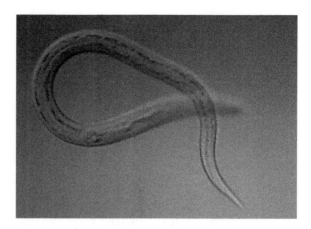

Figure 2.8 *Necator americanus* (hookworm) L3 infectious larva. (Photo courtesy of Jasper Lawrence, CC-BY-SA 4.0.)

As the larvae develop into adult worms, they initially cause a painful eosinophilic enteritis-like syndrome. Subsequently, the pain subsides and mature parasites proceed to feed on blood. Each adult hookworm has the ability to fasten deeply to the inner lining of the intestine and extract blood. The parasite lyses red blood cells and digests the hemoglobin component.[20] While feeding, the adult male and female hookworms mate and the female hookworm sheds thousands of eggs daily, which exit the body via the feces. In poor rural environments lacking adequate sanitation, either promiscuous defecation occurs or, in some societies, human feces are applied as fertilizer for crops (sometimes referred to as *night soil*). When feces are deposited on soil with adequate warmth and moisture, the eggs hatch and give rise to immature larvae that molt to become infective larvae.

People infected with hookworms become sick because of intestinal blood loss. The presence of as few as 25 *N. americanus* hookworms in the intestine is sufficient to cause about 1 ml of blood loss per day.[21] This amount of blood contains approximately 0.5 mg of iron, representing roughly a typical child's daily iron requirement. Therefore, hookworms essentially rob growing children of their daily iron and, as a result, cause iron deficiency anemia.[21] Higher hookworm loads cause even more blood loss and more-profound anemia. The disease resulting from chronic hookworm infection (sometimes referred to as hookworm disease) is long-standing iron deficiency anemia, which in children is responsible for growth retardation and intellectual and cognitive impairments. Because children tend to have low iron reserves from the outset, they are particularly vulnerable to hookworm-associated b-lood loss. Blood is also rich in protein, so that chronic blood loss can result in profound protein malnutrition, which is associated with edema of the face and limbs (Fig. 2.9). Many such children acquire a yellowish or sallow complexion; in several cultures, hookworm is known as the "yellow disease" or the "yellow puffy disease" (in Chinese, *huang zhong bing*, and in Brazilian Portuguese, *amarelao*). In antiquity, there are numerous references to the yellow disease, and there is an older term in

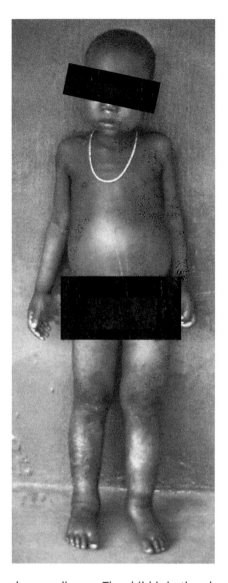

Figure 2.9 Severe hookworm disease. The child is both pale and edematous, thus reflecting severe loss of both iron and protein. (Image from CDC-PHIL [ID#5243]/CDC and the Family of Myron G. Schultz, DVM, MD, DCMT, 1986.)

the English medical literature, chlorosis, that refers to this condition.[18] Another unusual feature of chronic hookworm infection is pica, an appetite for consuming clay and other bulky substances. Referring to hookworm, Hippocrates described a syndrome in which "the skin is yellow, the intestine disturbed, and the person has an appetite for eating clay," and there are numerous references to clay eating in early Southern culture.[1] It has been suggested that eating of clay represents an effort to replace iron stores because of its high iron content.[1]

Hookworm is also an important health threat during pregnancy, and millions of pregnant women worldwide suffer from hookworm infection. One estimate indicates that almost 7 million pregnant women in sub-Saharan Africa (almost one-third of all pregnant women) are infected with hookworm.[22] Pregnant women typically have low iron reserves and are often iron deficient to begin with because of the iron demands of a growing fetus. There is a strong link between the added iron losses and anemia that result from hookworm and adverse maternal-fetal outcomes such as neonatal prematurity, low birth weight, and increased maternal mortality.[21,22] Among agricultural laborers, chronic hookworm iron deficiency results in impaired worker productivity and productive capacity. In the early part of the 20th century, the Brazilian writer Monteiro Lobato created the now famous character of Jeca Tatu, a laborer who is always lazy and lacking in energy until he is cured of his hookworm infection and then goes on to champion social causes (Fig. 2.10). The chronic disabilities associated with impaired child development, poor

Figure 2.10 Jeca Tatu. Image from Weise K. 1924. *Jéca Tatuzinho*. Monteiro Lobato, Brazil.

pregnancy outcome, and reduced worker productivity account for the observation that hookworm costs more healthy life years lost through disability annually than any other parasitic worm infection.[23]

Given that shoes do not protect against hookworm infection, what might be our options for controlling or preventing hookworm in developing countries? The sanitary disposal of human feces by increased use of latrines could under some circumstances dramatically reduce the prevalence of hookworm and other STH infections. However, the best evidence to date is that unless it is accompanied by substantial poverty reduction measures and urbanization, the isolated use of latrines has only modest impact on the transmission of hookworm or other STH infections.[17] Currently, the most effective approach to the control of STH infections is through deworming of large populations through mass drug administration of a specific anthelmintic with the ability to expel all three major parasite species. This approach is the first example that we will describe in which mass drug administration (frequently abbreviated as MDA) is used for the large-scale control or elimination of an NTD.

For the STHs, anthelmintic drugs belonging to the benzimidazole class (sometimes referred to as benzimidazole anthelmintics or BZAs) are primarily used in a single dose for purposes of mass deworming. The two major available BZAs are albendazole and mebendazole. Both drugs are available as low-cost generic products, and in some cases BZA donations are being organized through two programs housed at the Task Force for Global Health in Atlanta, including a Johnson & Johnson program for mebendazole donations and a GlaxoSmithKline program for albendazole. Because school-age children are particularly at risk for heavy STH infections with large numbers of worms, this group is the major one targeted for global deworming efforts. Frequent and periodic deworming of school-age children with BZAs has been shown to result in a number of pediatric health and nutritional benefits, including improvements in appetite, physical fitness, and physical growth, as well as improved iron status and reductions in anemia.[11,24] Deworming also produces neuropsychiatric progress, including positive intellectual and cognitive effects, such as improvements to short-term and long-term memory, problem solving, language, and cognition.[11,24] Michael Kremer and Ted Miguel, economists at Harvard University and the University of California, Berkeley, respectively, have recently confirmed the benefits of deworming in promoting educational advancement, while additional economic analyses conducted by Kremer and Miguel together with Sarah Baird and Joan Hicks, as well as Hoyt Bleakley of University of Chicago, suggest that these effects may also translate into economic benefits for the community.[10,11,24]

Every May, the world's ministers of health meet at the annual World Health Assembly, held at World Health Organization (WHO) headquarters in Geneva, Switzerland. At the 54th World Health Assembly in 2001, a resolution was adopted (resolution 54.19) that urged member nations to attain a minimum target of regular deworming of at least 75% and up to 100% of all at-risk school-age children. Since then, there has been heightened advocacy by the WHO and other international agencies for the administration of BZAs, typically a single dose of either albendazole or mebendazole, on a large scale. Increasingly, annual deworming is being practiced in schools because of the cost-effectiveness and efficiencies of having teachers rather than health care practitioners administer anthelmintic drugs.[25] This approach includes using schoolteachers who are specially trained to deliver the deworming tablets alongside health education messaging.[24] In many African and Asian countries, deworming is linked with school feeding programs sponsored by the World Food Programme (www.wfp.org) and nongovernmental organizations such as the Partnership for Child Development (www.imperial.ac.uk/partnership-for-child-development) and Deworm the World-Evidence Action (www.evidenceaction.org/dewormtheworld/).[24] Such interventions can be achieved for extremely low costs. For example, in Ghana and Tanzania, hundreds of thousands of children have been treated for as little as US$0.03 and $0.04 per capita.[24,25] In addition to the fact that the BZAs are often donated for free, another reason that the costs of school-based deworming are so low is that the excellent safety profile of a single dose of a BZA allows children to be treated regardless of whether they are infected with STHs. Instead, once it is established that the overall community prevalence of STH infections exceeds 50%, it no longer is necessary to conduct fecal examinations on each child. Authorities can then blanket the school with a single dose of either mebendazole or albendazole. This practice eliminates the high cost of bringing trained microscopists and laboratory equipment to the school. I believe that the advocacy efforts of two individuals, namely, Lorenzo Savioli at WHO and Don Bundy, now at the World Bank, were especially instrumental in promoting global deworming and advancing the agenda leading to resolution 54.19.[26]

Such efforts are not without controversy. A group based at the University of Liverpool and Liverpool School of Tropical Medicine, led by David Taylor-Robinson and Paul Garner, respectively, have reanalyzed much of the data linking educational and other benefits to deworming.[24] Many of their findings do not support earlier claims, thereby generating a scientific discussion or debate sometimes referred to as "worm wars." Recently, with Farhan Majid and Su Jin Kim, social scientists at the Baker Institute of Rice University, we have tried to resolve some of the controversy by pointing out the importance of considering the three major STH infections separately, given their differential susceptibility to deworming drugs, as well as examining the effects of heavy versus light infections.[24]

More than 600 million children in 50 countries received low-cost deworming in 2019.[27] While this number is impressive, it is still far short of the estimated 1 billion children who would need to be treated annually in order to meet the targets specified by World Health Assembly Resolution 54.19[27] (Fig. 2.11). Because many school-age children do not attend school in developing countries, as an alternative or complementary approach to school-based interventions, many children are being targeted worldwide through community-based interventions, such as child health days. In such programs, deworming is linked to vitamin A distribution as well as to some immunizations, such as measles vaccinations.[24] Child health days and other

Figure 2.11 Proportion of children (1 to 14 years of age) by country requiring preventive chemotherapy for soil-transmitted helminthiases, worldwide, 2010. (See http://gamapserver. who.int/mapLibrary/Files/Maps/Global_STH_2010.PNG [© 2011 WHO].)

community-based interventions are particularly suitable in regions of STH infection endemicity where preschool children, i.e., children under the age of five, also suffer from moderate and heavy infections. By some estimates, almost 200 million children have received vitamin A in more than 50 countries,[24] so that this mechanism provides an added opportunity to scale up deworming. Also, as pointed out earlier, in some developing countries pregnant women are at high risk for hookworm infection, and the WHO and other international agencies have therefore expanded their recommended targets to include this group in areas of high transmission.

Although for most school-based and community-based interventions a single dose of either mebendazole or albendazole is provided on an annual basis, in areas of intense transmission deworming may need to be conducted more frequently. STH reinfection can occur over a period of just a few months, so that sometimes two or three dewormings must take place in a single year. Currently, the WHO recommends two or three deworming treatments annually in areas of high prevalence (typically greater than 70% prevalence) or high intensity (where more than 10% of the population have moderate or heavy infections).

When frequent and periodic dewormings are required in order to control STH infections for large populations, there are concerns that STH parasites, like any other infectious agent, could over time become resistant to either mebendazole or albendazole. Indeed, BZA resistance is now widespread among intestinal helminth parasites of sheep and cattle in Australia, New Zealand, South America, South Africa, and elsewhere in the Southern Hemisphere.[28] The mechanisms by which BZA resistance occurs will be discussed later (in chapter 11). To date, there is no convincing evidence of the emergence of drug resistance to the BZAs used for human STH infections. However, a systematic review conducted by Jennifer Keiser and Juerg Utzinger from the Swiss Tropical and Public Health Institute revealed that single-dose mebendazole currently exhibits a cure rate for human hookworm infection of only 15%, with egg count reductions for *N. americanus* hookworm infections ranging from 0 to 68%.[28] Thus, while albendazole is still generally effective for *N. americanus* hookworm infection, single-dose mebendazole can no longer be considered a standard treatment for hookworm infection. In addition, a single dose of albendazole is also often inadequate for effectively treating trichuriasis, prompting efforts to explore new drugs, such as moxidectin or oxantel, either alone or in combination with albendazole (discussed in further detail in chapter 11).[28] Today, the high rate of drug failures for single-dose mebendazole and high rates of STH infection in areas of high transmission, coupled with emerging evidence of BZA resistance in animal nematodes, have led to international calls for increased monitoring of the effectiveness of the BZAs and for the development of new-generation STH drugs. A new global initiative known as "Starworms" is under way to search for drug resistance to human STH infections.[28]

Given the enormous health and educational benefits of deworming, I believe that we should try to do everything possible to scale up the use of BZAs in developing countries. Recently, through support of the Gates Foundation, an ambitious project known as the Deworm3 Project is under way. Led by Dr. Judd Walson at the University of Washington in Seattle and the British Natural History Museum, Deworm3 convenes experts to conduct operational research around the potential for deworming drugs to interrupt the transmission of STH infections.[29] In this context, they are going beyond the public health and humanitarian benefits of deworming children. Deworm3 aspires to determine if stepped-up measures to aggressively mass-treat human populations with deworming drugs could lead to the elimination of ascariasis, hookworm infection, or trichuriasis. Their activities include the potential for treating both adults and children under the premise that adults also release thousands of STH eggs into the environment, a consequence of transmission dynamics studies led by Sir Roy Anderson at Imperial College London. So far, the evidence that frequent and periodic deworming can interrupt transmission in areas where STH infections are highly endemic (and with lots of heavy infections) is not strong, but Deworm3 aspires to develop an evidence base for this concept.

One concern I have is the possibility that the drugs themselves may not be adequate for deworming, especially in a single dose for hookworm infection and trichuriasis. Unfortunately, the absence of a commercial market for such drugs has hampered a substantive research and development effort on this front. In the meantime, Deworm3 will look at novel drug combinations as highlighted above. As an alternative or complementary approach to STH control there has been a concerted effort to develop a recombinant anthelmintic vaccine, which would prevent reinfection following deworming.[30] In chapter 11, I will discuss the efforts of our nonprofit product development partnership, known as the Human Hookworm Vaccine Initiative, to develop a new hookworm vaccine or even pan-anthelmintic vaccines as important anti-poverty measures.

SUMMARY POINTS   **"The Unholy Trinity"**

- STH infections are caused by intestinal worms, with *Ascaris* roundworms, *Trichuris* whipworms, and hookworms being the most common.
- Ascariasis, hookworm infection, and trichuriasis are the world's most common NTDs.
- STH infections are highly prevalent in sub-Saharan Africa, Asia, and the Americas, especially in areas where rural poverty overlaps with tropical environments and adequate rainfall.
- Children typically exhibit heavier STH infections with higher worm burdens than do adults.
- The STHs live for years in the gastrointestinal tract.
- In children, chronic STH infections impair physical growth and development as well as cognition, memory, and school performance. Therefore, STHs produce educational deficits as well as ill health.

These poverty-promoting features probably result from parasite-induced malnutrition.

- Hookworms cause malnutrition by producing intestinal blood loss, which leads to iron deficiency anemia, especially in children and pregnant women with low underlying iron reserves. The DALYs lost to hookworm infection rank the highest among the STHs.
- Global control of STH infections currently focuses on morbidity reductions through frequent and periodic deworming with BZAs. School-based deworming is being frequently emphasized in order to target at-risk children.
- The Deworm3 Project is under way to evaluate whether the use of deworming drugs could be expanded to a point where STH transmission is halted. However, there are both theoretical and actual concerns about BZA drug resistance; a human hookworm vaccine is under development.

Notes

1. The Greek derivation of "helminth" is from Faust et al., 1970, p. 251. References to the historical documentation of human ascariasis and hookworm infection are found in Cox, 2002; Grove, 1990; and Sherman, 2006, p. 349–352. The GBD 2019 (Global Burden of Disease Collaborative Network, 2020) estimates that 909 million people live with intestinal nematode infections (http://www.healthdata.org/results/gbd_summaries/2019/intestinal-nematode-infections-level-3-cause).

2. Prevalence numbers are modified from Bethony et al., 2006.

3. The impact of STHs on child growth and development and cognition is summarized (with references) in Bethony et al., 2006; and Crompton and Nesheim, 2002. The Crompton and Nesheim reference also describes the nutritional basis of these deficits.

4. The results of the nationwide parasite survey are summarized in Hotez et al., 1997. The survey was later repeated (but on a smaller scale) between 2002 and 2004 and demonstrated

that the incidence had since decreased dramatically in areas of economic development. The new Chinese prevalence numbers are reported in Li et al., 2010.

5. These observations are summarized in Hotez, 2002.

6. These observations are reported in Fenghua et al., 1998.

7. From de Silva et al., 2003.

8. The concept of the worm index of economic development is reported in Hotez and Herricks, 2015, and later updated in Kang et al., 2018. Factors underlying the relationship between STH infection and poverty are described in Raso et al., 2005; Holland et al., 1988; and Hotez, 2008.

9. The exact mechanisms by which STHs impair growth and development are still poorly understood. An excellent review of the literature is found in Crompton and Nesheim, 2002.

10. Miguel and Kremer, 2004.

36 *Forgotten People, Forgotten Diseases*

11. The link between STHs and impaired cognition and memory has been studied extensively by D. A. P. Bundy and his colleagues. Two important papers include Nokes et al., 1992; and Sakti et al., 1999. However, the mechanisms by which worms impair cognition and memory are not well understood.

12. Bleakley, 2007.

13. The GBD 2019 study (Global Burden of Disease Collaborative Network, 2020) is found online at the Institute for Health Metrics and Evaluation website, www.healthdata.org, and then by going to the GBD Results Tool. The information for soil-transmitted helminth infections, reported as "Intestinal nematode infections," can also be found at the following link: http://www.healthdata.org/results/gbd_summaries/2019/intestinal-nematode-infections-level-3-cause.

14. Original maps of the distribution of all three STH infections can be found in de Silva et al., 2003.

15. An excellent historical account of hookworm in the United States is found in Ettling, 1981. A more global perspective can be found in Farley, 2004. The material on finding hookworm in Alabama in the United States is in McKenna et al., 2017.

16. Humphreys, 2001, p. 110–111.

17. A summary of the impact of sanitation and health education on STH infections is found in Asaolu and Ofoezie, 2003; and Ziegelbauer et al., 2012. A study showing the absence of a relationship between wearing shoes and avoiding hookworm infection is found in Bethony et al., 2002.

18. Descriptions of the hookworm life cycle and the clinical manifestations of hookworm are found in Hotez et al., 2005; and Hotez et al., 2004.

19. Studies showing a high hookworm prevalence and intensity in coastal regions (regions with sandy soils) are in Mabaso et al., 2003; and Mabaso et al., 2004.

20. The biochemical mechanism by which hookworms digest blood is described in Williamson et al., 2004.

21. Some of these calculations can be found in Stoltzfus et al., 1997a; and Stoltzfus et al., 1997b. They are summarized in Crompton, 2000; and Crompton and Nesheim, 2002. The link between moderate and heavy hookworm infections and intestinal blood loss leading to iron deficiency anemia was recently confirmed in a systematic review; see Smith and Brooker, 2010.

22. The global importance of hookworm in pregnancy is highlighted in Bundy et al., 1995; and Christian et al., 2004. A recent systematic review confirmed the link between hookworm and anemia in pregnancy and provided the estimate of the number of hookworm-infected pregnant women in sub-Saharan Africa; see Brooker et al., 2008.

23. Data are summarized in Hotez et al., 2006b.

24. Information on the control of STH infections through deworming is summarized in Albonico et al., 2006; Brooker et al., 2004; and Crompton and Nesheim, 2002. The educational and economic benefits of deworming are discussed in Miguel and Kremer, 2004; Baird et al., 2016; and Bleakley, 2007. Some of these studies operate under the auspices of the innovative Abdul Latif Jameel Poverty Action Lab, based at Massachusetts Institute of Technology (www.povertyactionlab.org), and the Partnership for Child Development in London (http://www.imperial.ac.uk/partnership-for-child-development). Information on the "worm wars" can be found in Taylor-Robinson et al., 2019; and Majid et al., 2019.

25. Hotez et al., 2006a; and World Bank, 2003.

26. Savioli et al., 1992.

27. Estimates of the scope of global deworming and other NTD interventions are in World Health Organization, 2020.

28. The information about resistance is summarized in Brooker et al., 2004. The systematic review of the efficacy of drugs for STH infections is found in Keiser and Utzinger, 2008. Additional information about increasing the efficacy of deworming for trichuriasis by introducing new drugs is in Keller et al., 2020. The Starworms project is in Vlaminck et al., 2020.

29. This information is summarized in https://www.nhm.ac.uk/our-science/our-work/sustainability/deworm3/objectives.html.

30. Information about the Human Hookworm Vaccine Initiative is summarized in Hotez et al., 2016; and Bartsch et al., 2016.

References

Albonico M, Montresor A, Crompton DWT, Savioli L. 2006. Intervention for the control of soil-transmitted helminthiasis in the community. *Adv Parasitol* **61**:311–348.

Asaolu SO, Ofoezie IE. 2003. The role of health education and sanitation in the control of helminth infections. *Acta Trop* **86**:283–294.

Baird S, Hicks JH, Kremer M, Miguel E. 2016. Worms at Work: Long-run Impacts of a Child Health Investment. *Q J Econ* **131**:1637–1680.

Bartsch SM, Hotez PJ, Hertenstein DL, Diemert DJ, Zapf KM, Bottazzi ME, Bethony JM, Brown ST, Lee BY. 2016. Modeling the economic and epidemiologic impact of hookworm vaccine and mass drug administration (MDA) in Brazil, a high transmission setting. *Vaccine* **34**:2197–2206.

Bethony J, Brooker S, Albonico M, Geiger SM, Loukas A, Diemert D, Hotez PJ. 2006. Soil-transmitted helminth infections: ascariasis, trichuriasis, and hookworm. *Lancet* **367**:1521–1532.

Bethony J, Chen J, Lin S, Xiao S, Zhan B, Li S, Xue H, Xing F, Humphries D, Yan W, Chen G, Foster V, Hawdon JM, Hotez PJ. 2002. Emerging patterns of hookworm infection: influence of aging on the intensity of *Necator* infection in Hainan Province, People's Republic of China. *Clin Infect Dis* **35**:1336–1344.

Bleakley H. 2007. Disease and development: evidence from hookworm eradication in the American South. *Q J Econ* **122**:73–117.

Brooker S, Bethony J, Hotez PJ. 2004. Human hookworm infection in the 21st century. *Adv Parasitol* **58**:197–288.

Brooker S, Hotez PJ, Bundy DA. 2008. Hookworm-related anaemia among pregnant women: a systematic review. *PLoS Negl Trop Dis* **2**:e291.

Bundy DA, Chan MS, Savioli L. 1995. Hookworm infection in pregnancy. *Trans R Soc Trop Med Hyg* **89**:521–522.

Christian P, Khatry SK, West KP Jr. 2004. Antenatal anthelmintic treatment, birthweight, and infant survival in rural Nepal. *Lancet* **364**:981–983.

Cox FE. 2002. History of human parasitology. *Clin Microbiol Rev* **15:**595–612.

Crompton DW. 2000. The public health importance of hookworm disease. *Parasitology* **121**(Suppl)**:**S39–S50.

Crompton DW. 2001. *Ascaris* and ascariasis. *Adv Parasitol* **48:**285–375.

Crompton DW, Nesheim MC. 2002. Nutritional impact of intestinal helminthiasis during the human life cycle. *Annu Rev Nutr* **22:**35–59.

de Silva NR, Brooker S, Hotez PJ, Montresor A, Engels D, Savioli L. 2003. Soil-transmitted helminth infections: updating the global picture. *Trends Parasit* **19:**547–551.

Despommier DD, Gwadz RW, Hotez PJ, Knirsch C. 2006. *Parasitic Diseases*, 5th ed. Apple Tree Productions, New York, NY.

Ettling J. 1981. *The Germ of Laziness: Rockefeller Philanthropy and Public Health in the New South.* Harvard University Press, Cambridge, MA.

Farley J. 2004. *To Cast Out Disease: a History of the International Health Division of the Rockefeller Foundation (1913–1951).* Oxford University Press, Oxford, United Kingdom.

Faust EC, Russell PF, Jung RC. 1970. *Craig and Faust's Clinical Parasitology*, 8th ed. Lea & Febiger, Philadelphia, PA.

Fenghua S, Zhongxing W, Yixing Q, Hangqun C, Haichou X, Hainan R, Shuhua X, Bin Z, Hawdon JM, Zheng F, Hotez PJ. 1998. Epidemiology of human intestinal nematode infections in Wujiang and Pizhou counties, Jiangsu Province, China. *Southeast Asian J Trop Med Public Health* **29:**605–610.

Global Burden of Disease Collaborative Network. 2020. Global Burden of Disease Study 2019 (GBD 2019). GBD Results Tool. Institute for Health Metrics and Evaluation, Seattle, WA.

Grove DI. 1990. *A History of Human Helminthology.* CAB International, Wallingford, United Kingdom.

Holland CV, Taren DL, Crompton DW, Nesheim MC, Sanjur D, Barbeau I, Tucker K, Tiffany J, Rivera G. 1988. Intestinal helminthiases in relation to the socioeconomic environment of Panamanian children. *Soc Sci Med* **26:**209–213.

Hotez PJ. 2002. China's hookworms. *China Q* **172:**1029–1041

Hotez P. 2008. Hookworm and poverty. *Ann N Y Acad Sci* **1136:**38–44.

Hotez PJ, Bethony J, Bottazzi ME, Brooker S, Buss P. 2005. Hookworm: "the great infection of mankind". *PLoS Med* **2:**e67.

Hotez PJ, Brooker S, Bethony JM, Bottazzi ME, Loukas A, Xiao S. 2004. Hookworm infection. *N Engl J Med* **351:**799–807.

Hotez PJ, Bundy DA, Beegle K, Brooker S, Drake L, de Silva N, Montresor A, Engels D, Jukes M, Chitsulo L, Chow J, Laxminarayan R, Michaud CM, Bethony J, Correa-Oliveira R, Xiao SH, Fenwick A, Savioli L. 2006a. Helminth infections: soil-transmitted helminth infections and schistosomiasis, p 467–482. *In* Jamison DT, Breman JG, Measham AR, Alleyne G, Claeson M, Evans DB, Prabhat J, Mills A, Musgrove P (ed), *Disease Control Priorities in Developing Countries*, 2nd ed. Oxford University Press, Oxford, United Kingdom.

Hotez PJ, Zheng F, Long-qi X, Ming-gang C, Shu-hua X, Shu-xian L, Blair D, McManus DP, Davis GM. 1997. Emerging and reemerging helminthiases and the public health of China. *Emerg Infect Dis* **3:**303–310.

Hotez PJ, Herricks JR. 2015. Helminth elimination in the pursuit of sustainable development goals: a "worm index" for human development. *PLoS Negl Trop Dis* **9:**e0003618.

Hotez PJ, Molyneux DH, Fenwick A, Ottesen E, Ehrlich Sachs S, Sachs JD. 2006b. Incorporating a rapid-impact package for neglected tropical diseases with programs for HIV/AIDS, tuberculosis, and malaria. *PLoS Med* **3:**e102.

Hotez PJ, Strych U, Lustigman S, Bottazzi ME. 2016. Human anthelminthic vaccines: rationale and challenges. *Vaccine* **34:**3549–3555.

Humphreys M. 2001. *Malaria, Poverty, Race, and Public Health in the United States.* The Johns Hopkins University Press, Baltimore, MD.

Kang S, Damania A, Majid MF, Hotez PJ. 2018. Extending the global worm index and its links to human development and child education. *PLoS Negl Trop Dis* **12:**e0006322.

Keiser J, Utzinger J. 2008. Efficacy of current drugs against soil-transmitted helminth infections: systematic review and meta-analysis. *JAMA* **299:**1937–1948.

Keller L, Palmeirim MS, Ame SM, Ali SM, Puchkov M, Huwyler J, Hattendorf J, Keiser J. 2020. Efficacy and safety of ascending dosages of moxidectin and moxidectin-albendazole against *Trichuris trichiura* in adolescents: a randomized controlled trial. *Clin Infect Dis* **70:**1193–1201.

Li T, He S, Zhao H, Zhao G, Zhu XQ. 2010. Major trends in human parasitic diseases in China. *Trends Parasit* **26:**264–270.

Mabaso ML, Appleton CC, Hughes JC, Gouws E. 2003. The effect of soil type and climate on hookworm (*Necator americanus*) distribution in KwaZulu-Natal, South Africa. *Trop Med Int Health* **8:**722–727.

Mabaso ML, Appleton CC, Hughes JC, Gouws E. 2004. Hookworm (*Necator americanus*) transmission in inland areas of sandy soils in KwaZulu-Natal, South Africa. *Trop Med Int Health* **9:**471–476.

Majid MF, Kang SJ, Hotez PJ. 2019. Resolving "worm wars": an extended comparison review of findings from key economics and epidemiological studies. *PLoS Negl Trop Dis* **13:**e0006940.

McKenna ML, McAtee S, Bryan PE, Jeun R, Ward T, Kraus J, Bottazzi ME, Hotez PJ, Flowers CC, Mejia R. 2017. Human intestinal parasite burden and poor sanitation in rural Alabama. *Am J Trop Med Hyg* **97:**1623–1628.

Miguel E, Kremer M. 2004. Worms: identifying impacts on education and health in the presence of treatment externalities. *Econometrica* **72:**159–217.

Nokes C, Grantham-McGregor SM, Sawyer AW, Cooper ES, Bundy DA. 1992. Parasitic helminth infection and cognitive function in school children. *Proc Biol Sci* **247:**77–81.

Raso G, Utzinger J, Silué KD, Ouattara M, Yapi A, Toty A, Matthys B, Vounatsou P, Tanner M, N'Goran EK. 2005. Disparities in parasitic infections, perceived ill health and access to health care among poorer and less poor schoolchildren of rural Côte d'Ivoire. *Trop Med Int Health* **10**:42–57.

Sakti H, Nokes C, Hertanto WS, Hendratno S, Hall A, Bundy DA, Satoto. 1999. Evidence for an association between hookworm infection and cognitive function in Indonesian school children. *Trop Med Int Health* **4**:322–334.

Savioli L, Bundy D, Tomkins A. 1992. Intestinal parasitic infections: a soluble public health problem. *Trans R Soc Trop Med Hyg* **86**:353–354.

Sherman IW. 2006. *The Power of Plagues.* ASM Press, Washington, DC.

Smith JL, Brooker S. 2010. Impact of hookworm infection and deworming on anaemia in non-pregnant populations: a systematic review. *Trop Med Int Health* **15**:776–795.

Stoll NR. 1962. On endemic hookworm, where do we stand today? *Exp Parasitol* **12**:241–252.

Stoltzfus RJ, Chwaya HM, Tielsch JM, Schulze KJ, Albonico M, Savioli L. 1997a. Epidemiology of iron deficiency anemia in Zanzibari schoolchildren: the importance of hookworms. *Am J Clin Nutr* **65**:153–159.

Stoltzfus RJ, Dreyfuss ML, Chwaya HM, Albonico M, Albonico M. 1997b. Hookworm control as a strategy to prevent iron deficiency. *Nutr Rev* **55**:223–232.

Taylor-Robinson DC, Maayan N, Donegan S, Chaplin M, Garner P. 2019. Public health deworming programmes for soil-transmitted helminths in children living in endemic areas. *Cochrane Database Syst Rev* **9**:CD000371.

Vlaminck J, Cools P, Albonico M, Ame S, Chanthapaseuth T, Viengxay V, Do Trung D, Osei-Atweneboana MY, Asuming-Brempong E, Jahirul Karim M, Al Kawsar A, Keiser J, Khieu V, Faye B, Turate I, Mbonigaba JB, Ruijeni N, Shema E, Luciañez A, Santiago Nicholls R, Jamsheed M, Mikhailova A, Montresor A, Mupfasoni D, Yajima A, Ngina Mwinzi P, Gilleard J, Prichard RK, Verweij JJ, Vercruysse J, Levecke B. 2020. Piloting a surveillance system to monitor the global patterns of drug efficacy and the emergence of anthelmintic resistance in soil-transmitted helminth control programs: a Starworms study protocol. *Gates Open Res* **4**:28.

Williamson AL, Lecchi P, Turk BE, Choe Y, Hotez PJ, McKerrow JH, Cantley LC, Sajid M, Craik CS, Loukas A. 2004. A multi-enzyme cascade of hemoglobin proteolysis in the intestine of blood-feeding hookworms. *J Biol Chem* **279**:35950–35957.

World Health Organization. 2020. Schistosomiasis and soil-transmitted helminthiases: numbers of people treated in 2019. *Wkly Epidemiol Rec* **95**:629–640.

Ziegelbauer K, Speich B, Mäusezahl D, Bos R, Keiser J, Utzinger J. 2012. Effect of sanitation on soil-transmitted helminth infection: systematic review and meta-analysis. *PLoS Med* **9**:e1001162.

3

Schistosomiasis (Snail Fever) and the Food-borne Trematodiases

Welcome the sunrise, come under the stars, work from dusk to daybreak
Our strength is boundless, our enthusiasm is redder than fire . . .
Be the river like a sea, drained clean it shall be . . .
Empty the rivers to wipe out the snails, resolutely to fight the big-belly disease.

WEI WEN-PO, 1958

Schistosomiasis

Schistosomiasis is a waterborne parasitic worm infection affecting 140 million people in low- and middle-income countries (LMIC)s.[1] More than 90% of the cases occur in sub-Saharan Africa, and today schistosomiasis is one of the most common parasitic infections on the African continent. Additional cases also occur in the Americas (primarily in Brazil and Venezuela) and the Middle East (Yemen and Iraq). A unique form of Asian schistosomiasis transmitted from animals (water buffalo and pigs) is still a significant illness in Southeast Asia, especially in the Philippines, and it still remains endemic to a few areas of China (Fig. 3.1).[1] Unlike the soil-transmitted helminths, which are nematodes, the schistosomes are a type of flatworm, also known as a *trematode* or *fluke*. As adult worms, schistosomes live in the bloodstream (and are known also as *blood flukes*), where they release eggs armed with a spine that produce serious inflammation and disease either in the urogenital tract or in the intestine and liver, depending on the particular species of parasite. Humans acquire schistosomiasis by direct contact with the larval stages (known as *cercariae*) that swim in freshwater. Prior to becoming cercariae, the immature developing and reproducing forms of these parasites spend a part of their life history living in various species of aquatic snails.

Forgotten People, Forgotten Diseases: The Neglected Tropical Diseases and Their Impact on Global Health and Development, Third Edition. Peter J. Hotez.
© 2022 American Society for Microbiology. DOI: 10.1128/9781683673903.ch03

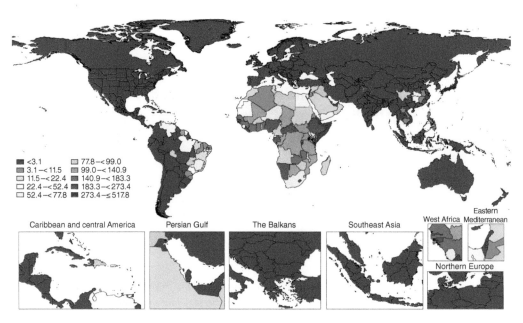

Figure 3.1 Distribution of schistosomiasis, worldwide. (From Global Burden of Disease Collaborative Network, 2020 [http://www.healthdata.org/results/gbd_summaries/2019/schistosomiasis-level-3-cause].)

Few other infectious diseases have influenced history more than schistosomiasis has. Joshua's curse and the abandonment of Jericho have been attributed to the disease.[2] Schistosomiasis was well known to the ancient Egyptians, and it was a major scourge of Napoleon's army during its disastrous military campaign in Egypt between 1798 and 1801.[2] However, some of the most dramatic examples of the schistosome's historical impact draw from postrevolutionary China during the second half of the 20th century. Shortly after the Chinese revolution in 1949, Mao Zedong (Mao Tse-tung) began planning a massive amphibious assault to bring Taiwan (then known as Formosa) under Communist control. While undergoing rigorous water training around the eastern tributaries of the Yangtze River, tens of thousands of People's Liberation Army (PLA) troops were exposed to the infectious cercariae of *Schistosoma japonicum*, the parasitic larvae that live in the water and are shed by snails living in the wet mud along the riverbanks. Within weeks, tens of thousands of soldiers experienced the early acute phase of schistosomiasis, a condition known as Katayama syndrome, which can last for weeks and is characterized by fever, extreme fatigue, muscle pains, and coughing.[3] As a result, the launch of the amphibious attack was delayed just long enough for the U.S. Seventh Fleet to enter the Strait of Formosa and abort a Communist takeover. These events are described in a 1959 *Harper's Magazine* article titled "The Blood Fluke That Saved Formosa."[4]

Undoubtedly, the derailment of the PLA by a blood fluke transmitted by a snail made a deep impression on the Chinese Communist Party leadership. Beginning in 1955 on the order of Chairman Mao, a special nine-man committee on schistosomiasis was established and a seven-year plan for the eradication of the disease was launched.[5] In 1958, beginning with the Great Leap Forward, millions of peasants were mobilized to the Yangtze River valley, where they drained the marshes and buried the "devil snails" under dirt or, in some cases, removed the snails individually with sticks.[6] The quotation in the opening of this chapter is from a 1958 paper written by Wei Wen-po (the second-in-command of the nine-man committee) that appeared in the *Chinese Medical Journal* under the title "The People's Boundless Energy during the Current Leap Forward."[7] Later, chemical agents to kill the snails were applied. Such low-technology interventions directed at snail control had an important impact and helped to reduce the overall prevalence of schistosomiasis from 12 million or more cases before the revolution to approximately 1.6 million cases by the mid-1980s.[6] Mao himself wrote a poem about these successes entitled "Farewell to the God of Plague." Later, there were renewed concerns that the completion of the massive Three Gorges Dam across the Yangtze River in 2006 might establish new ecological niches for snails to thrive and reproduce. This could cause a rebound in the number of cases of schistosomiasis, just as other dam projects on the Nile at Aswan and the Senegal and Volta Rivers in sub-Saharan Africa have caused massive increases in snail populations and reemergence of the disease.[8] However, a more recent assessment determined that the Three Gorges Dam may have altered water levels in unique ways to actually lower schistosomiasis transmission.[8] This, together with stepped-up direct control measures, is showing promise for eliminating schistosomiasis as a public health threat in China.

In Egypt and elsewhere in East Africa, schistosomiasis has had an equally important historical impact. Schistosome eggs have even been recovered from Egyptian mummies dating from the 20th dynasty, around 1000 BCE.[2] During the first half of the 20th century, more than half of the populations in some rural villages of the Nile Delta were infected with either *Schistosoma haematobium* or *Schistosoma mansoni*. Both forms of schistosomiasis were considered the scourge of the *fellaheen*, Egypt's peasant agricultural laborers.[9] It was said that hematuria, the bloody urine that results from *S. haematobium* infection, was so common among Egyptian children and adolescent boys that it was considered a form of male menstruation.[9] The initial efforts to control schistosomiasis in Egypt and in neighboring Sudan (where the Gezira Scheme, one of the world's largest irrigation projects, is located near the confluence of the Blue Nile and White Nile) were organized by the British. Their approach included multiple injections of toxic antimony compounds to cure the disease in humans, together with widespread use of molluscicides, snail-destroying chemical agents distributed in the environment.[10] During the 1960s, Bayluscide (known generically as niclosamide), a molluscicide developed by Bayer, was widely used but later largely abandoned because of price increases combined with a realization that it caused significant damage to fish and other wildlife. Subsequently, during the 1970s, newly discovered drugs, such as ambilhar and hycanthone, were used in large-scale treatment of schistosomiasis, but they too fell into disuse because of their toxicity and some unexplained deaths.[10]

Complicating Egypt's schistosomiasis problem was the fact that many schistosome-infected individuals also contracted hepatitis C when contaminated needles and equipment were used to administer tartar emetic, an older and injectable antiparasitic drug that is now seldom used.[10] Patients with schistosome and hepatitis C coinfections often suffered from a severe, progressive form of fibrotic liver disease.[10] In 1977, one of Egypt's most beloved and celebrated singers and artists, Abdel Halim Hafez, died of liver complications from schistosomiasis at the age of 48. He was revered throughout the Arab world, and known as the "King of Emotions and Feelings."[10] At his funeral march, girls and women were said to have committed suicide by jumping from their balconies.[10] Finally, during the 1990s, widespread use of the anthelmintic drug praziquantel in a 14-year-long mass drug administration (MDA) campaign, supported in part by the World Bank and the U.S. Agency for International Development (USAID), reduced the overall prevalence of schistosomiasis in Egypt to less than 10% of the population.[10] In China, a 10-year World Bank initiative also supporting MDA of praziquantel resulted in similar dramatic reductions of schistosomiasis in the Yangtze River valley, recently hastened by the Three Gorges Dam completion.[10]

Despite the successes in China and Egypt, today schistosomiasis still rivals hookworm infection as among the most important helminth infections of humans. Urogenital schistosomiasis caused by *S. haematobium* is responsible for approximately 63% of the cases worldwide, while an intestinal and hepatic form caused by *S. mansoni* accounts for another 35% (Table 3.1).[11] Less than 1% of the global burden of schistosomiasis results from the Cold War warrior *S. japonicum*, the major cause of Asian schistosomiasis.[1] Several other minor species make up the remaining 1% of the cases. Almost all of the people infected with schistosomiasis live in Africa, with up to 30 African nations each harboring hundreds of thousands or more cases (Fig. 3.1).[1] Most of these cases are caused by *S. haematobium*, followed by *S. mansoni*. Outside Africa, only the nations of Brazil and possibly Iraq or Yemen have equivalent numbers of cases of schistosomiasis.[1] In the Americas, *S. mansoni* infection was likely introduced by a flourishing slave trade with sub-Saharan Africa that began in the 1600s.[12]

Humans contract schistosomiasis through freshwater contact with free-swimming cercariae (Fig. 3.2). Therefore, poor rural populations whose everyday

Table 3.1 The major human schistosomes

Species	Length as adult	Disease	Percentage of cases worldwide	Major geographic locations
Schistosoma haematobium	10–20 mm (0.4–0.8 in.)	Urogenital schistosomiasis	63%	Africa and Middle East
Schistosoma mansoni	6–17 mm (0.2–0.7 in.)	Intestinal and hepatic schistosomiasis	35%	Africa, Middle East, and the Americas
Schistosoma japonicum and *Schistosoma mekongi*	12–26 mm (0.5–1.0 in.)	Intestinal and hepatic schistosomiasis	1%	China, the Philippines, and Southeast Asia

activities involve fishing, bathing, or swimming in schistosome-contaminated waters or working in agricultural areas irrigated by contaminated waters are at the highest risk of infection.[10,13] By some estimates, almost 800 million people in developing countries live in proximity to either irrigated agricultural fields or dam reservoirs, where the risk of acquiring schistosomiasis is the highest.[1]

Schistosome cercariae have a forked tail that allows them to swim and ultimately to directly penetrate human skin. Following skin penetration, the cercariae lose their tail and undergo a number of biochemical changes that allow them to resist attack by the human immune system. These larval schistosomes (also known as *schistosomulae*) migrate through the lungs and over a period of approximately 1 to 2 months make their way to the portal vein of the liver, where they mature into adult male and female schistosomes.[13] The paired worms ultimately migrate to their final destination, which for *S. haematobium*, the cause of urogenital schistosomiasis, is the small veins that drain the bladder and other pelvic organs, while *S. mansoni* and *S. japonicum* live in the mesenteric veins that drain the intestine.[13] While living in the blood vessels, the adult schistosomes feed on blood, breaking down the hemoglobin components by using enzymes similar to those found in hookworms. Through evolution, the adult male and female schistosomes living

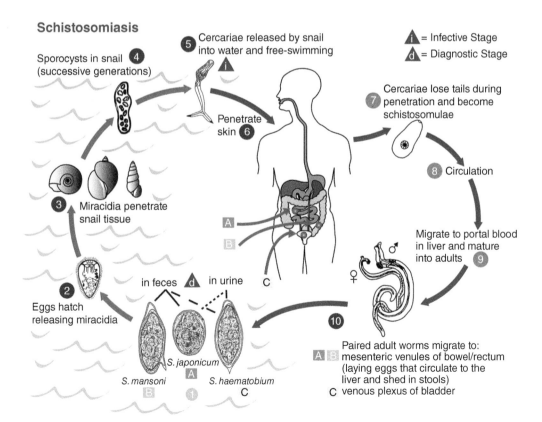

Figure 3.2 Life cycle of human schistosomes. (Modified from CDC-PHIL [ID#3417]/Alexander de Silva, PhD/Melanie Moser.)

in the bloodstream have developed remarkable mechanisms for masking their identity, including the accretion of host molecules on their surface. In this way, the schistosomes avoid attack by antibodies and cells of the human immune system.

The female worms subsequently produce hundreds of eggs daily.[13] In order to continue the schistosome life cycle, the eggs ultimately require a mechanism to exit from the body. In the case of hookworm infection and other soil-transmitted helminth infections in which the parasites live in the gastrointestinal tract, the feces provide a straightforward path for the eggs to escape into the environment. In contrast, schistosome eggs have a more formidable challenge because they are present in the human blood vessels. As shown in Fig. 3.3, schistosome eggs are equipped with an ominous-looking spine that permits them to bore their way through the blood vessels and then into either the bladder or intestine from the outside. Through a combination of mechanical boring and the release of tissue-dissolving enzymes, the eggs gain access to either the lumen of the bladder or the intestine; they exit the body in urine or feces, respectively. When deposited in freshwater, the eggs live for about a week. They hatch and give rise to free-swimming ciliated forms known as *miracidia*, which seek out a suitable snail species. Upon entry into the appropriate snail, each miracidium can give rise to multiple progeny through asexual reproduction. Eventually, these progeny develop into cercariae that exit the snail.[13]

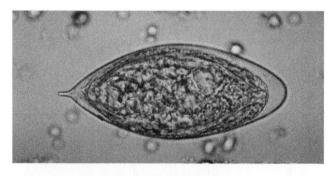

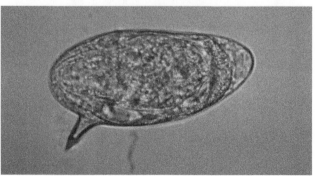

Figure 3.3 Spined eggs of *S. haematobium* (top) and *S. mansoni* (bottom). (Images from CDC PHIL [ID#4843/CDC/Dr. D.S. Martin, 1966 and 4841/CDC, 1979].)

Unfortunately, egg migration through human body tissues is not an efficient process, so many eggs become trapped in either the bladder and reproductive organs (*S. haematobium*) or intestine and liver (*S. mansoni* and *S. japonicum*). The trapped eggs cause mechanical damage and the rupture of small blood vessels, which lead to bleeding and the appearance of blood in either urine or feces. The eggs also trigger an inflammatory response composed of masses of human white cells and other host-derived components (known as *granulomas*), which can obstruct urine or blood flow. Because schistosomes can live for years in the small veins of the bladder and intestine, their constant release of eggs is associated with chronic blood loss causing anemia, as well as damage to the bladder, kidneys, and reproductive organs (*S. haematobium*) or to the intestine and liver (*S. mansoni*). The combination of long-standing anemia, inflammation, and target organ damage causes growth impairments, undernutrition, and cognitive delays in children, as well as chronic abdominal pain, exercise intolerance, poor school performance, and reduced work capacity.[13,14] Anemia and chronic inflammation also partially account for the developmental delays occurring in chronic pediatric soil-transmitted helminth infections.

Sub-Saharan Africa bears the overwhelming burden of disease caused by the schistosomes and almost all of the cases of *S. haematobium* infection. Figure 3.4 shows several children, each holding a cup of his or her reddened urine (hematuria; schistosomiasis is sometimes known locally as "red-water fever," as well as "snail fever"). Endemic hematuria from schistosomiasis was first recorded by a Western physician in 1798, by J. Renoult, a French army surgeon who accompanied Napoleon on his invasion of Egypt.[2] Just as hookworm causes chronic blood loss in the intestine, leading to anemia, the chronic blood loss resulting from *S. haematobium* egg deposition in the bladder is a significant cause of anemia in Africa, particularly among adolescent children, who on average harbor larger numbers of schistosomes than do any other age group.[13,14] We saw previously how chronic intestinal blood loss and anemia resulting from hookworm were associated with physical and mental delays in children. For urogenital schistosomiasis, the anemia results not only from urinary blood loss but also from other factors, including chronic inflammation. These processes also contribute to inhibition of physical and mental growth for the child. Other important contributors to the morbidity of urogenital schistosomiasis are the inflammatory granulomas that develop in the bladder. Severe bladder wall pathology occurs in an estimated 18 million people in Africa.[11] When the bladder granulomas coalesce, they can obstruct urine flow and cause distension of the ureter and kidneys. This condition is known as hydronephrosis and occurs in up to 20 million people in Africa as a result of *S. haematobium* infection.[11] Long-standing hydronephrosis can lead to renal failure; this progression probably accounts for a significant number of the thousands of deaths annually from schistosomiasis.[11] Another major consequence of chronic *S. haematobium* infection is its ability to predispose people to acquiring an unusual form of bladder cancer. Whereas most bladder carcinomas in the industrialized world are adenocarcinomas, *S. haematobium*

Figure 3.4 Children in Niger with hematuria. (Photo courtesy of Juerg Utzinger, Swiss Tropical Institute.)

infection is associated with a unique squamous cell carcinoma of the bladder. It is conjectured that the schistosome granulomas in the bladder may increase the exposure of the bladder epithelium to environmental carcinogens.[13] Alternatively, the recent sequencing of the *S. haematobium* genome has also allowed scientists to search for carcinogenic molecules released by schistosome eggs.[15] Overall, however, most investigators believe that these molecules by themselves do not cause cancer but require a second host component such as smoking or nitrosamines in the diet, or possibly genetic factors, or even second infections such as with human papillomavirus.[15]

Still another important component of the morbidity of *S. haematobium* infection is involvement of the female reproductive tract. Up to 75% of women with this form of schistosomiasis develop fibrotic lesions known as "sandy patches" in the vulva, vagina, cervix, and uterus.[16] Such lesions and sandy patches are associated with genital bleeding, pain on intercourse, infertility, and even clinical depression.[16] Given that *S. haematobium* infection potentially affects roughly two-thirds of the more than 100 million schistosome-infected individuals, this complication, known as female genital schistosomiasis (FGS) or female urogenital schistosomiasis (FUGS), may actually be one of the most common gynecologic conditions on the African continent.[16] Beyond the direct clinical effects are the social determinants of FGS. They include, for example, stigmatization when FGS is mistaken as a sexually transmitted infection. Adolescent girls may be accused of promiscuity by family members or even health workers,[16] or the condition may become a source of marital discord. It has also been linked to clinical depression. These are among the ways that schistosomiasis, like many neglected tropical diseases (NTDs), disproportionately affects girls and women. Moreover, FGS lesions have been linked to a 3- and 4-fold increase in the transmission of HIV/AIDS in Zimbabwe and Tanzania, respectively, and presumably elsewhere in sub-Saharan Africa.[16] From these and similar studies, there is considerable interest in looking at schistosomiasis control and prevention efforts in terms of their impact on HIV transmission in rural areas of Africa.[16] The opportunity of exploring schistosomiasis elimination efforts as a back-door strategy for HIV/AIDS control is examined further in chapter 10.

S. mansoni is also a significant cause of intestinal and liver pathology in sub-Saharan Africa and Brazil. The presence of eggs and granulomas in the intestinal wall, usually of the large intestine and rectum, is associated with bleeding and diarrhea, as well as with loss of appetite, while liver granulomas can cause inflammation and liver enlargement (hence the term "big-belly disease").[13] Chronic schistosomiasis of the liver can progress to fibrosis, splenic enlargement, and bleeding from the esophagus. Occasionally, severe bleeding can result in death. An estimated 8.5 million cases of liver disease result from S. mansoni infection in sub-Saharan Africa.[11]

Although snail control was the major tool used to effect an almost 10-fold reduction in the prevalence of S. japonicum infection in China, this approach has not been an efficient or very effective means for controlling either S. haematobium or S. mansoni in Africa and elsewhere. Failed efforts to control snails environmentally reflect the unique biology of the snail intermediate hosts of S. haematobium and S. mansoni as well as the epidemiology of these forms of schistosomiasis. In addition, the environmental toxicities of molluscicides preclude their constant and widespread use. Even in China, snail control has not been front and center in terms of eliminating the last remaining cases in the Yangtze River valley.[6] Instead, a key approach for Asian schistosomiasis in China has relied on animal control. Whereas schistosomiasis in Africa, the Middle East, and Brazil is almost exclusively an infection of humans, Asian schistosomiasis has important animal reservoirs harboring the infection. They include water buffalo and pigs, which shed millions of schistosome eggs into the environment. Accordingly, the Chinese government has been gradually replacing water buffalo with tractors as a means to interrupt schistosomiasis transmission. This approach has been especially effective when combined with mass treatment with praziquantel.

We saw previously (in chapter 2) how MDA of benzimidazole anthelmintics is beginning to have an impact on the global burden of disease caused by soil-transmitted helminth infections. Today, the most effective means of controlling schistosome infections is MDA of praziquantel to affected and at-risk human populations. Developed at Bayer in Germany, praziquantel was shown to be effective against all forms of schistosomiasis in multicenter trials, as evidenced by schistosome egg reductions or outright cures.[10] Treatment of schistosome-infected children with praziquantel results in a number of health benefits similar to those experienced by children with soil-transmitted helminth infections who are treated with either albendazole or mebendazole, including improvements in growth and physical fitness, as well as reductions of anemia. In addition, multiple treatments with praziquantel can sometimes reverse both urinary tract and liver pathology, particularly if children receive these treatments early in their lives. Building on these breakthrough medical observations were the low-cost synthesis and production methods for manufacturing generic praziquantel by the Korean pharmaceutical company Shin Poong. These manufacturing improvements have dramatically reduced the price of the drug, possibly by as much as 90%.[10] As a result, by the 1990s, it became possible to launch praziquantel mass treatment programs for schistosomiasis in several middle-income countries, including Brazil, China, Egypt, Morocco, the Philippines, Saudi Arabia, and Tunisia, as well as Puerto Rico.[10,13,17] Several of these initiatives were funded by the World Bank. Praziquantel MDA in these countries has been largely successful, particularly in Morocco and Puerto Rico, where the disease was eliminated (zero transmission) partly because of parallel poverty reduction measures that were enacted.[10]

Despite the promise that praziquantel offers for controlling human schistosomiasis, up until recently it has not been made widely available.[17] Previously (in chapter 2), we learned how in 2001, the 54th World Health Assembly adopted a resolution calling for treatment with a benzimidazole anthelmintic drug of at least 75% of school-age children at risk of acquiring schistosomiasis and soil-transmitted helminth infections by 2010. Based on the experiences and successes of MDA of praziquantel, the resolution was written to include this drug along with either albendazole or mebendazole for the treatment of school-age children. In many regions of the developing world, especially in sub-Saharan Africa and Brazil, it is common for children to be polyparasitized with both soil-transmitted helminths (especially hookworm) and schistosomes,[18] so they could benefit from combination therapy with a benzimidazole anthelmintic and praziquantel. Unfortunately, by the 2010 proposed target date the World Health Organization (WHO) estimated that only about 33 million of the 236 million children at risk for schistosomiasis actually received praziquantel in that year, just a little over 10%.[19] The low drug coverage results from the absence of political will in expanding MDA, together with limited availability of praziquantel. In response, in January 2012, the Bill & Melinda Gates Foundation partnered with the major pharmaceutical companies, the World Bank, the WHO and other international agencies, and key nongovernmental organizations in a London Declaration to reaffirm support for control of NTDs through MDA.[20] The announcement coincided with a commitment by the German company Merck KGaA to dramatically ramp up praziquantel donations. I have met on several occasions the Merck KGaA CEO, Stefan Oschmann, who has a deep personal commitment to this project. He is an enlightened CEO who is passionate about global health. These activities are also linked to the formation of the Global Schistosomiasis Alliance (GSA) for raising awareness and promoting the availability of a pediatric formulation of praziquantel to also treat younger children.[20]

A key implementing partner that provides technical assistance to the African health ministries in order to facilitate MDA of praziquantel (usually together with albendazole for soil-transmitted helminth coinfections) is the Schistosomiasis Control Initiative (SCI), established by Alan Fenwick. It was based initially at Imperial College London but now exists as a free-standing organization in London, where it is headed by Dr. Wendy Harrison (https://schistosomiasiscontrolinitiative.org/). Previously, Fenwick had acquired wide-ranging experience in large-scale control of schistosomiasis through his leadership in World Bank- and USAID-funded projects in Egypt and Sudan. Working in collaboration with ministries of health in almost a dozen sub-Saharan African countries as well as Yemen, together with the WHO, the SCI facilitates once-yearly praziquantel treatments using donated praziquantel from Merck KGaA. The SCI operates through algorithms based on the schistosomiasis prevalence determined in local school surveys. The number of praziquantel tablets administered to each individual is based on an easy-to-use height pole that approximates a dose of 40 mg of drug per kg of body weight, and the delivery and distribution of praziquantel are carried out in health centers, in schools, or through community-based drug distributors.[10]

To date, the SCI and its partners have made some impressive public health gains, including some reductions in either the prevalence or intensity of schistosomiasis in several African nations.[21] By the end of 2018, the SCI estimated, more than 200 million people had been treated in Africa and the Middle East. Such successes and the prospects of expanded MDA campaigns and ongoing donations of praziquantel have led some investigators to suggest that global schistosomiasis elimination is a potentially attainable goal.[22] Indeed, in 2012 the World Health Assembly adopted a resolution that affirmed the feasibility of elimination.[22] Here the term "elimination" refers to reduction in the prevalence of human schistosomiasis to the point where transmission has been interrupted, but unlike eradication, elimination may require ongoing implementation of public health control measures. In 2019, the WHO estimated that 86 million school-age children received praziquantel.[22]

There are a number of important but still unanswered questions about the impact of praziquantel MDA on sustainable poverty reduction in developing countries. Current population-based approaches with praziquantel are focused on reducing the morbidity of this condition in areas where high-intensity infections and serious urogenital, intestinal, and liver morbidities are present. However, it is unclear whether mass treatment will also reduce transmission and prevent reinfection in the community.[23] This uncertainty is a serious concern, since ongoing low levels of schistosomiasis reinfection could lead to subtle but persistent clinical problems, including anemia, growth stunting, and diminished productive capacities.[23] In addition, there are potential worries that frequent and periodic use of praziquantel could lead to anthelmintic drug resistance. These issues have prompted some nascent efforts to develop alternative or complementary control tools for schistosomiasis, just as they have for human hookworm infection. They include exploration of new drug development, including the antimalarial artemisinin artemether, which also exhibits antischistosomal properties.[24] Still another approach has been efforts to develop a recombinant protein vaccine for schistosomiasis, such that several candidate vaccines have entered into phase 1 and 2 clinical trials, including a vaccine developed at our Texas Children's Center for Vaccine Development in Houston (discussed in chapter 11).[25] However, for now the major approach to schistosomiasis control still relies on MDA of praziquantel or of praziquantel and albendazole or mebendazole in areas of schistosomiasis and soil-transmitted helminth coinfection. We will further explore (in chapter 10) how these treatments might link with MDA for other NTDs, including lymphatic filariasis, onchocerciasis, and trachoma.

Food-borne Trematodiases

There is a group of trematode (also known as fluke) infections related to schistosomiasis that are of widespread public importance. Whereas schistosomiasis is transmitted by direct contact from cercariae released by snails in freshwater, these fluke infections are acquired by ingesting the encysted larval stages found in contaminated food, such as fish (liver or bile duct fluke) or freshwater crabs (lung fluke). For that reason, this group of infections is referred to as food-borne trematodiases. Approximately 34 million people are infected with these flukes globally, with most of the infections found in East Asia and a remaining focus in Ecuador, Peru, and Bolivia in the Americas. Among the most devastating food-borne trematodiases are a group of liver fluke infections—clonorchiasis in China and Korea, opisthorchiasis in Thailand and Laos, and a second form of opisthorchiasis in Siberia and Kazakhstan. Chronic infections with clonorchiasis and opisthorchiasis lead to an unusual form of bile-duct carcinoma, now a major cause of cancer in Asia.[26] Some exciting work is under way to uncover the molecular basis of carcinogenesis resulting from liver fluke infection. It results from consuming raw fish, usually in the form of an Asian fish paste contaminated with the larval stages, known as encysted metacercariae. Paragonimiasis is another major fluke disease, causing chronic lung infection and hemoptysis (coughing of blood) in East Asia and the Americas. It results from consuming freshwater crab contaminated with metacercariae. In South America, locally prepared ceviche is a cause. There is also a form in the south-central United States resulting from accidental or intentional ingestion of raw crawdads.

SUMMARY POINTS   **Schistosomiasis**

- Significant impact on the modern history of China and Egypt
- One of the most common and clinically important human helminth infections
- Schistosomiasis is a waterborne infectious disease transmitted by a snail intermediate host.
- Approximately 140 million cases worldwide, with more than 90% of the infections in Africa

- Children and adolescents are at highest risk. All forms of chronic schistosomiasis are associated with anemia, undernutrition and growth impairments, poor school performance, and reduced productive capacity.
- Urogenital schistosomiasis caused by *S. haematobium* accounts for approximately two-thirds of the schistosome infections in Africa. This form is responsible

for hematuria, female genital schistosomiasis, and squamous cell carcinoma of the bladder. Genital lesions increase the risk of HIV/AIDS transmission.

- Intestinal and liver schistosomiasis caused by S. mansoni accounts for approximately one-third of the cases in Africa and all of the cases of schistosomiasis in Brazil. This form is responsible for bloody

diarrhea, abdominal pain, and liver involvement (hepatomegaly and fibrosis).

- The major approach to control of schistosomiasis today is mass drug administration of praziquantel. It has led to the near elimination of schistosomiasis in some middle-income countries and to significant morbidity reductions in some sub-Saharan African countries.

Notes

1. The 140 million case number is from the GBD 2019 (Global Burden of Disease Collaborative Network, 2020); see http://www.healthdata.org/results/gbd_summaries/2019/schistosomiasis-level-3-cause. However, not everyone agrees with this assessment, with Charles King (King, 2010) determining that current estimates do not consider individuals who go undiagnosed with schistosomiasis because of variabilities in parasite egg excretion.

2. Historical accounts of schistosomiasis are found in Hulse, 1971; Hotez et al., 2006; Cox, 2002; Fenwick et al., 2006; and Grove, 1990.

3. Ross et al., 2007.

4. This incident is described in Kernan, 1959. Because the article is not easy to find, readers may also wish to read a summary of this information in Farley, 1991.

5. Information is contained in Lampton, 1974.

6. Information is found in Horn, 1969, p. 94–106; Hotez, 2002; Utzinger et al., 2005; and Farley, 1991, p. 201–215.

7. Wei, 1958, quoted in Farley, 1991, p. 206.

8. The impact of dam construction projects on the emergence of human schistosomiasis is described in Fenwick, 2006; Hotez et al., 1997; and Zhou et al., 2016.

9. Farley, 1991, p. 45–54, 188–200.

10. Information is contained in Fenwick et al., 2006. Information about schistosomiasis and hepatitis C coinfections in Egypt may be found in Rao et al., 2002; El-Sabah et al., 2011; and Sanghvi et al., 2013. Information about Abdel Halim Hafez can be found in Hotez and Kassem, 2016.

11. There is not consensus in the scientific community regarding the 280,000 deaths estimated by Van der Werf et al., 2003. For example, the GBD 2019 (Global Burden of Disease Collaborative Network, 2020) estimates that 11,500 people die annually; see www.healthdata.org, GBD Results Tool (http://www.healthdata.org/results/gbd_summaries/2019/schistosomiasis-level-3-cause). The significant differences may reflect the accounting by the GBD 2019 to attribute schistosomiasis deaths in a separate category linked to chronic renal disease or liver disease.

12. Information about the link between schistosomiasis and other NTDs and slavery from Africa is found in Lammie et al., 2007.

13. Gryseels et al., 2006.

14. The morbid effects of long-standing schistosomiasis are described in King et al., 2005; King and Dangerfield-Cha, 2008; and King, 2010.

15. Information about the relationship between S. haematobium eggs and bladder cancer can be found in Mitreva, 2012; Fu et al., 2012; and Ishida and Hseih, 2018.

16. Information about female urogenital schistosomiasis and its links with HIV/AIDS can be found in Kjetland et al., 2012; and Hotez et al., 2019b.

17. Fenwick and Webster, 2006; and Hotez et al., 2010.

18. Papers on schistosomiasis and hookworm coinfections include Raso et al., 2006; and Fleming et al., 2006.

19. Information about global coverage for schistosomiasis with praziquantel can be found in World Health Organization, 2012.

20. Information about the London Declaration can be found at https://www.who.int/neglected_diseases/London_Declaration_NTDs.pdf and in Hotez, 2012. Information about the Global Schistosomiasis Alliance and praziquantel for pediatric use can be found in Faust et al., 2020.

21. The results obtained by the SCI through MDA with praziquantel have been reported extensively in the peer-reviewed literature. Some recent examples include Landouré et al., 2012; Leslie et al., 2011; and Oshish et al., 2011.

22. The concept and possibilities of using praziquantel to interrupt schistosomiasis transmission and the possibility of disease elimination have been put forward in several published papers; see Rollinson et al., 2013; and Hotez, 2011. The World Health Assembly resolution on schistosomiasis elimination can be found at www.who.int/neglected_diseases/Schistosomiasis_wha65/en/. Information about progress to date for schistosomiasis treatment can be found in World Health Organization, 2020.

23. King et al., 2006.

24. Utzinger et al., 2001.

25. A recent paper on the development of a schistosomiasis vaccine is Hotez et al., 2019a.

26. Information on the global disease burden of the food-borne trematodiases is found at http://www.healthdata.org/results/gbd_summaries/2019/food-borne-trematodiases-level-3-cause and in Qian et al., 2016; and Sripa et al., 2018.

References

Colley DG, Secor WE. 2007. A schistosomiasis research agenda. *PLoS Negl Trop Dis* **1:**e32.

Cox FE. 2002. History of human parasitology. *Clin Microbiol Rev* **15:**595–612.

El-Sabah AA, El-Metwally MT, Abozinadah NY. 2011. Hepatitis C and B virus in schistosomiasis patients on oral or parenteral treatment. *J Egypt Soc Parasitol* **41:**307–314.

Farley J. 1991. *Bilharzia: a History of Imperial Tropical Medicine.* Cambridge University Press, Cambridge, United Kingdom.

Faust CL, Osakunor DNM, Downs JA, Kayuni S, Stothard JR, Lamberton PHL, Reinhard-Rupp J, Rollinson D. 2020. Schistosomiasis control: leave no age group behind. *Trends Parasitol* **36:**582–591.

Fenwick A. 2006. Waterborne infectious diseases—could they be consigned to history? *Science* **313:**1077–1081.

Fenwick A, Rollinson D, Southgate V. 2006. Implementation of human schistosomiasis control: challenges and prospects. *Adv Parasitol* **61:**567–622.

Fenwick A, Webster JP. 2006. Schistosomiasis: challenges for control, treatment and drug resistance. *Curr Opin Infect Dis* **19:**577–582.

Fleming FM, Brooker S, Geiger SM, Caldas IR, Correa-Oliveira R, Hotez PJ, Bethony JM. 2006. Synergistic associations between hookworm and other helminth species in a rural community in Brazil. *Trop Med Int Health* **11:**56–64.

Fu CL, Odegaard JI, Herbert DR, Hsieh MH. 2012. A novel mouse model of *Schistosoma haematobium* egg-induced immunopathology. *PLoS Pathog* **8:**e1002605.

Global Burden of Disease Collaborative Network. 2020. Global Burden of Disease Study 2019 (GBD 2019). GBD Results Tool. Institute for Health Metrics and Evaluation, Seattle, WA.

Grove DI. 1990. *A History of Human Helminthology.* CAB International, Wallingford, United Kingdom.

Gryseels B, Polman K, Clerinx J, Kestens L. 2006. Human schistosomiasis. *Lancet* **368:**1106–1118.

Horn JS. 1969. *Away with All Pests: an English Surgeon in People's China,* 1954–1969. Monthly Review Press, New York, NY.

Hotez PJ. 2002. China's hookworms. *China Q* **172:**1029–1041.

Hotez P. 2011. Enlarging the "Audacious Goal": elimination of the world's high prevalence neglected tropical diseases. *Vaccine* **29**(Suppl 4):D104–D110.

Hotez P. 2012. The London Declaration: a tipping point for the world's poor. *Huffington Post.* January 30, 2012.

Hotez PJ, Bottazzi ME, Bethony J, Diemert DD. 2019a. Advancing the development of a human schistosomiasis vaccine. *Trends Parasitol* **35:**104–108.

Hotez PJ, Engels D, Gyapong M, Ducker C, Malecela MN. 2019b. Female genital schistosomiasis. *N Engl J Med* **381:**2493–2495.

Hotez PJ, Engels D, Fenwick A, Savioli L. 2010. Africa is desperate for praziquantel. *Lancet* **376:**496–498.

Hotez PJ, Zheng F, Long-qi X, Ming-gang C, Shu-hua X, Shu-xian L, Blair D, McManus DP, Davis GM. 1997. Emerging and reemerging helminthiases and the public health of China. *Emerg Infect Dis* **3:**303–310.

Hotez PJ, Kassem M. 2016. Egypt: its artists, intellectuals, and neglected tropical diseases. *PLoS Negl Trop Dis* **10:**e0005072.

Hotez P, Ottesen E, Fenwick A, Molyneux D. 2006. The neglected tropical diseases: the ancient afflictions of stigma and poverty and the prospects for their control and elimination. *Adv Exp Med Biol* **582:**23–33.

Hulse EV. 1971. Joshua's curse and the abandonment of ancient Jericho: schistosomiasis as a possible medical explanation. *Med Hist* **15:**376–386.

Ishida K, Hsieh MH. 2018. Understanding urogenital schistosomiasis-related bladder cancer: an update. *Front Med (Lausanne)* **5:**223.

Kernan FA. 1959. The blood fluke that saved Formosa. *Harper's Magazine* 1959(April):45–47.

King CH. 2010. Parasites and poverty: the case of schistosomiasis. *Acta Trop* **113:**95–104.

King CH, Dickman K, Tisch DJ. 2005. Reassessment of the cost of chronic helmintic infection: a meta-analysis of disability-related outcomes in endemic schistosomiasis. *Lancet* **365:**1561–1569.

King CH, Sturrock RF, Kariuki HC, Hamburger J. 2006. Transmission control for schistosomiasis—why it matters now. *Trends Parasitol* **22:**575–582.

Kjetland EF, Leutscher PD, Ndhlovu PD. 2012. A review of female genital schistosomiasis. *Trends Parasitol* **28:**58–65.

Lammie PJ, Lindo JF, Secor WE, Vasquez J, Ault SK, Eberhard ML. 2007. Eliminating lymphatic filariasis, onchocerciasis, and schistosomiasis from the Americas: breaking a historical legacy of slavery. *PLoS Negl Trop Dis* **1:**e71.

Lampton DM. 1974. Health policy during the Great Leap Forward. *China Q* **60:**668–698.

Landouré A, Dembélé R, Goita S, Kané M, Tuinsma M, Sacko M, Toubali E, French MD, Keita AD, Fenwick A, Traoré MS, Zhang Y. 2012. Significantly reduced intensity of infection but persistent prevalence of schistosomiasis in a highly endemic region in Mali after repeated treatment. *PLoS Negl Trop Dis* **6:**e1774.

Leslie J, Garba A, Oliva EB, Barkire A, Tinni AA, Djibo A, Mounkaila I, Fenwick A. 2011. Schistosomiasis and soil-transmitted helminth control in Niger: cost effectiveness of school based and community distributed mass drug administration [corrected]. *PLoS Negl Trop Dis* **5:**e1326.

Mitreva M. 2012. The genome of a blood fluke associated with human cancer. *Nat Genet* **44:**116–118.

Oshish A, AlKohlani A, Hamed A, Kamel N, AlSoofi A, Farouk H, Ben-Ismail R, Gabrielli AF, Fenwick A, French MD. 2011. Towards nationwide control of schistosomiasis in Yemen: a pilot project to expand treatment to the whole community. *Trans R Soc Trop Med Hyg* **105:**617–627.

Qian MB, Utzinger J, Keiser J, Zhou XN. 2016. Clonorchiasis. *Lancet* **387:**800–810.

Rao MR, Naficy AB, Darwish MA, Darwish NM, Schisterman E, Clemens JD, Edelman R. 2002. Further evidence for association of hepatitis C infection with parenteral schistosomiasis treatment in Egypt. *BMC Infect Dis* **2:**29.

Raso G, Vounatsou P, Singer BH, N'Goran EK, Tanner M, Utzinger J. 2006. An integrated approach for risk profiling and spatial prediction of *Schistosoma mansoni*-hookworm coinfection. *Proc Natl Acad Sci USA* **103:**6934–6939.

Rollinson D, Knopp S, Levitz S, Stothard JR, Tchuem Tchuenté LA, Garba A, Mohammed KA, Schur N, Person B, Colley DG, Utzinger J. 2013. Time to set the agenda for schistosomiasis elimination. *Acta Trop* **128:**423–440.

Ross AG, Vickers D, Olds GR, Shah SM, McManus DP. 2007. Katayama syndrome. *Lancet Infect Dis* **7:**218–224.

Sanghvi MM, Hotez PJ, Fenwick A. 2013. Neglected tropical diseases as a cause of chronic liver disease: the case of schistosomiasis and hepatitis C co-infections in Egypt. *Liver Int* **33:**165–168.

Sripa B, Tangkawattana S, Brindley PJ. 2018. Update on pathogenesis of opisthorchiasis and cholangiocarcinoma. *Adv Parasitol* **102:**97–113.

Utzinger J, Xiao S, N'Goran EK, Bergquist R, Tanner M. 2001. The potential of artemether for the control of schistosomiasis. *Int J Parasitol* **31:**1549–1562.

Utzinger J, Zhou XN, Chen MG, Bergquist R. 2005. Conquering schistosomiasis in China: the long march. *Acta Trop* **96:**69–96.

Werf MJ, de Vlas SJ, Brooker S, Looman CW, Nagelkerke NJ, Habbema JD, Engels D. 2003. Quantification of clinical morbidity associated with schistosome infection in sub-Saharan Africa. *Acta Trop* **86:**125–139.

Wei WP. 1958. The people's boundless energy during the current leap forward. I. New victories on the antischistosomiasis front. *Chin Med J* **77:**107–110.

World Health Organization. 2012. Integrated preventive chemotherapy for neglected tropical diseases: estimation of the number of interventions required and delivered, 2009–2010. *Wkly Epidemiol Rec* **87:**17–27.

World Health Organization. 2020. Schistosomiasis and soil-transmitted helminthiases: numbers of people treated in 2019. *Wkly Epidemiol Rec* **95:**629–640.

Zhou YB, Liang S, Chen Y, Jiang QW. 2016. The Three Gorges Dam: does it accelerate or delay the progress towards eliminating transmission of schistosomiasis in China? *Infect Dis Poverty* **5:**63.

Elephantiasis: Lymphatic Filariasis, Endemic Nonfilarial Elephantiasis (Podoconiosis), and Dracunculiasis (Guinea Worm)

Moving from the jungle was a native with elephantiasis . . . pushing a rude wheelbarrow before him. In the barrow rested his scrotum, a monstrous growth that . . . weighed more than 70 pounds and tied him a prisoner to his barrow.

JAMES MICHENER, *TALES OF THE SOUTH PACIFIC*

Elephantiasis is the most severe and dramatic complication of lymphatic filariasis (LF), a chronic infection caused primarily by the filarial parasite *Wuchereria bancrofti*. However, there is now increasing attention being paid to podoconiosis, a related condition that clinically resembles elephantiasis but results from noninfectious causes. Dracunculiasis is another chronic infection caused by a filaria-like parasite, *Dracunculus medinensis*, also known as the *guinea worm*. Although both LF and dracunculiasis are still important parasitic infections in the developing world, we are much farther along in our global efforts to control these ancient scourges than we are with the soil-transmitted helminth (STH) infections or schistosomiasis. For LF, mass drug administration (MDA) is progressing to the point where we believe that there is a real possibility that this infection could one day be eliminated. We are on the verge of eradicating dracunculiasis. Our use of the terms "elimination" and "eradication" is deliberate. *Elimination* refers to reduction of the prevalence of an infection to the point where transmission has been interrupted but ongoing public health control measures must still be maintained. *Eradication* refers to the disappearance of naturally occurring infection to the point where public health intervention measures can be halted. To date, smallpox is the only disease ever to be eradicated.

Although filariae are nematodes, they bear little resemblance to the STHs described in chapter 2. As adult worms, the filariae grow to enormous lengths (sometimes a yard or more) in various body tissues such as the lymphatics,

Forgotten People, Forgotten Diseases: The Neglected Tropical Diseases and Their Impact on Global Health and Development, Third Edition. Peter J. Hotez.
© 2022 American Society for Microbiology. DOI: 10.1128/9781683673903.ch04

genitals, or subcutaneous tissues, instead of in the gastrointestinal tract. More-over, they are transmitted by the bites of arthropod vectors rather than through soil contamination. Table 4.1 lists the major filarial parasites of humans, the diseases that they cause, their geographic prevalence and location, and their arthropod vectors. I include dog heartworm on this list, because many pet owners in North America are already familiar with this parasite.

This chapter will consider LF (and podoconiosis) and dracunculiasis, while onchocerciasis, another filarial infection, will be discussed together with trachoma (in chapter 5) as major causes of blindness.

LF

Ancient records of LF include a statue with swollen limbs of the Egyptian pharaoh Mentuhotep II from 2000 BCE and descriptions by the Persian phy-sician Avicenna (Ibn Sina, 981–1037), who, like the Greek and Roman physi-cians before him, distinguished this disease from leprosy.[1] Today, most of the world's 72 million LF cases occur among the poorest people living in India, Southeast Asia, and sub-Saharan Africa, although large numbers of cases still occur on the Pacific islands and in some tropical areas of the Americas (espe-cially Haiti, Guyana, and northeastern Brazil) (Fig. 4.1).[2]

The adult *W. bancrofti* worms that cause 90% of the cases of LF (the other infections are caused by worms of the genus *Brugia*) live in human lym-phatics (vessel-like structures that parallel and feed into the bloodstream) mostly located in the inguinal region (groin) and genitals. An adult female worm grows to up to 4 in. in length and resembles a coiled-up piece of string or angel-hair pasta. How and why the adult worms develop in the lymphatics versus other sites is not known. However, since the lymphatics are a rich source of antibodies and cells of the immune system, it is all the more remarkable that the adult filarial worms could survive for up several years in what is arguably one of the most inhospitable places for foreign pathogens imaginable.

After mating, the female filarial worm, like the STHs and the schistosomes, produces a large number of embryos contained within eggshells (Fig. 4.2). However, the embryos of *W. bancrofti* do not take an egg-like shape but instead take an elongated shape, resembling tiny worms measuring roughly 0.25 mm in length. These so-called *microfilariae* migrate to the bloodstream, where ultimately they can be ingested by an appropriate species of female mosquito feeding on blood. It was Sir Patrick Manson, often considered the father of modern tropical medicine, who in 1877 first elucidated some of the fundamental aspects of the life cycle of *W. bancrofti*, including its transmission via mosquitoes.[1]

Once in the mosquito, the microfilariae develop, penetrate the stomach wall, and after about 10 to 12 days develop into larval stages that are capable of infecting humans the next time that the mosquito bites. In sub-Saharan Africa and elsewhere, the same mosquito species that transmit malaria also transmit LF. However, *W. bancrofti* is far less selective, so several different genera of mosquitoes are capable of hosting infective larvae and transmitting the infec-tion. The infective larvae that migrate to the mouth parts of the mosquito can

Table 4.1 Major filariae or filaria-like parasites of humans

Disease	Alternative name	Estimated global prevalence (no. of cases); location	Parasite	Location of adult parasite in humans	Length as adults (male and female)	Arthropod vector
Lymphatic filariasis	Elephantiasis	72 million[a]; Africa, South Asia, East Asia and Pacific, and the Americas	Wuchereria bancrofti (major); Brugia malayi and Brugia timori (minor)	Lymphatics, genitals	4–10 cm (1.5–4 in.)	Mosquito
Onchocerciasis	River blindness	19 million[a]; Africa	Onchocerca volvulus	Subcutaneous tissues	2–50 cm (1–20 in.)	Blackfly
Loiasis	African eye worm	Prevalence unknown, 14 million people live in areas of high transmission[b]; Africa	Loa loa	Subcutaneous tissues	3–7 cm (1–3 in.)	Tabanid fly
Dracunculiasis	Guinea worm	Near eradication; Africa	Dracunculus medinensis	Subcutaneous tissues	4–100 cm (1.5–40 in.)	Copepod
Heartworm	Dog heartworm	Rare parasite of humans	Dirofilaria immitis	Heart	12–30 cm (5–12 in.)	Mosquito

[a] Prevalence of lymphatic filariasis and onchocerciasis from Global Burden of Disease Collaborative Network, 2020 (http://www.healthdata.org/results/gbd_summaries/2019/lymphatic-filariasis-level-3-cause and http://www.healthdata.org/results/gbd_summaries/2019/onchocerciasis-level-3-cause).

[b] Data from Zouré et al., 2011.

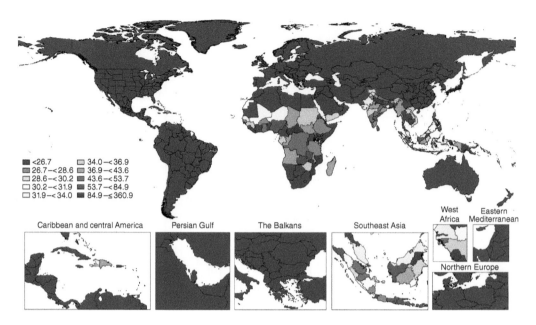

Figure 4.1 Prevalence of LF, worldwide. (From Global Burden of Disease Collaborative Network, 2020 [http://www.healthdata.org/results/gbd_summaries/2019/lymphatic-filariasis-level-3-cause].)

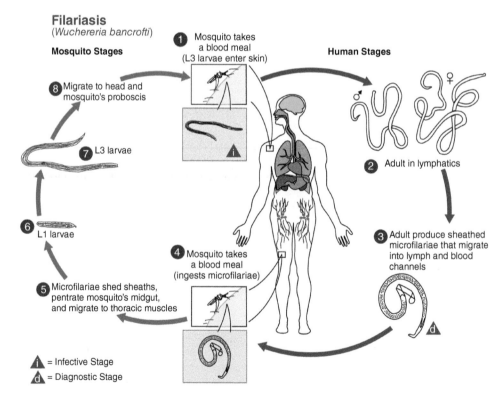

Figure 4.2 Life cycle of *W. bancrofti*. (Modified from CDC-PHIL [ID#3425]/Alexander de Silva, PhD/Melanie Moser.)

penetrate human skin. Possibly this presentation occurs through a puncture site generated by the insect. Then, over a period of months, the larvae migrate to their final destination in the lymphatics, where they grow into adult worms. The adult worms can live for extraordinarily long periods, often for 8 years or more.[3] This life span is amazing for an invertebrate but not unusual for parasitic worms. Why parasitic worms tend to live so much longer than their free-living counterparts is still a mystery.

In fact, there are a number of mysteries about the life history of *W. bancrofti* and the disease that it causes. Among them is the curious observation that over a 24-h period, the microfilariae tend to accumulate in the blood only at certain hours. In most of the world, *W. bancrofti* microfilariae appear in the victim's circulation at night, when the mosquitoes are biting the most. In some cases, the timing is very precise so that microfilariae appear between 10 p.m. and 4 a.m. Given that older methods for diagnosing LF required sampling the blood to look for microfilariae under a microscope, establishing a diagnosis of LF used to require operating clinics in the dead of night.[4] Like so many neglected tropical diseases (NTDs), LF is understudied and we still do not understand the physiological basis by which the microfilariae exhibit the unusual circadian rhythm known as *nocturnal periodicity*. Of interest is the observation that it is sometimes possible to reverse nocturnal periodicity by altering the sleep-wake cycle of an infected individual; this observation suggests that somehow the human host trains the microfilariae to keep up with the human schedule. Yet another mystery is how the adult *W. bancrofti* worms can survive for 8 years or more in a part of the body where they are practically bathed in host antibody and immune attacking cells. The immune evasion tactics by the adult worms are still not well understood, but there are some data to indicate that they somehow can manipulate the human immune system in order to make the host tolerant of their presence.[5]

We also do not understand well the sequence of events by which a person infected with LF can progress from having no symptoms to having horrible and disfiguring elephantiasis. Adding to the complexity is the observation that some patients can remain asymptomatic and never develop further complications. To our best knowledge, the pathologic sequence of events leading to disfigurement is initiated when the adult *W. bancrofti* worms dilate the lymphatic vessels where they reside.[6] In some cases, this vessel dilation phenomenon (known as *lymphangiectasia*) occurs because the adult worms cause lymphatic obstruction, but blockage is not always a requirement. Lymphangiectasia often occurs silently, or in some cases it can be seen by applying an ultrasound probe to the skin surface in the region of the groin. When ultrasonography is done, it is frequently possible to see worms writhing within the dilated lymphatic vessels. This "filarial dance sign" is one of the first indications that someone has acquired a new filarial infection, and it is one of the first signs of infection in children and young adolescents as they grow up in a part of the world where LF is endemic.

Late in the lymphangiectasia phase of the infection, the filarial worms reach the upper limit of their life span and start to die. In a village where LF is endemic, parasite death typically starts to occur by the time infected children are well into adolescence. However, unlike many infections with which we are familiar—such as bacterial and viral infections, in which it is a desired goal for either our host immune response or antimicrobial agents administered by a physician to kill the offending pathogens—the death of adult filarial worms in the lymphatics of adolescents is not necessarily a good thing. Instead, it is believed that the dead and dying worms lose their immunological masking properties, which results in a stimulus for the human immune response to release a cascade of cytokine inflammatory mediators. For the patient, the resulting cytokine storm produces fever and swelling of the lymphatics. Concurrently, secondary bacterial infections occur, with the parasites themselves contributing some of the bacteria. Therefore, adding to this host inflammatory component of the pathologic sequence of LF is the fact that within the parasites themselves are symbiotic bacteria of the genus *Wolbachia*. These are required for parasite survival, and the death of the parasites might contribute to the release of *Wolbachia* organisms that contribute to the disease (chapters 5 and 11 contain additional details). Our current thinking about LF is that the new inflammation against the dead parasites and the accompanying bacterial infections combine to initiate the first steps in a disease process lasting years, one subsequently leading to the characteristic lymph and genital deformities of elephantiasis.[6,7] The long time required for this series of events to unfold largely explains why elephantiasis usually only occurs in adulthood.

The health and socioeconomic impact of LF for the poorest people of India, Africa, and elsewhere is huge. The initial filarial fevers of adolescents are associated with enlarged, warm, and painful lymph nodes, and males experience tenderness and swelling in their scrotum, which is sometimes associated with *hydrocele*, an accumulation of fluid in the sac surrounding the testicles. Some patients develop *lymphedema*, i.e., edema and swelling of the legs and scrotum, with thickening and loss of skin elasticity. As lymphedema progresses, the limb and overlying skin begin to resemble an elephant's leg. The affected leg can become ulcerated and ooze foul-smelling pus.[6,7] Thus, LF exhibits a wide spectrum of disease, ranging from asymptomatic cases to filarial fevers and swellings to lymphedema and elephantiasis. Of the roughly 100 million people with LF, approximately 40% suffer from hydrocele, lymphedema, and other evidence of severe LF disease.[8]

While the health impact of these pathologic changes is fairly obvious, like many NTDs, LF has equally serious economic and social effects. Although death from LF is rare, the disease is nevertheless devastating because it strikes otherwise young and productive adults. In males, LF is a particular threat for agricultural laborers, usually of extremely modest means, who either are incapacitated with filarial fevers or are forced to give up their jobs because of their hydrocele or lymphedema. Myrtle Perera, a social worker from Sri Lanka, has extensively studied young coconut pickers who are no longer able to climb trees because of their hydroceles.[9] In one case, a worker commented that his hydrocele was as big as the coconuts he was picking.[9] Similarly, in Haiti, women with lymphedema frequently cannot participate in market trading, their major form of economic activity, while in northern Ghana, the incapacitating episodes of acute lymphangitis often reach their peak in rainy season, ordinarily a time of maximal agricultural activity.[9] Throughout the regions of the developing world where LF is endemic, it is a common practice for afflicted workers to change their trade to one that requires fewer physically demanding tasks in order to accommodate their disabilities.[9]

K. D. Ramaiah of India has conducted economic analyses of the impact of LF pathology on worker productivity and has identified a huge effect: the annual loss of US$842 million, almost 1% of India's GNP (gross national product).[9,10] In Orissa State, India, chronic LF patients lose a total of 68 days of work per year, equivalent to 19% of the total working time of the year, and spend 2% of their annual wages on treatment.[9,10] Another important impact, although one less easy to quantify, is the stigma of having LF. The examples of the man and his ashamed daughter and the woman who lost her job because of LF's stigmatizing aspects were illustrated in chapter 1. In addition, the stigma of hydrocele and elephantiasis is today causing millions of young men to hide their condition for years and prevent them from seeking medical attention.

Figure 4.3 presents pictures of Stephen J. Hawking and Mao Zedong. They are connected in a very interesting way by LF and the prospects for controlling it. While almost everyone has heard of Professor Hawking, the great cosmologist and father of the Big Bang theory, far fewer know that his father, Frank Hawking, was a parasitologist who studied LF and traveled widely in order to develop new approaches to its control. During the 1950s and 1960s, the senior Hawking made important contributions to the development and testing

Figure 4.3 What do Stephen Hawking and Mao Zedong have in common? (Hawking photo courtesy of NASA, Mao Zedong image from IISH / Stefan R. Landsberger / Private Collection, https://chineseposters.net/.)

of drugs that target *W. bancrofti*, including a compound known as diethylcar-bamazine, often abbreviated as DEC.[11] Hawking and others demonstrated that DEC was highly effective against the microfilarial stages of *W. bancrofti* but much less so against the adult stages. DEC is now widely administered for the treatment of LF worldwide. But because all of the pathology resulting from LF is caused by adult worms and not by the microfilariae, what then would be the rationale for administering DEC?

During World War II, LF was a major health problem for U.S. troops fighting in the South Pacific. It was estimated that approximately 38,300 U.S. naval personnel were exposed during the war, about one-third of whom developed filarial fevers.[12] These concerns led groups of American and other scientists to take a greater interest in the problem of LF and possible approaches for controlling it. Shortly after the end of World War II, John F. Kessel of UCLA and his colleagues working in Tahiti began looking at the impact of mass administration of DEC to infected populations. They demonstrated that by administering a drug that lowers the number of microfilariae among an entire village, it would be possible to interrupt the transmission of LF even if it did not directly reduce the immediate suffering from the disease.[13] To make this process more efficient, Frank Hawking pioneered the concept of fortifying dietary salt with DEC.[14] DEC is a highly stable compound, and even cooking does not destroy the drug. Therefore, instead of iodizing the salt, Hawking proposed DEC'ing it, and he began administering the medicated salt in food in order to reduce microfilarial loads among the population. The DEC-fortified salt study provided an important proof of concept that this approach could be an important weapon in the fight against endemic LF. However, it took the Chinese to apply it on an almost unimaginable scale.

While Hawking was working in Brazil, on the other side of the world, the People's Republic of China under the leadership of Mao was suffering some of the highest rates of LF. Indeed, the life cycle of LF was discovered by Sir Patrick Manson in the city of Amoy (renamed Xiamen) in Fujian Province on China's coast. Manson's landmark study demonstrated for the first time that a human infection could be transmitted by a mosquito vector. This observation laid the foundation for subsequent studies in India by Sir Ronald Ross, who demonstrated that mosquitoes transmitted malaria. In 1976 in Shandong Province, also on the coast of China (roughly 100 years following Manson's discoveries), DEC-medicated salt was administered to almost 40,000 people for 6 months, with a resulting dramatic reduction in the prevalence of LF as determined by the numbers of microfilariae in blood.[15] Studies such as this one were followed with efforts to scale up LF control for all of China's major provinces at risk for the disease. As a result of blanket coverage with DEC-fortified salt, today the People's Republic of China represents one of the first large nations to have successfully eliminated LF. Subsequently, widespread use of DEC-fortified salt resulted in the elimination or near elimination of *W. bancrofti* infection from Brazil, Japan, Tanzania, and Taiwan,[16] while mass administration of the drug as tablets has over a period of 5 years resulted in the elimination of LF from Egypt, one of the first nations to eliminate LF through this approach.[17] In all,

LF has been eliminated in more than 20 countries through MDA, including highly populated nations such as China, Cambodia, Thailand, and Vietnam in East Asia; several South Pacific islands; and Egypt and Malawi in Africa.[17] In addition, there is optimism that LF might be eliminated soon in nations such as Bangladesh and the Dominican Republic.[17] For some of these countries, mosquito control was a critical adjunct for LF elimination efforts, while in the case of the Solomon Islands, vector control was sufficient for LF elimination.[18] In most of sub-Saharan Africa, the drug ivermectin is used in place of DEC because of its better safety profile in patients who are coinfected with onchocerciasis (the widespread use of ivermectin in Africa is discussed in chapter 5).

A number of factors, both biological and social, provide great optimism that LF elimination efforts through MDA could be extended worldwide.[19] They include the following. (i) A single dose of either DEC or ivermectin can dramatically reduce the number of blood-circulating microfilariae for up to 12 months, and a highly effective and simple-to-use diagnostic test for LF is available.[20] (ii) Humans are the only significant reservoir of infection caused by *W. bancrofti*, and there are no concerns about the need to target other animal species with DEC or ivermectin. (iii) The infection process, which occurs through mosquito bites, is not nearly as efficient as it is for malaria (another mosquito-borne infection) and therefore is more easily interrupted through vector control methods (including bed nets). (iv) In addition, we have the proof of concept that it has been possible to eliminate LF in some countries. Based on these assumptions, in 1997 the 50th World Health Assembly (WHA) adopted a resolution to eliminate LF as a public health problem by 2020 (www.who.int/lymphatic_filariasis/resources/WHA_50%2029.pdf). The major components of the LF elimination strategy are (i) to interrupt LF transmission by using one of two single-dose, two-drug treatment regimens, either DEC plus albendazole or ivermectin plus albendazole (albendazole is added to forestall the emergence of drug resistance), with once-yearly administration of these two drug regimens in order to achieve four to six annual consecutive rounds of high coverage; and (ii) to reduce the existing morbidity of LF, particularly filarial fevers, hydrocele, and lymphedema and/or elephantiasis, with an emphasis on hygiene, skin care, and simple surgery.[19] Thus, MDA for LF of either ivermectin and albendazole or DEC and albendazole joins other mass treatments for STH infections (using albendazole or mebendazole) and schistosomiasis (using praziquantel) as major approaches to global NTD control. In a more recent advance, teams from Case Western Reserve University, led by Christopher King and James Kazura, and Washington University in St. Louis, led by Gary Weil, found that a three-drug regimen comprising ivermectin, DEC, and albendazole was even more effective than two-drug regimens in terms of sustaining clearance of blood microfilariae.[19] As an alternative approach, Mark Taylor and his colleagues at the Liverpool School of Tropical Medicine are exploring antibiotics that target *Wolbachia* bacterial endosymbionts, highlighted above, as novel therapies for *W. bancrofti* infection.[7]

In response to the WHA resolution, the health ministers from all of the countries in the Pacific region agreed to meet with the World Health

Organization (WHO) Western Pacific Regional Office in order to establish PacELF for the elimination of LF in the region through MDA, with each country adopting its own such strategy.[18]

Then in 2000, the Global Alliance to Eliminate LF (GAELF) was formed to support a global program—the Global Programme to Eliminate LF (GPELF)—for helping to ensure global elimination, now targeted for 2030 under the auspices of a new WHO Roadmap. Among the areas of emphasis, GAELF is working to facilitate establishment of programs in Africa and parts of Asia where infrastructure is lacking. GAELF also has as a major goal the alleviation of the physical, social, and economic hardships in individuals who have LF-induced disabilities. The Secretariat of GAELF is based at the Liverpool School of Tropical Medicine. The work is being facilitated by a number of operational research activities together with large-scale donations of ivermectin by Merck & Co., albendazole by GlaxoSmithKline, and DEC by the Japanese company Eisai, which affirmed its philanthropic commitment at the January 2012 meeting to announce the London Declaration for NTDs.

Global LF control and elimination efforts represent one of the world's largest public health programs. According to the WHO, more than 8 billion treatments have been administered to almost 1 billion people since 2000.[19] In 2019 alone, more than 500 million people received MDA, representing almost two-thirds of the global population at risk of LF.[19] Currently, the three-drug regimen is used regularly in the regions where LF is endemic, with plans to evaluate the impact of this approach in terms of elimination targets. By 2030, the GPELF and WHO hope that 80% of the world's countries of LF endemicity will meet elimination targets, and ultimately a situation in which no one requires MDA for LF.[19] In parallel, the WHO emphasizes the importance of ancillary care for patients suffering from the clinical effects of LF. This is an important point, since neither the two-drug nor three-drug regimen administered once annually likely destroys the adult *W. bancrofti* worms. Instead, the MDA approach is often referred to by the WHO as preventive treatment or preventive chemotherapy, meaning that it can interrupt transmission by reducing the number of microfilariae circulating in blood. Therefore, it is still necessary to provide medical or surgical management for treating hydrocele and lymphedema.

In addition to the enormous health gains realized from LF control and elimination efforts, an economic assessment by Eric Ottesen and colleagues of the first 8 years of the GPELF estimated that the program produced US$21.8 billion of direct economic benefits.[19]

Even though we have excellent tools in hand and the biological and social features of LF are favorable for elimination, it is not yet clear that at our current pace we will meet the WHA elimination targets. Additional hurdles include the fact that most of the 750 million living in extreme poverty who require access to treatment are at the bottom of the economic scale and therefore often politically voiceless.[19] Adding even further to these social and political barriers to MDA is the stigmatizing element of LF, which means that many people with the disease are hidden from view. Stigma prevents accurate assessments of disease burden and access to essential medicines. The very good news is that the GAELF has had an extraordinary track record of success, which includes the impressive accomplishment of eliminating LF in more than 20 countries.[21] Very few public health programs in place today can match that record of achievement. Later, we will see how, through linking LF control with other NTD control efforts, programs to integrate controls for multiple NTDs could increase the efficiency of ongoing LF elimination efforts as a means for achieving ambitious elimination targets set by the WHA.

Podoconiosis (Endemic Nonfilarial Elephantiasis)

Podoconiosis is another type of lymphedema found in the tropics, and bears some clinical resemblance to elephantiasis caused by LF. It primarily affects the feet and lower legs and is associated with marked swelling and disfigurement.[21] However, unlike LF, podoconiosis is a noninfectious condition believed to be caused by a host response, possibly genetically determined, to small mineral particles found in soil composed of volcanic deposits. Although it is not yet officially listed by the WHO as an NTD, podoconiosis has many of the same features, including a geographic distribution in the tropics, especially in the sub-Saharan country of Ethiopia. It also highly stigmatizing, and linked to poverty and involvement in agricultural pursuits. Although prevention relies on the straightforward measure of wearing footwear to prevent soil-particle exposure and covering the earthen floors of simple dwellings, it remains a significant public health problem. Professor Gail Davey at Brighton and Sussex Medical School in the United Kingdom leads a lot of the work on podoconiosis internationally and has been a strong advocate for people afflicted by the condition. This includes raising awareness about podoconiosis, as well as its pathogenesis and prevention. The true global burden of the disease outside areas of high endemicity such as Ethiopia and Cameroon is not well established. However, it is worth noting that new evidence reveals a higher mortality rate among podoconiosis sufferers than the general population living in the same area or region, although the basis for this finding is unknown.[21] Still another important aspect is understanding the social science surrounding the stigma associated with podoconiosis and related disfiguring conditions such as LF and *Onchocerca* skin disease (see chapter 5), and Davey has helped to create a foundation for this purpose known as 5S.[21]

Dracunculiasis (Guinea Worm Infection)

Except for just over a few dozen known remaining cases, mostly in Chad and Ethiopia, but possibly elsewhere in sub-Saharan Africa, human guinea worm infection has been practically eradicated from the planet.[22] Guinea worm is thought to have been the Biblical "fiery red serpent" that attacked Israelites in the desert after their exodus from Egypt, with therapies describing methods for extracting it found in Egyptian medical papyri.[1] By 1986, it was estimated that approximately 3.5 million cases of guinea worm infection occurred among the poorest people in Africa, the Middle East, and Asia. However, through an extraordinary program of public health control and advocacy, the Carter Center in cooperation with the U.S. Centers for Disease Control and

Prevention (CDC), UNICEF, and the WHO has led a 35-year-long global dracunculiasis eradication program, which has so far resulted in more than a 99% reduction in incidence of the disease, with 192 countries and territories certified free of dracunculiasis transmission.[22]

The guinea worm, *D. medinensis*, is not a true filarial worm like *W. bancrofti*, but they have a number of similarities. The adult worms grow to up to a yard in length (females are longer than males) and inhabit the tissues of the legs and feet, just under the skin. Occasionally, they are located in other body parts. The adult female has evolved an interesting way of getting her larval offspring into the environment by producing substances that create a blister under the skin. The blister is under pressure, and when an individual immerses his leg in water, the blister ruptures. During this process, the female adult worm discharges a milky-white substance containing thousands of immature *D. medinensis* larvae. Larval discharge and release can continue for days. The larvae survive in freshwater and develop further after being swallowed by a small crustacean called a copepod. Humans then become infected when they swallow unfiltered or unboiled water containing copepods (Fig. 4.4). After ingestion, the copepod is digested by gastric acid, but the infective guinea worm larvae survive this insult. The larvae released in the

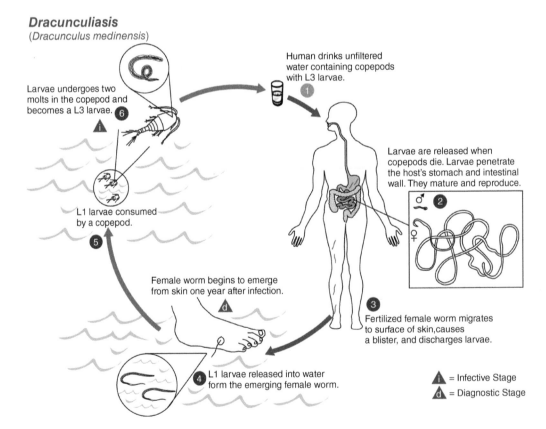

Dracunculiasis
(*Dracunculus medinensis*)

Larvae undergoes two molts in the copepod and becomes a L3 larvae. **6** **i**

L1 larvae consumed by a copepod. **5**

Female worm begins to emerge from skin one year after infection. **d**

L1 larvae released into water form the emerging female worm. **4**

Human drinks unfiltered water containing copepods with L3 larvae. **1**

Larvae are released when copepods die. Larvae penetrate the host's stomach and intestinal wall. They mature and reproduce.

♂ **2** ♀

Fertilized female worm migrates to surface of skin, causes a blister, and discharges larvae. **3**

i = Infective Stage
d = Diagnostic Stage

Figure 4.4 Life cycle of the guinea worm. (Modified from CDC-PHIL [ID#3391]/CDC/ Alexander J. de Silva, PhD; Melanie Moser, 2002.)

human gastrointestinal tract penetrate the human gut and over a period of several months make their way to the connective tissue of the legs, where they sexually mature and mate.

Almost all of the symptoms of guinea worm infection result from the presence of the adult worm residing in the legs (Fig. 4.5). The blister produced by female worms is painful and can cause a burning sensation. These processes can go on for 2 or 3 months, during which time the individual is often incapacitated and unable to work.[22] In some cases, the pain can continue for 1 to 2 years. In the resource-poor settings where guinea worm occurs, it is common for the ulcer that results from the ruptured blister to become infected with bacteria; both secondary bacterial infections and sometimes even tetanus

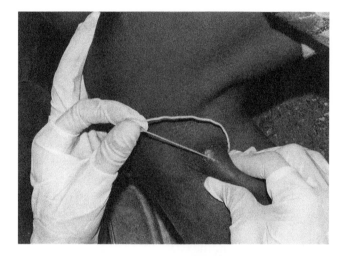

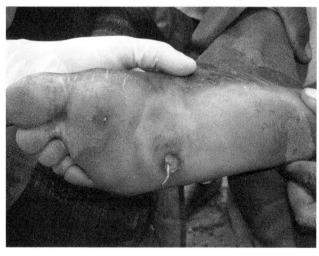

Figure 4.5 Clinical infection with guinea worm. (Top) Subcutaneous emergence of a female guinea worm from a sufferer's lower left leg, just distal to the lateral left knee. (© 2001 The Carter Center.) (Bottom) Subcutaneous emergence of a female guinea worm from the plantar surface, i.e., sole, of a patient's right foot. (© 2003 The Carter Center/E. Wolfe.)

are well-known complications. When the worm is located near joints (commonly the ankle joint), the inflammation can cause arthritis, which can be severe at times.

Prior to large-scale control efforts, it was common for entire villages to be affected simultaneously. When guinea worm epidemics struck during a harvest or planting season, the agricultural productivity of the village would be severely impaired.[22] For this reason, guinea worm infection is sometimes known as the "empty granary" disease. These losses have been quantified in several regions of West Africa.[23] Therefore, the economic impact of guinea worm infection is almost as serious as its health impact. Moreover, in children, guinea worm infection is also associated with malnutrition and school absenteeism,[23] so that, like the STH infections and schistosomiasis, guinea worm has an important educational impact as well.

There is no straightforward cure for dracunculiasis. Even today, a major approach to treating guinea worm is to use the ancient practice of slowly extracting the worm onto a small stick and then twisting the stick a bit each day. It is believed that the sign of the caduceus, the universal symbol of medicine depicting a snake wrapped around a staff, probably evolved from what was undoubtedly one of the first treatments ever for an infectious disease! Unfortunately, no anthelmintic drug is effective against guinea worm.

The successes in guinea worm control over the past 35 years represent one of the more interesting stories in the field of the NTDs.[24] In the early 1980s, there was great excitement about disease eradication following the successful eradication of smallpox through widespread vaccination. The last naturally occurring case of smallpox was recorded in Somalia in 1977, and the world was certified free of smallpox in 1981. Although there is no vaccine for guinea worm, the CDC suggested dracunculiasis as a potential new target for eradication because of its relatively low prevalence compared to that of other helminth infections such as hookworm infection and schistosomiasis. Additional reasons were the observations that guinea worm could be prevented by health education and straightforward behavioral changes, such as filtering water through cloth to remove copepods and avoiding public water supplies during active infection. Proof of concept that such measures were effective included previously successful eradication efforts in Uzbekistan, Iran, and some states in India, such as Tamil Nadu and Gujarat.[25] In 1986, two landmark events occurred. First, the 39th WHA adopted a resolution to eliminate dracunculiasis, and second, former U.S. President Jimmy Carter, 6 years after his defeat for reelection, began to champion efforts to eradicate the guinea worm. His advocacy began in Pakistan, where he persuaded General Mohammad Zia ul-Haq to step up eradication efforts, and then subsequently in sub-Saharan Africa, where he worked with the leaders of Ghana, Mali, and Nigeria to embark on national control programs. In 1993, Pakistan became the first country to eradicate guinea worm, and by 1995 to 1996, guinea worm eradication efforts were under way in every country where dracunculiasis was endemic. These efforts were then linked with a new WHA resolution (WHA 50.35), which urged all "Member States, international and nongovernmental organizations and appropriate entities to continue to ensure political support and the availability of much-needed resources for completion of eradication of dracunculiasis as quickly as technically feasible and for the 'International Commission for the Certification of Dracunculiasis Eradication' and its work."[26] The three major elements of these proposed eradication efforts included (i) provision of safe water through the creation of deep wells (as a source of water free from copepods), filtering of water, and application of larvicidal agents (primarily the agent Temephos [Abate; American Cyanamid]) that destroy copepods; (ii) health education; and (iii) case containment and management.[24] Among the important new tools in this effort was a new and more efficient nylon cloth filter that was developed by DuPont specifically for guinea worm filtration.[24]

Over the last few decades, the Guinea Worm Eradication Program (GWEP)—previously under the direction of Ernesto Ruiz-Tiben, and since taken over by Adam Weiss (together with the advice of Donald Hopkins)—has been an unqualified success. Up until just a few years ago, only 1,000 or so cases of guinea worm infection remained on the planet, of which almost all were in South Sudan, where a long-standing civil war has thwarted public health interventions over the last 25 years.[27] Even during the worst years of fighting in war-torn Sudan, former President Carter made surprising gains in fighting guinea worm. In 1995, he negotiated a 4-month-long "Guinea Worm Cease-Fire," with the Carter Center facilitating coordination of eradication activities from both sides, using offices in both Khartoum (the Sudanese capital) and Nairobi.[28] The cease-fire demonstrated that it is possible to continue public health interventions in areas of conflict (in chapter 7, we will see how wars in Angola, the Democratic Republic of the Congo, and Sudan enabled the recrudescence of human African trypanosomiasis). More recently, however, cases have been reported in Chad, where it was noted that dogs and possibly other animal reservoirs were also infected.[22] Finding guinea worm in dogs raises concerns that final eradication may be more daunting than previously believed since it will require additional measures directed jointly at both humans and animals, sometimes referred to as a One Health approach.[22]

Conclusions

The control of LF and guinea worm infection, like so many NTD control programs, represents extremely low-cost and cost-efficient measures. For LF, ivermectin, DEC, and albendazole are donated free of charge. Therefore, LF MDA can be provided for less than US$1 per person (www.filariasis.org). Similarly, the estimated cost of the GWEP between 1987 and 1998 was US$87.5 million, or approximately US$5 to US$8 per prevented case.[29] Since 2000, the Gates Foundation has provided additional and generous support to continue these activities. The Carter Center is working furiously to ensure that it meets targets for guinea worm eradication. According to one estimate of the cost of the GWEP between 1986 and 2030, a 44-year program will cost approximately US$11 per case averted.[29]

In their own ways, LF MDA and the GWEP have been successful.[30] For the GWEP, major reasons for success include the availability of simple interventions together with President Carter's high-level advocacy and excellent coordination by the Carter Center and its partners. For LF, wherever elimination efforts have been aggressively pursued, the outcome has been extremely favorable. However, because the magnitude of the LF elimination problem is so much greater than the problem facing the GWEP, the WHO and other organizations estimate that about two-thirds or fewer of the 850 billion people who live in areas at risk for infection have so far received MDA of either DEC or ivermectin. Therefore, we could be looking at several more years before LF elimination is under way in most of the developing world, especially in Africa. Later (in chapter 10), we will see how linking LF control with other NTD MDA programs can catalyze greater efficiencies and increase coverage.

SUMMARY POINTS Elephantiasis

LF

- LF occurs in an estimated 72 million people in the world's poorest countries, where approximately 800 million to 900 million people are at risk for acquiring the infection. Most of the world's cases occur in South Asia, Southeast Asia, and sub-Saharan Africa, with some tropical regions of the Americas (especially Haiti and northeastern Brazil) accounting for the remainder.

- Approximately 90% of the world's LF cases are caused by the nematode parasite *W. bancrofti*. The adult worms live for 8 years in the lymphatics. They produce microfilariae that enter the bloodstream and are transmitted person to person through a mosquito vector.

- The morbidity and pathology of LF result from adult *W. bancrofti* worms in the lymphatics, which cause dilation (lymphangiectasia), obstruction, lymphedema, and, in some cases, elephantiasis. Bacterial endosymbionts of the genus *Wolbachia* may also contribute to the pathology of LF.

- The chronic morbidity and disfigurement resulting from LF exact a huge economic toll, at least US$1 billion in losses annually to the Indian economy. Much of the loss is from reduced worker productivity. In addition, the stigma of LF has an enormous but still poorly quantified social impact.

- Transmission of LF in poor rural communities can be interrupted by mass drug administration of anthelmintic drugs that target the microfilariae. These drugs include DEC and ivermectin, together with albendazole. Widespread use of DEC in national programs has resulted in the elimination or near elimination of LF in more than 20 countries including Brazil, China, Egypt, Japan, and Taiwan. In some cases, it has been brought about through use of DEC-fortified salt. Recently, a three-drug regimen comprising DEC, ivermectin, and albendazole has been shown to result in more-sustained suppression and is being used increasingly for elimination efforts.

- Through years of administering DEC or ivermectin, coverage of about two-thirds of people at risk worldwide has been achieved. Greater efficiencies in coverage might be achieved through integrated NTD control. A new WHO Roadmap calls for the elimination of LF by the year 2030.

Podoconiosis

- This condition is also known as endemic nonfilarial elephantiasis. Although it is not officially listed by the WHO as an NTD, podoconiosis has many of the features of the NTDs, including disabling and chronic features accompanying poverty.

- It clinically resembles LF in some aspects, but podoconiosis is not caused by an infectious agent. Instead, it represents a host inflammatory response to particles found in volcanic soils. Major areas where it occurs include Ethiopia and Cameroon, although podoconiosis is also found outside Africa.

- Podoconiosis is a highly stigmatizing condition, similar to LF. Prevention includes adequate provision of footwear and hygiene.

Guinea worm

- Dracunculiasis is caused by the guinea worm, *D. medinensis*. Prior to global eradication efforts that began over 35 years ago, an estimated 3.5 million people were infected, primarily in Africa and the Middle East and on the Indian subcontinent.

- The adult *D. medinensis* worm grows to almost 1 yard in length and lives in the subcutaneous tissues of the legs and feet (as well as elsewhere). The adult female worm produces a blister that ruptures in water, with the release of immature *D. medinensis* larvae. The larvae are ingested by copepods. Human infection occurs after ingestion of the copepods.

- Stepped-up measures for guinea worm control began in 1986, with former President Carter and the Carter Center spearheading many of the initiatives. The major tools of the GWEP include provision of safe water, health education, and case containment. In 1993, Pakistan became the first country to eradicate guinea worm infection.

- Today, only a few dozen cases of guinea worm infection remain, mostly in Chad and possibly Ethiopia and other Sahel countries. There is optimism that guinea worm will be eradicated in the coming years. However, a new concern, the finding of guinea worm infection in dogs, suggests there is an emerging important animal reservoir of the infection. Worries are that this might thwart current elimination strategies and require shifting to a One Health approach integrating human and animal health.

Notes

1. Historical accounts of LF and dracunculiasis are found in Cox, 2002; and Hotez et al., 2006.

2. Information on the prevalence of LF and the estimates of people in these developing regions at risk of acquiring the disease are found in the GBD 2019 (Global Burden of Disease Collaborative Network, 2020) (see http://www.healthdata.org/results/gbd_summaries/2019/lymphatic-filariasis-level-3-cause); Cromwell et al., 2020; Ottesen, 2006; and Chu et al., 2010.

3. These numbers are quoted from Rajan, 2005.

4. Today, diagnosing LF can be done by sampling the blood at any time of the day and chemically detecting the presence of *W. bancrofti* antigen through use of a card known as an ICT card or test, the development of which was pioneered by Gary Weil at Washington University in St. Louis.

5. This is discussed in Babu et al., 2006; and Nutman and Kumaraswami, 2001.

6. A good description of the pathologic sequence of events in patients with LF is found in Addiss and Brady, 2007.

7. There are several excellent published clinical descriptions of LF and elephantiasis. The one I like best is in Simonsen, 2002. Information about *Wolbachia* can be found in Taylor et al., 2019.

8. Figures are from Michael et al., 1996.

9. The productivity losses from LF are described in Perera et al., 2007 and on the website of the Lymphatic Filariasis Support Center of the Task Force for Global Health (https://www.gaelf.org/).

10. The economic impact of LF has been reported in at least three papers: Ramaiah et al., 2000a; Ramaiah et al., 2000b; and Babu et al., 2002.

11. Some of the senior Hawking's work is summarized in Hawking et al., 1950; Hawking, 1955; and Hawking, 1958.

12. Coggeshall, 1945, quoted in Rajan, 2005.

13. Kessel et al., 1953.

14. Hawking and Marques, 1967.

15. Zhong and Zhen, 1991.

16. Houston, 2000.

17. Ramzy et al., 2006; and Hotez, 2011. See World Health Organization, 2020 for an updated list.

18. Ichimori et al., 2007.

19. Information on progress in the treatment and elimination of LF can be found on the GAELF website (www.filariasis.org) and in World Health Organization, 2020; Chu et al., 2010; and Ottesen, 2006. Information about the new three-drug regimen for sustained clearance can be found in King et al., 2018; and Weil et al., 2020.

20. The diagnostic test was developed in the laboratory of Gary Weil at Washington University in St. Louis; see Weil et al., 1997.

21. Information about podoconiosis is found at https://www.who.int/lymphatic_filariasis/epidemiology/podoconiosis/en/ and in Deribe et al., 2020; Masraf et al., 2020; and Zaman et al., 2020.

22. Global guinea worm eradication efforts are summarized in Hopkins et al., 2005; and Ruiz-Tiben and Hopkins, 2006 and at www.who.int/dracunculiasis/en/index.html. The current status is reported in Molyneux et al., 2020; and Liu et al., 2020.

23. Cairncross et al., 2002.

24. The account here of the modern history of guinea worm eradication is summarized from Levine and the What Works Working Group, 2004, p. 91–104. The book contains a number of interesting histories of successes in NTD control and in global health.

25. Historical background information on dracunculiasis can be found at www.who.int/dracunculiasis/background/en/index.html.

26. http://whqlibdoc.who.int/hq/2008/WHO_HTM_NTD_PCT_2008.1_eng.pdf.

27. Carter Center, 2006, p. 30.

28. Hopkins and Withers, 2002. The article is reprinted at www.cartercenter.org/news/documents/doc1255.html.

29. The economic figures are from Kim et al., 1997; and Fitzpatrick et al., 2017.

30. McNeil, 2006; and Barry, 2006.

References

Addiss DG, Brady MA. 2007. Morbidity management in the Global Programme to Eliminate Lymphatic Filariasis: a review of the scientific literature. *Filaria J* **6:**2.

Babu BV, Nayak AN, Dhal K, Acharya AS, Jangid PK, Mallick G. 2002. The economic loss due to treatment costs and work loss to individuals with chronic lymphatic filariasis in rural communities of Orissa, India. *Acta Trop* **82:**31–38.

Babu S, Blauvelt CP, Kumaraswami V, Nutman TB. 2006. Regulatory networks induced by live parasites impair both Th1 and Th2 pathways in patent lymphatic filariasis: implications for parasite persistence. *J Immunol* **176:**3248–3256.

Barry M. 2006. Slaying little dragons: lessons from the dracunculiasis eradication program. *Am J Trop Med Hyg* **75:**1–2.

Bockarie MJ, Molyneux DH. 2009. The end of lymphatic filariasis? *BMJ* **338:**b1686.

Cairncross S, Muller R, Zagaria N. 2002. Dracunculiasis (guinea worm disease) and the eradication initiative. *Clin Microbiol Rev* **15:**223–246.

Carter Center. 2006. *25th Anniversary Annual Report: 2005–2006.* Carter Center, Atlanta, GA.

Chu BK, Hooper PJ, Bradley MH, McFarland DA, Ottesen EA. 2010. The economic benefits resulting from the first 8 years of the Global Programme to Eliminate Lymphatic Filariasis (2000-2007). *PLoS Negl Trop Dis* **4:**e708.

Coggeshall LT. 1945. Malaria and filariasis in the returning servicemen. *Am J Trop Med Hyg* **25:**177–196.

Cox FE. 2002. History of human parasitology. *Clin Microbiol Rev* **15:**595–612.

Cromwell EA, et al, Local Burden of Disease 2019 Neglected Tropical Diseases Collaborators. 2020. The global distribution of lymphatic filariasis, 2000-18: a geospatial analysis. *Lancet Glob Health* **8:**e1186–e1194.

Deribe K, Mackenzie CD, Newport MJ, Argaw D, Molyneux DH, Davey G. 2020. Podoconiosis: key priorities for research and implementation. *Trans R Soc Trop Med Hyg* **114:**889–895.

Fitzpatrick C, Sankara DP, Agua JF, Jonnalagedda L, Rumi F, Weiss A, Braden M, Ruiz-Tiben E, Kruse N, Braband K, Biswas G. 2017. The cost-effectiveness of an eradication programme in the end game: evidence from guinea worm disease. *PLoS Negl Trop Dis* **11:**e0005922.

Global Burden of Disease Collaborative Network. 2020. Global Burden of Disease Study 2019 (GBD 2019). GBD Results Tool. Institute for Health Metrics and Evaluation, Seattle, WA.

Hawking F. 1955. The chemotherapy of filarial infections. *Pharmacol Rev* **7:**279–299.

Hawking F. 1958. Filariasis. *Sci Am* **199:**94–101.

Hawking F, Marques RJ. 1967. Control of Bancroftian filariasis by cooking salt medicated with diethylcarbamazine. *Bull World Health Organ* **37:**405–414.

Hawking F, Sewell P, Thurston JP. 1950. The mode of action of hetrazan on filarial worms. *Br J Pharmacol Chemother* **5:**217–238.

Hopkins DR, Ruiz-Tiben E, Downs P, Withers PC Jr, Maguire JH. 2005. Dracunculiasis eradication: the final inch. *Am J Trop Med Hyg* **73:**669–675.

Hopkins DR, Withers PC Jr. 2002. Sudan's war and eradication of dracunculiasis. *Lancet* **360**(Suppl)**:**s21–s22.

Hotez P. 2011. Enlarging the "Audacious Goal": elimination of the world's high prevalence neglected tropical diseases. *Vaccine* **29**(Suppl 4)**:**D104–D110.

Hotez P, Ottesen E, Fenwick A, Molyneux D. 2006. The neglected tropical diseases: the ancient afflictions of stigma and poverty and the prospects for their control and elimination. *Adv Exp Med Biol* **582:**23–33.

Houston R. 2000. Salt fortified with diethylcarbamazine (DEC) as an effective intervention for lymphatic filariasis, with lessons learned from salt iodization programmes. *Parasitology* **121**(Suppl)**:**S161–S173.

Ichimori K, Graves PM, Crump A. 2007. Lymphatic filariasis elimination in the Pacific: PacELF replicating Japanese success. *Trends Parasitol* **23:**36–40.

Kessel JF, Thooris GC, Bambridge B. 1953. The use of diethylcarbamazine (hetrazan or notezine) in Tahiti as an aid in the control of filariasis. *Am J Trop Med Hyg* **2:**1050–1061.

Kim A, Tandon A, Ruiz-Tiben E.1997. *Cost-Benefit Analysis of the Global Dracunculiasis Eradication Campaign. Policy Research Working Paper* 1835. Africa Human Development Department, World Bank, Washington, DC.

King CL, Suamani J, Sanuku N, Cheng YC, Satofan S, Mancuso B, Goss CW, Robinson LJ, Siba PM, Weil GJ, Kazura JW. 2018. A trial of a triple-drug treatment for lymphatic filariasis. *N Engl J Med* **379:**1801–1810.

Levine R, What Works Working Group. 2004. *Millions Saved: Proven Successes in Global Health.* Center for Global Development, Washington, DC.

Liu EW, Sircar AD, Matchanga K, Mahamat AM, Ngarhor N, Ouakou PT, Zirimwabagabo H, Ruiz-Tiben E, Sankara D,

Wiegand R, Roy SL. 2020. Investigation of dracunculiasis transmission among humans, Chad, 2013-2017. *Am J Trop Med Hyg* **104:**724–730.

Masraf H, Azemeraw T, Molla M, Jones CI, Bremner S, Ngari M, Berkley JA, Kivaya E, Fegan G, Tamiru A, Kelemework A, Lang T, Newport MJ, Davey G. 2020. Excess mortality among people with podoconiosis: secondary analysis of two Ethiopian cohorts. *Trans R Soc Trop Med Hyg* **114:**1035–1037.

McNeil DG Jr. 2006. Dose of tenacity wears down an ancient horror. *The New York Times* 2006 (March 26).

Michael E, Bundy DA, Grenfell BT. 1996. Re-assessing the global prevalence and distribution of lymphatic filariasis. *Parasitology* **112:**409–428.

Molyneux DH, Eberhard ML, Cleaveland S, Addey R, Guiguemd ÈRT, Kumar A, Magnussen P, Breman JG. 2000. Certifying guinea worm eradication: current challenges. *Lancet* **396:** 1857–1860.

Nutman TB, Kumaraswami V. 2001. Regulation of the immune response in lymphatic filariasis: perspectives on acute and chronic infection with *Wuchereria bancrofti* in South India. *Parasite Immunol* **23:**389–399.

Ottesen EA. 2006. Lymphatic filariasis: treatment, control and elimination. *Adv Parasitol* **61:**395–441.

Perera M, Whitehead M, Molyneux D, Weerasooriya M, Gunatilleke G. 2007. Neglected patients with a neglected disease? A qualitative study of lymphatic filariasis. *PLoS Negl Trop Dis* **1:**e128.

Rajan TV. 2005. Natural course of lymphatic filariasis: insights from epidemiology, experimental human infections, and clinical observations. *Am J Trop Med Hyg* **73:**995–998.

Ramaiah KD, Das PK, Michael E, Guyatt H. 2000a. The economic burden of lymphatic filariasis in India. *Parasitol Today* **16:**251–253.

Ramaiah KD, Radhamani MP, John KR, Evans DB, Guyatt H, Joseph A, Datta M, Vanamail P. 2000b. The impact of lymphatic filariasis on labour inputs in southern India: results of a multi-site study. *Ann Trop Med Parasit* **94:**353–364.

Ramzy RM, El Setouhy M, Helmy H, Ahmed ES, Abd Elaziz KM, Farid HA, Shannon WD, Weil GJ. 2006. Effect of yearly mass drug administration with diethylcarbamazine and albendazole on bancroftian filariasis in Egypt: a comprehensive assessment. *Lancet* **367:**992–999.

Richards FO, Ruiz-Tiben E, Hopkins DR. 2011. Dracunculiasis eradication and the legacy of the smallpox campaign: what's new and innovative? What's old and principled? *Vaccine* **29**(Suppl 4): D86–D90.

Ruiz-Tiben E, Hopkins DR. 2006. Dracunculiasis (guinea worm disease) eradication. *Adv Parasitol* **61:**275–309.

Simonsen PE. 2002. Filariases, p 1487–1526. *In* Cook GC, Zumla AI (ed), *Manson's Tropical Diseases*, 21st ed. W B Saunders, New York, NY.

Taylor MJ, von Geldern TW, Ford L, Hübner MP, Marsh K, Johnston KL, Sjoberg HT, Specht S, Pionnier N, Tyrer HE, Clare RH, Cook DAN, Murphy E, Steven A, Archer J, Bloemker D, Lenz F, Koschel M, Ehrens A, Metuge HM, Chunda VC, Ndongmo Chounna PW, Njouendou AJ, Fombad FF, Carr R, Morton HE, Aljayyoussi G, Hoerauf A, Wanji S, Kempf DJ, Turner JD, Ward SA. 2019. Preclinical development of an oral anti-*Wolbachia* macrolide drug for the treatment of lymphatic filariasis and onchocerciasis. *Sci Transl Med* **11:**eaau2086.

Weil GJ, Jacobson JA, King JD. 2020. A triple-drug treatment regimen to accelerate elimination of lymphatic filariasis: from conception to delivery. *Int Health* **13**(Supplement_1):S60–S64.

Weil GJ, Lammie PJ, Weiss N. 1997. The ICT Filariasis Test: A rapid-format antigen test for diagnosis of bancroftian filariasis. *Parasitol Today* **13:**401–404.

World Health Organization. 2020. Global programme to eliminate lymphatic filariasis: progress report, 2019. *Wkly Epidemiol Rec* **95:**509–524.

Zaman S, Nahar P, MacGregor H, Barker T, Bayisenge J, Callow C, Fairhead J, Fahal A, Hounsome N, Roemer-Mahler A, Mugume P, Tadele G, Davey G. 2020. Severely stigmatised skin neglected tropical diseases: a protocol for social science engagement. *Trans R Soc Trop Med Hyg* **114:**1013–1020.

Zhong CH, Zhen TM. 1991. Control and surveillance of filariasis in Shandong. *Chin Med J (Engl)* **104:**179–185.

Zouré HG, Wanji S, Noma M, Amazigo UV, Diggle PJ, Tekle AH, Remme JH. 2011. The geographic distribution of *Loa loa* in Africa: results of large-scale implementation of the Rapid Assessment Procedure for Loiasis (RAPLOA). *PLoS Negl Trop Dis* **5:**e1210.

5

The Blinding Neglected Tropical Diseases: Onchocerciasis (River Blindness) and Trachoma

I remember vividly those days in what was then called Upper Volta and is now Burkina Faso. . . . We had heard, before and during the visit, about the terrible disease called river blindness, and some had suggested that the World Bank should play a role in doing something about it. We could hardly pronounce the name of the disease, much less spell onchocerciasis, but were horrified by what we heard about it. Literally millions of people were at risk of a fate that could be worse than death in that society and time.

ROBERT S. MCNAMARA[1]

An estimated 216 million people in the world suffer from moderate or severe visual impairment, and 36 million people are blind.[2] Sadly, 90% of the world's blind people live in low- and middle-income countries (LMICs), and the majority (about two-thirds or more) of the cases of blindness are either preventable or curable.[3]

A startling example of how blindness disproportionately affects vulnerable and impoverished people in developing countries is a recent analysis of visual loss in post-conflict South Sudan.[4] Between 1983 and 2005, the Second Sudanese Civil War killed an estimated 1.9 million civilians, one of the highest death tolls since World War II, and forced the relocation of another 4 million people.[4] Because of its strategic location near oil fields, the Mankien district of Sudan was particularly affected by the conflict (Fig. 5.1).[4] A team from Cambridge University found that 4% of the people in Mankien villages were blind and that almost double that number had poor vision. Therefore, people living under the harsh conditions of South Sudan are 6 to 7 times more likely to become blind than are those living in the rest of the world. Indeed, the World Health Organization (WHO) considers blindness a severe public health problem in a community when the local prevalence is 1% or greater[5]; this region of Sudan exceeds even this level by 4-fold. Blindness in such

Forgotten People, Forgotten Diseases: The Neglected Tropical Diseases and Their Impact on Global Health and Development, Third Edition. Peter J. Hotez.
© 2022 American Society for Microbiology. DOI: 10.1128/9781683673903.ch05

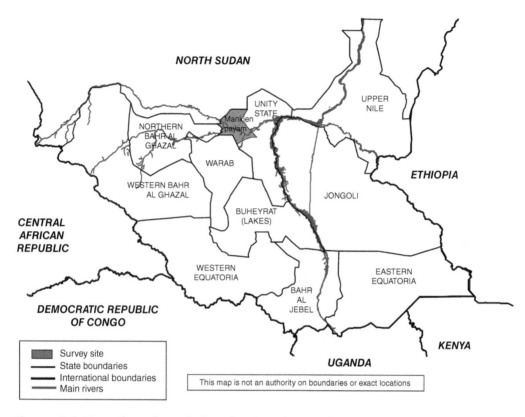

Figure 5.1 Map of southern Sudan showing the Mankien study site. (Reprinted from Ngondi J, et al. 2006. Prevalence and causes of blindness and low vision in southern Sudan. *PLoS Med* 3:e477. © 2006 Ngondi et al., CC BY 4.0)

resource-poor settings is especially onerous because of the absence of the support services found in the industrialized world. Not surprisingly, blind people who live in settings of rural poverty are at much higher risk of death and injury, possibly as much as 5-fold.[6]

Here we will consider the impact of blindness that results from two neglected tropical diseases (NTDs) that are entirely preventable with simple, safe, and effective preventive chemotherapy. Onchocerciasis is a parasitic helminth infection occurring in an estimated 19 million people in West and Central Africa. Trachoma is a bacterial infection causing visual impairment in approximately 2 million persons.

Onchocerciasis

Onchocerciasis is also known as river blindness because the *Simulium* black-flies that transmit this filarial infection breed along fast-flowing streams. The infection is caused by *Onchocerca volvulus*, a filarial worm with similarities to the parasites that cause lymphatic filariasis (LF) (chapter 4). The *Onchocerca* adult worms resemble long pieces of spaghetti up to 20 in. in length, which coil up in fibrous nodules that form under the skin. Both the adult female *Wuchereria bancrofti* worms that cause LF and the adult female *O. volvulus* worms also produce microscopic microfilariae. However, unlike the *W. bancrofti* micro-filariae, the *Onchocerca* microfilariae do not enter the bloodstream in large numbers. Instead, they mostly migrate in the skin, where they cause intense itching and disfigurement. However, from the skin some of the microfilariae also reach the eye, where they form small opacities. Over a period of years, sometimes decades, these opacities begin to coalesce and eventually block out light.[7] This progression is the basis by which *O. volvulus* causes blindness or visual impairment in approximately 1.15 million people.[8] Although onchocer-ciasis is best known as a cause of eye disease and blindness, the manifestations that result from the presence of microfilariae in the skin are almost as severe. These cutaneous manifestations are both highly disfiguring and stigmatizing,[9] and the associated itching has been alleged to be so intense that it can even prompt suicide. Yet another component of the onchocerciasis disease burden is emerging evidence for its links to seizures and epilepsy.

A majority of the estimated 19 million people infected with onchocerci-asis are impoverished subsistence farmers and their families living in West, Central, and parts of East Africa (Fig. 5.2). The disease is widely distributed in these regions, where it takes two major forms, a savanna form that is primarily associated with blindness and a forest form in which skin disease is a more prominent feature.[9] In some savanna communities, there is hyperendemic transmission of onchocerciasis, and blindness can be found in 10% or more of poor rural villages, while in the forest zone of West Africa, severe *Onchocerca* skin disease (OSD) afflicts millions of people. Outside of Africa, the at-risk population includes small foci remaining in Yemen in the Middle East and

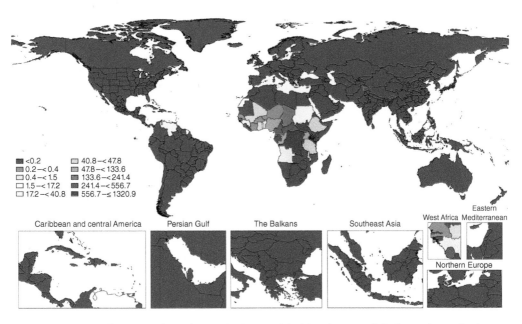

Figure 5.2 Distribution of onchocerciasis, worldwide. (From Global Burden of Disease Collaborative Network, 2020 [http://www.healthdata.org/results/gbd_summaries/2019/onchocerciasis-level-3-cause].)

in a tropical region of the Americas—the Amazon region of Venezuela and in neighboring areas of Brazil. In the Amazon, the disease disproportionately affects the Yanomami indigenous populations.

Humans become infected with onchocerciasis when *Simulium* blackflies, which breed along rivers or streams, deposit *O. volvulus* larvae while biting and feeding on blood (Fig. 5.3). Thus, whereas hookworm infective larvae enter human skin from the soil, both *O. volvulus* and *W. bancrofti* infective larvae gain entry into humans during the bite of an insect vector, a mosquito in the case of *W. bancrofti* and a blackfly for *O. volvulus*. Over the course of approximately 1 year, the larvae then migrate to different areas under the skin of the body (subcutaneous tissues) and develop into adult worms. The fibrous nodules created by the adult worms lie under the skin, typically located near bony prominences, especially the hip. Encased in these nodules, the adult worms can live for a decade or more.[9]

Like *W. bancrofti* and *Dracunculus medinensis*, adult female *O. volvulus* worms grow to enormous lengths, up to 20 in. or more. After fertilization, the female worm produces millions of microfilariae (about 500 to 1,500 microfilariae per day), which then migrate out of the nodules and travel in the skin over all aspects of the body. In some regions of Africa where river blindness is endemic, the density of microfilariae in the skin can be enormous. Some estimates suggest that a human can harbor as many as 100 million microfilariae,

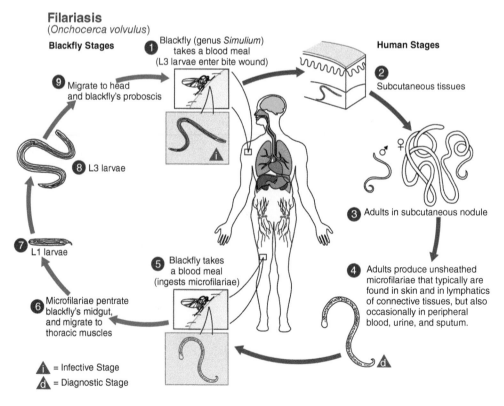

Figure 5.3 Life cycle of *O. volvulus*. (Modified from CDC-PHIL [ID#3413]/Alexander de Silva, PhD; Melanie Moser, 2002.)

with some regions of the skin having as many as 2,000 microfilariae per mg of skin.[9] A few of these microfilariae travel to the eye after spreading directly from the adjacent skin. The microfilariae residing in the skin and subcutaneous tissues develop further when they are taken up by blackflies as the insects are feeding on blood. Onchocerciasis is then transmitted to another person when the microfilariae develop into infective larvae (a period of about 6 to 12 days) and are deposited onto the skin during a subsequent insect bite.

Onchocerciasis is a debilitating and disabling disease because of resulting disfigurement and blindness. Disfigurement results primarily from OSD, an inflammatory condition that develops in response to the enormous numbers of microfilariae in the skin. In West Africa, OSD is sometimes known locally as craw craw. Contributing to the severity of OSD is the fact that the microfilariae themselves contain bacterial organisms (of the genus *Wolbachia*). The bacteria are in a sense parasites of the *Onchocerca* parasites and are referred to as *endosymbionts*, a concept first introduced (in chapter 4) when reporting on lymphatic filariasis. It is thought that as microfilariae die in the skin, the released endosymbiotic bacteria serve to incite additional host inflammatory responses.[10] Later (in chapter 11), we will see how the *Wolbachia* bacterial endosymbionts may be targeted as a novel therapeutic approach for the treatment of onchocerciasis.[10] Most commonly, OSD occurs on the legs, but it is not at all unusual for the entire body to be involved. Over a period of years, the chronic inflammatory reactions in the skin cause it to lose its elasticity and firmness so that the skin takes on a loose and aged appearance.[9] When OSD occurs in the groin, the skin often hangs down in folds, a condition sometimes known as "hanging groin."[2,9] Loss of skin pigmentation also results; this can occur in patches, imparting a leopard-like appearance.[10] Simultaneously, multiple nodules under the skin (commonly along the bony prominences of the hip, although hip nodules are more common in Africa than in the Americas) arise because of the presence of adult worms. In Africa's rain forest belt, OSD can be found in 50% or more of the villages.[9,11] In these regions, the stigma surrounding OSD is huge, and it is associated with severe ostracism and psychosocial aspects.[12,13] For instance, in eastern Nigeria, OSD is sometimes known by a moniker that translates as "that which prevents a woman from getting married," and the mistaken belief that this condition results from poor hygiene leads to even further isolation and even reduction in infant breastfeeding.[12,13] Affected men have diminished economic prospects from farming and other rural occupations.[12,13]

Onchocerca eye disease also results from the presence of microfilariae in the skin. Each time a microfilaria enters the eye and dies, it combines with cells from the human immune system to form tiny opacities in the cornea. Over a period of years, sometimes decades, these opacities merge and scar to the point where they block out light. In addition, the major nerve that connects the eye with the brain, the optic nerve, becomes affected; this exacerbates the processes leading to blindness.[9] In a typical West or Central African community affected by onchocerciasis, blindness commonly begins to occur in the 4th decade of life and then continues to increase. Thus, in the hardest-hit African communities, individuals who ordinarily are parents and heads of households, typically the family breadwinners, are now rendered blind. Frequently, they are led about their villages by their children or grandchildren and therefore no longer contribute appreciably to the economic well-being of their farming communities. The economic consequences of this scenario are absolutely devastating. When village blindness reaches epidemic proportions, there are too few people who can tend the fields, resulting in food shortages and the abandonment of homelands in otherwise fertile river valleys.[12] When subsistence farmers are forced to migrate into highland regions with poor soils, an entire village can be thrown into poverty.[13] Today in sub-Saharan Africa, we may still find rural savanna villages where 10% or more of the adult population is blind from onchocerciasis.[9] Still another aspect of the condition's impact is the increasing realization that onchocerciasis can cause epilepsy. This includes a unique pediatric form known as "nodding syndrome" associated with repeat head bobbing or nodding accompanied by impairments in neurocognition.[11]

In the foyer of the World Bank headquarters in Washington, DC, stands a remarkable bronze statue of a child leading a man blinded by onchocerciasis, presumably his father or grandfather, by a stick (Fig. 5.4). Throughout history

Figure 5.4 A bronze study of a blind man led by a young boy (R. T. Wallen). Reproductions of this statue are located at the World Bank and at the WHO. (Image courtesy of R.T. Wallen under license CC BY-SA 4.0.)

in sub-Saharan Africa, this tragic scene has been replayed millions of times. To understand the World Bank–onchocerciasis connection, we need to turn the clock back to 1972, when then World Bank president Robert McNamara witnessed firsthand the devastation wrought by onchocerciasis in West Africa. Although he is better known for escalating the Vietnam War in his role as secretary of defense in the Johnson administration, it is not widely appreciated that McNamara also helped to launch one of the world's most successful health and humanitarian programs. Today, the impact of his early efforts to improve the health of the world's poorest people is still widely felt. In 1972, while touring drought-stricken regions of Burkina Faso (then called Upper Volta) in West Africa, McNamara and his wife were overcome by the sight of villages affected by river blindness, particularly the large numbers of middle-aged men and women led by sticks.[14] In the 1970s, it was estimated that more than 60% of some savanna populations were infected with *O. volvulus*, with approximately one-half of the males over the age of 40 years being blind.[13] While visiting the capital, Ouagadougou, he met with scientists from ORSTROM (Office de Recherche Scientifique et Technique Outre-Mer), a scientific arm of the French overseas development agency, who informed McNamara that if blackflies could be controlled for as long as adult *O. volvulus* worms live in the human body (10 to 20 years), then in theory transmission of the disease could be interrupted. In response, McNamara held a summit in Paris, which ultimately led to the establishment of the Onchocerciasis Control Program (OCP), a unique partnership of the World Bank with the WHO, the Food and Agriculture Organization, and the United Nations Development Programme.

Launched in 1974, the OCP followed on the heels of earlier attempts to control populations of the blackfly vector in Ghana and in Francophone West Africa (including Burkina Faso, Ivory Coast, and Mali) led by ORSTROM and other organizations.[16] During its first 14 years of existence, the major focus of the OCP was to interrupt transmission of onchocerciasis by controlling the *Simulium* insect vector. Blackfly populations were controlled by frequent and periodic use of Temephos, an organophosphorus insecticide that destroys *Simulium* larvae. Earlier (in chapter 4), we saw how Temephos was applied to water in order to destroy copepod larvae for the control of guinea worm infection. For onchocerciasis, Temephos was delivered either through aerial spraying (helicopters or fixed-wing aircraft) or by ground application. The rationale for aerial spraying was to reduce blackfly populations in order to stop transmission and then sustain zero transmission for 2 decades or more, a time equivalent to the life of *O. volvulus* in human tissues.[12] Ultimately, the OCP used more environmentally friendly larvicides, often rotating them periodically to defeat emerging insect resistance. An important component of the OCP was an emphasis on parallel research to continually identify new larvicides.[15] This practice is a useful lesson for 21st-century large-scale efforts to control malaria with bed nets and indoor residual spraying.

In its first few years of existence, the OCP concentrated larvicidal spraying in seven West African countries—Benin, Burkina Faso, Côte d'Ivoire, Ghana, Mali, Niger, and Togo—where in many regions more than 60% of village populations were infected, with blindness rates that exceeded 10%. Despite this intense concentration of disease and misery, in most of the regions of these countries where onchocerciasis was endemic, the OCP was successful in eliminating onchocerciasis, meaning that transmission of the disease was reduced to zero.[15] Because blackflies do not respect international borders, an added key component of the OCP was international cooperation among the seven countries.[12] Ultimately, four additional countries—Guinea, Guinea-Bissau, Senegal, and Sierra Leone—were targeted for aerial spraying after OCP scientists discovered that blackflies were traveling up to 600 km away from the major targeted areas.[12] The achievements of the OCP are impressive. By the time it ceased operations in 2002, it was estimated that the OCP had prevented approximately 600,000 cases of blindness, while 18 million children born in the OCP area were freed from risk of river blindness and 25 million hectares of arable land (which could feed 17 million people) were made available.[15] Even today, decades after larvicidal spraying first began, many regions of West Africa remain onchocerciasis free. Moreover, these efforts were extremely cost-effective; some estimates indicate that the OCP achieved its goals for less than US$1 per person.[15]

The OCP was successful, but not entirely, because of insecticides and their widespread use. Roughly at the halfway mark in the program, a second control tool was introduced, which greatly enhanced the progress of the OCP and still provides a basis for onchocerciasis control programs today. During the 1970s, Merck & Co. scientists led by William Campbell co-discovered a new drug with anthelmintic properties. The precursor of ivermectin (later given the trade name Mectizan) originated from a microorganism (*Streptomyces avermitilis*) living in soil samples obtained from a Japanese golf course![16] Unlike human parasites, veterinary parasites represent a rather large commercial market because of their adverse health impact on livestock. What was so amazing about Merck's role in the story of human onchocerciasis was that after developing ivermectin for veterinary use, the company did not close the door on developing it for a human disease with no commercial value. Instead, Merck & Co. scientists, together with the WHO, began a number of clinical trials in regions of Africa and the Americas where onchocerciasis was endemic, which demonstrated that a single dose of ivermectin was both safe and effective in reducing the density of microfilariae living in human tissue.[16] This work ultimately was recognized with the awarding of the 2015 Nobel Prize in Physiology or Medicine to Dr. William Campbell who worked at Merck & Co., together with Satoshi Ōmura who co-discovered the molecule working at the famed Kitasato Institute in Japan. In efforts headed by then Merck & Co. CEO Roy Vagelos, the company in 1987 worked successfully with William Foege, then the Carter Center's executive director, to establish a unique public-private partnership for donating Mectizan free of charge to anyone who needed the drug for as long as it was needed.[16] The Mectizan Donation Program has since 1988 provided hundreds of millions of treatments for onchocerciasis, as well as for LF, and currently approximately 400 million treatments are being approved annually (www.mectizan.org).[7]

A key to the success of the OCP was providing poor people with oncho-cerciasis with access to Mectizan. Access was accomplished through an innovative use of community-based drug distributors. The early experience with Mectizan in the OCP was that well-trained and expensive mobile health teams would enter villages where onchocerciasis was endemic, only to find that villagers were often away, tending to their fields or hunting.[12] The excellent safety profile of Mectizan, however, meant that it was both safe and practical for the mobile teams to leave the drug in the village and allow community-based workers to administer it on their own.[13] Eventually, these steps led to the creation of a network of community drug distributors, today known as CDTI (community-directed treatment with ivermectin), which has since revolutionized access to this essential medicine. At the same time, many nongovernmental development organizations began to coordinate their activities through CDTI. In the future, it is likely that CDTI and CDTI-like programs could provide a mechanism for large-scale interventions against several NTDs, as well as for antimalaria measures, vitamin A distribution, and pediatric vaccinations.[12] Through the activities of the OCP and annual or semiannual treatments lasting more than a decade, onchocerciasis has been eliminated in areas of Mali and Senegal in which it was hyperendemic.

Within a few years after CDTI began, it was clear that the OCP was becoming an important success story, but because the OCP did not cover the disease in Central or East Africa (or in the Americas), there was a need for additional partnerships to cover these regions. The African Programme for Onchocerciasis Control (APOC) built on OCP successes to reach 19 additional African countries—Angola, Burundi, Cameroon, the Central African Republic, Chad, the Democratic Republic of the Congo, the Republic of the Congo, Equatorial Guinea, Ethiopia, Gabon, Kenya, Liberia, Malawi, Mozambique, Nigeria, Rwanda, Sudan, Tanzania, and Uganda—with an initial aim to treat 75 million people annually with Mectizan, but ultimately scaling to 90 million treatments annually.[12,15,17] CDTI also became a cornerstone of APOC, with ivermectin distributed annually in areas where the prevalence of villagers with *Onchocerca* nodules exceeds 20%. Monitoring and evaluation efforts have determined that CDTI has achieved enormous therapeutic successes in the target APOC countries.[12] Through APOC-sponsored CDTI, approximately 76 million people living in over 100,000 villages were receiving regular treatments from hundreds of thousands of community distributors.[7] This achievement was highly cost-effective, estimated at US$6.50 per disability-adjusted life year.[12,15] For much of its existence, APOC was headed by a charismatic and highly effective African woman, Uche Amazigo, who managed the program from offices in Ouagadougou, Burkina Faso, the birthplace of onchocerciasis control. Later, it came under the direction of Paul-Samson Lusamba-Dikassa as part of AFRO, the African regional offices of the WHO and the World Bank. As APOC was integrated into Africa's primary health care system, it became a model for distributing other interventions. Particularly exciting was the observation that CDTI greatly enhances the efficiency of other health interventions. For example, a study conducted by Frank O. Richards of the Carter Center has shown that by piggybacking onto CDTI, antimalaria bed net distribution increased 9-fold. As a result, several nongovernmental development organizations, including the Carter Center and Helen Keller International, are now linking CDTI with malaria control initiatives.[12] Later (in chapter 10), we will see how this is just the beginning of an effort to link multiple NTD and malaria control efforts. Meanwhile, in the Americas, the Onchocerciasis Elimination Programme for the Americas (OEPA) focused on eliminating the disease in the Western Hemisphere, leaving only a single area on the Brazil-Venezuela border as the last remaining focus of ongoing transmission.

In 1995, APOC was initially launched as a 12-year program, but has since been extended to a 20-year time frame finally ending in 2015. It was estimated that the community-directed drug distributors working through APOC delivered more than 500 million treatments in almost 150,000 African communities; as a result, the prevalence of onchocerciasis in the 19 APOC countries has been reduced by approximately 73%.[12] Additional estimates indicate that up to 17 million disability-adjusted life years were averted.[12] However, based on the premise that annual treatments would need to be continued for as long as the adult *O. volvulus* parasites live in the body, it is widely accepted that the activities of APOC or an equivalent program would likely need to be sustained for much longer to make a significant impact on the African elimination of onchocerciasis.[7,17] Based in part on the demonstration of the feasibility of onchocerciasis elimination after 15 to 17 years of mass drug administration (MDA) in Mali and Senegal,[15] some investigators initially proposed long-distance targets for onchocerciasis elimination.[17]

In 2016, APOC was reorganized and its remit extended to include additional NTDs for mass treatment. Under the new Expanded Special Project for Elimination of Neglected Tropical Diseases (ESPEN), a private-public partnership between WHO-AFRO, African member states, and key NTD partners, efforts are under way to move toward eliminating not only onchocerciasis but at least four other NTDs.[17] Dr. Maria Rebollo Polo, a public health physician, serves as the ESPEN inaugural director. The pace of treatment through ESPEN has continued in the tradition of APOC. For instance, the WHO estimates that in 2019 approximately 70% of the more than 200 million people who require preventive treatments or MDA for onchocerciasis received donated ivermectin.

Efforts are under way to examine in detail the potential for finally eliminating onchocerciasis—a "Stamp Out Oncho" campaign—with a possible target date for 2030. However, many stars will need to align for this to happen. Among them is the potential replacement of ivermectin with a related drug known as moxidectin, which appears to be superior in terms of suppressing the number of microfilariae for longer periods.[19] However, this would require public donations of the drug. Another is shifting from annual or biannual treatments to possibly quarterly interventions, or also reexamining treatments in areas where another parasitic infection known as loiasis (infection with the filarial parasite *Loa loa*) occurs.[19] The issue with *L. loa* is that some

individuals with heavy infections with this parasite can experience toxic neurologic responses if they receive a dose of ivermectin or the equivalent.

Yet another challenge to global efforts to control onchocerciasis is the
observation that *O. volvulus* resistance to ivermectin (and possibly moxidectin as well, given that it is from a similar drug class) may have developed in
some specific regions of Ghana and possibly other OCP countries.[19] Whether
or not resistance has truly developed is considered quite controversial, but if
it is confirmed, presumably it was brought about because of repeated MDA
with Mectizan over a period of 2 decades. There is no evidence that resistance
will become widespread as it did for the antimalarial drug chloroquine. In any
case, without the benefit of having additional backup drugs in hand, we have
no choice but to continue to aggressively pursue CDTI in most of the regions
where onchocerciasis is endemic. Identification of possible emerging drug
resistance to ivermectin, however, points out the need for constant vigilance
in developing and testing new control tools, which could include new drug
development. Later (in chapter 11), we will examine the prospects of new
classes of anthelmintic drugs for onchocerciasis, which are based on targeting
the parasite's *Wolbachia* bacterial endosymbionts, and even the possibility that
an anti-*Onchocerca* vaccine could be developed.

Trachoma

Trachoma is an infection caused by the bacterium *Chlamydia trachomatis*
causing blindness or visual impairment in approximately 1.9 million people.[20]
It is still likely the most important cause of infectious blindness worldwide,
accounting for up to 3% of all the world's cases of blindness.[21] The disease is
highly prevalent in remote and rural areas of the African Sahel. The Sahel
refers to ancient lands bridging the Sahara Desert to the north and the African
savanna to the south, stretching across the African continent. In addition, trachoma is still common in Pakistan and Afghanistan in Central Asia; Myanmar
and Papua New Guinea in Southeast Asia, and Honduras in Latin America.[20]
Like the other NTDs, trachoma is an infection that occurs only in the poorest
people of the world. The infection is spread from person to person on dirty
hands and clothing, but it is also carried by flies, which can carry the *Chlamydia*
bacteria from the eye discharges of one person to another. For that reason, the
disease often strikes entire families. Moreover, women are affected 3 times
more than men. According to the nonprofit International Trachoma Initiative
(ITI), based at the Task Force for Global Health in Atlanta, GA, when a woman
who runs the household can no longer work, the burden falls on the daughter,
who is forced to leave school, in turn losing her opportunity for education.[21]
Such an impact on women and the close association between trachoma and
filth and flies ensure a tight link between the disease and poverty. Therefore,
trachoma, like onchocerciasis, is an excellent example of an NTD that not only
occurs in the setting of poverty but helps to promote it as well. An estimated
US$5.3 billion is lost annually from the global trachoma disease burden.[22]

At the turn of the 20th century, trachoma was a major reason why poor
European immigrants were prevented from entering the United States. Back

then, it was a common practice for U.S. Public Health Service physicians assigned to Ellis Island to screen potential immigrants by turning up their eyelids and looking for signs of the disease. During the early part of the 20th century, trachoma was a common disease in the United States, particularly on Native American reservations, where crowding, poverty, and lack of clean water combined to create conditions favorable for transmission.[23] In 1913, President Woodrow Wilson signed antitrachoma legislation making the control of this infection a priority for the United States. As a result of improved housing and standards of living, trachoma is no longer a major public health threat,[23] although the infection may still occur sporadically among the Navajo and other Native American tribes in the West.[24] Today, the disease principally occurs throughout the poorest developing countries of Africa, Central Asia, the Middle East, India, and Southeast Asia (Fig. 5.5). Trachoma is also found in Australian regions that are home to aboriginal populations.

Unlike many NTDs discussed previously, which depend on adequate moisture and rainfall to ensure survival of parasitic larval stages in the soil, in water, or in arthropod vectors, trachoma more often occurs in dry and dusty environments, particularly those without adequate sanitation and with large populations of flies. The environmental risk factors of dryness and filth for trachoma have been summarized by the six D's (dryness, dust, dirt, dung, discharge, and density [overcrowding]) or the five F's (flies, feces, faces, fingers, and fomites).[20,21] Dirty and unwashed faces, as well as human or animal feces on the ground, attract flies that spread the disease.[20,21] Other important risk factors include lack of adequate access to clean water, which requires villagers

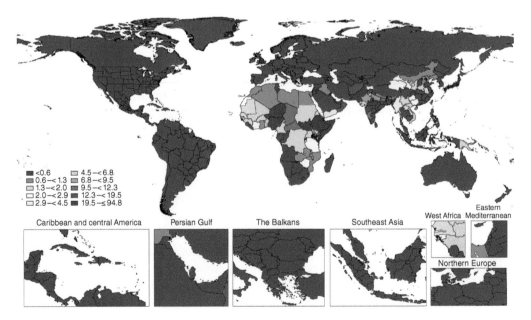

Figure 5.5 Distribution of trachoma, worldwide. (From Global Burden of Disease Collaborative Network, 2020 [http://www.healthdata.org/results/gbd_summaries/2019/trachoma-level-3-cause].)

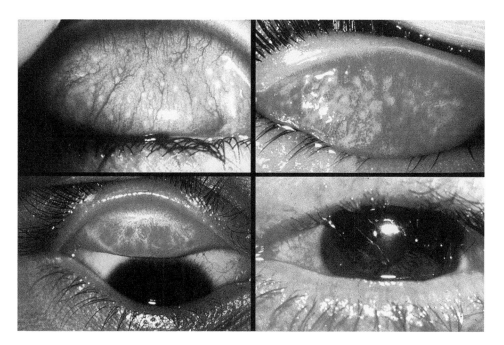

Figure 5.6 Trachoma patient—a fifth-grade student—in Silti Zone, Ethiopia. (Image by Daniel Ostermayer, MD, Deputy Editor and Head of Software Development, WikEM, Assistant Professor Emergency Medicine, UT-Houston, CC-BY-SA 4.0.)

to walk long distances for their water supply. This in turn lessens the likelihood of adequate hygiene.[20,21] The absence of education, especially maternal education, is another risk factor.

Trachoma is a chronic infection that usually begins in childhood, sometimes with redness (pinkeye), itching, and pain[20,25] (Fig. 5.6). The disease and blindness result from recurrent and multiple infections with *C. trachomatis*, which occur over a period of 10 to 20 years. During this time, recurrent infection leads to scarring of the eyelids. When the scarred eyelids turn inward, the lashes scratch the surface of the eyeball, leading to inflammation and scarring of the cornea (an ordinarily transparent part of the eye) and ultimately blindness. These later stages are known as trachomatous trichiasis.[25] Trachomatous trichiasis is often a painful condition.

Because these pathologic processes are insidious and occur over a period of years or even decades, and because they occur only among poor people usually in remote and rural areas, trachoma has been often overlooked as a serious public health problem. Adding to the neglect of trachoma is the commonness of this condition in areas made inaccessible by conflict or post-conflict turmoil. However, over the last several years, the WHO, together with the ITI, has developed an innovative strategy of trachoma management and prevention that goes by the acronym SAFE.[20,21,26]

- S is for simple surgery, for the immediate relief of trichiasis to prevent corneal scarring. In many countries, health professionals other than physicians, such as nurses and specialized ophthalmic assistants, can be trained to perform this simple operation, which involves making a slit in the eyelid and peeling back a portion to prevent further scarring.

- A is for antibiotics. The major antibiotic used is a single oral dose of azithromycin, marketed by Pfizer as the drug Zithromax. Zithromax has largely replaced tetracycline, which had to be administered topically for a period of 4 to 6 weeks.

- F is for facial hygiene. Children with clean faces have a reduced likelihood of developing severe trachoma, especially when face washing is combined with administration of Zithromax. Face washing can also help to interrupt transmission.

- E is for environmental improvement. This component relies on improving access to clean water and improving sanitation and latrines to reduce fly populations. Combined with health education, environmental change is critical for reducing transmission.

One of the most important breakthroughs in the global control of trachoma was the discovery that a single dose of azithromycin was as effective as 4 to 6 weeks of topical tetracycline for the treatment of trachoma.[25,26] Azithromycin bears some structural resemblance to tetracycline, but its chemistry is modified so that it achieves greater penetration inside human cells. Since *C. trachomatis* lives inside cells (it is an intracellular bacterium), this finding afforded an opportunity to achieve similar antibiotic efficacy with fewer doses. Although Zithromax was developed primarily as the Z-PAK for the treatment of outpatients with pneumonia and other community-acquired infections, during the 1990s, Pfizer, together with the Edna McConnell Clark Foundation, conducted clinical trials showing that a single dose of the drug was also effective for treating trachoma. In 1998, Pfizer and the Clark Foundation established the ITI, which included Pfizer's donation of US$60 million worth of azithromycin.[26] Given that Zithromax was one of Pfizer's best-selling products and that there was a risk that the donated drug would be sold in developing countries on the black market,[27] the donation of this product is considered an extraordinary gesture of corporate philanthropy and one that rivals Merck's Mectizan Donation Program. Formerly under the leadership of Joe Cook, Jacob Kumaresan, and Danny Haddad, the ITI is currently under the direction of Paul Emerson, formerly of the Carter Center, and PJ Hooper, who serves as deputy director. The ITI works closely with the Alliance for the Global Elimination of Trachoma based at the WHO.[26]

Through the activities of the WHO Alliance, the ITI, and other organizations dedicated to fighting blindness, such as Helen Keller International, the WHO has established guidelines to administer azithromycin en masse in affected communities where the prevalence of trachoma exceeds 20%. In some cases, MDA is used when the prevalence exceeds 5%, although targeted treatment of specific groups is sometimes warranted in these settings.[27] In 1999, Morocco became the first country to embark on a comprehensive program of trachoma control, which used both the SAFE strategy and single-dose azithromycin.[26] Prior to the control campaign, an estimated 5.4% of the population was affected, with all of the cases in five southeastern rural provinces.[26] To ensure success, Morocco's National Blindness Control Program created a comprehensive partnership comprising five government ministries including health and education; international organizations such as the ITI, Helen Keller, UNICEF, and the WHO; bilateral and multilateral agencies; and local nongovernmental organizations.[27] Their efforts included mobile surgical units staffed by physicians and nurses to perform tens of thousands of surgeries and administer several million doses of Zithromax, health education and media blitzes, and the construction of new latrines and wells for rural communities.[27] In 2005, Morocco became the first developing country to successfully interrupt the transmission of trachoma through the SAFE strategy and to seek certification that it had eliminated the disease. In November 2006, the king of Morocco hosted a high-profile celebration to mark the end of mass distribution of antibiotics.[26]

In 2020, the WHO determined that trachoma has been eliminated in nine countries—Cambodia, China, Ghana, Islamic Republic of Iran, Lao People's Democratic Republic, Mexico, Morocco, Nepal, and Oman—while Gambia, Iraq, Myanmar, and Togo may soon achieve elimination.[27] Today, the ITI is supporting trachoma control programs in 40 countries in Africa, the Middle East, and Asia, in addition to Colombia and Guatemala in Latin America.[22] Moreover, the WHO Alliance estimates that the number of people living with trachomatous trichiasis has fallen from 7.6 million people in 2002 to 2 million in 2020. Still another benefit noted from mass treatment with azithromycin is the finding that overall child mortality may diminish when the drug is administered to large populations where trachoma is endemic. These studies were led by the MORDOR (Macrolides Oraux pour Réduire les Décès avec un Oeil sur la Résistance) group working in Africa. The mechanisms of this important global health observation are unknown but may be related to collateral reductions or treatments of bacterial respiratory or bacterial pathogens, thereby reducing cases of bacterial pneumonias or diarrheal infections, respectively.

SUMMARY POINTS   The Blinding NTDs

Blindness
- 90% of the world's blind people live in low- and middle-income countries.
- The majority of the cases of blindness are either preventable or curable.

Onchocerciasis
- Onchocerciasis is also known as river blindness because the *Simulium* blackflies that transmit this filarial infection breed along fast-flowing streams.
- The infection is caused by *Onchocerca volvulus*, a filarial worm with similarities to the parasites that cause LF.
- An estimated 19 million people are infected with onchocerciasis, which is responsible for visual impairment or blindness in approximately 1.15 million people.

- Onchocerciasis also produces a serious, debilitating, and stigmatizing skin disease (OSD), and is emerging as an important cause of epilepsy, especially in children who experience debilitating nodding syndrome.
- The OCP was launched in 1974 following the advocacy of then World Bank president Robert McNamara. Over the next 30 years, the OCP helped to facilitate the elimination of onchocerciasis in large regions of 11 West African countries.
- The cornerstone of the OCP was initially insecticidal spraying, but this was gradually replaced by annual treatments with ivermectin. Since 1988, ivermectin has been donated by Merck through its Mectizan Donation Program.
- Later, onchocerciasis control continued through the activities of APOC and the OEPA. The cornerstone of APOC is CDTI. In 2016, APOC transitioned to ESPEN with an expanded remit that includes mass drug administration or preventive treatments for several NTDs.
- Through these activities, elimination targets for onchocerciasis have been proposed with a 10-year time horizon. However, the details of exactly how future onchocerciasis elimination activities will be conducted must still be worked out, and may depend on introducing newer anthelmintic drugs such as moxidectin, or even an onchocerciasis vaccine.

Trachoma

- Trachoma is an infection caused by the bacterium *C. trachomatis*; its chronic trachomatous trichiasis form causes visual impairment in approximately 2 million people, especially in the Sahel region of Africa and in Central Asia.
- It is the most important cause of infectious blindness worldwide.
- Trachoma spreads from person to person on dirty hands and clothing. The disease is also carried by flies that can carry the *Chlamydia* bacteria from the eye discharges of one person to another.
- Women are affected 3 times more than are men.
- The environmental risk factors of dryness and filth for trachoma have been summarized by the six D's (dryness, dust, dirt, dung, discharge, and density [overcrowding]) or the five F's (flies, feces, faces, fingers, and fomites).
- An innovative strategy of trachoma management and prevention goes by the acronym SAFE—surgery, antibiotics, face washing, and environmental control.
- A breakthrough in the global control of trachoma was the discovery that a single dose of azithromycin was as effective as 4 to 6 weeks of topical tetracycline. Pfizer, in association with the ITI, is providing large-scale donations of azithromycin (Zithromax) in approximately 40 countries where trachoma is endemic. These activities are led by the WHO Alliance for the Global Elimination of Trachoma.
- In 2006, Morocco became the first country to eliminate trachoma using the SAFE strategy and single-dose azithromycin. Since then, trachoma has been eliminated in multiple countries.
- It has been further noted that annual treatments with azithromycin produce collateral benefits that include reductions in overall child mortality. However, the scientific basis of this observation is still not well understood.

Notes

1. Quotation from Robert S. McNamara, former president of the World Bank, from http://documents1.worldbank.org/curated/en/100751468202452180/text/315690WP0Box349497B01PUBLIC1.txt

2. World Health Organization, 2012a; and Bourne et al., 2017.

3. In developing countries, the leading causes of pediatric blindness are vitamin A deficiency, congenital cataract, and ophthalmia neonatorum, an infectious disease; in adults, the leading causes of blindness are cataract, trachoma, and chronic glaucoma (McGavin, 2002).

4. Two papers on infectious and other causes of blindness as a consequence of the war are Buchan, 2006; and Ngondi et al., 2006.

5. The WHO Prevention of Blindness team works with member states to prevent blindness and restore sight. Its global target is to reduce the prevalence of blindness to less than 0.5% in all countries or less than 1% in any particular country (https://www.who.int/health-topics/blindness-and-vision-loss#tab=tab_1).

6. Courtright et al., 1997.

7. Basáñez et al., 2006; https://mectizan.org/what/ and http://www.who.int/apoc/en.

8. The figure of 1.15 million with blindness or visual impairment is from https://www.who.int/news-room/fact-sheets/detail/onchocerciasis.

9. Two excellent reviews of onchocerciasis are Rodríguez-Pérez et al., 2011; and Brattig et al., 2020. Information about onchocerciasis among the Yanomami indigenous populations is in Sauerbrey et al., 2018. Information about onchocerciasis in Yemen is in Al-Kubati et al., 2018. Clinical descriptions of onchocerciasis are found in Simonsen, 2002.

10. A description of these *Wolbachia* bacterial endoparasites and their potential as an onchocerciasis treatment target can be found in Taylor et al., 2019.

11. Kale, 1998. Summaries of onchocerciasis epilepsy and nodding syndrome can be found in Gumisiriza et al., 2020; and Johnson et al., 2020.

12. Information on the accomplishments of APOC can be found in Amazigo et al., 2006; and is updated at the APOC website: www.who.int/apoc/en/index.html. The information about 17 million DALYs averted based on past estimates and projections is reported in Coffeng et al., 2013.

13. Papers on stigma and OSD include Amazigo and Obikeze, 1992; and Vlassoff et al., 2000.

14. The story of Robert McNamara's interest in onchocerciasis can be found in Cobb, 2001.

15. Excellent descriptions of the modern history of programs for onchocerciasis control can be found in Boatin and Richard, 2006; and Levine and the What Works Working Group, 2004, p. 57–64. Descriptions of elimination efforts in Mali and Senegal can be found in Diawara et al., 2009; and Cupp et al., 2011.

16. The history of ivermectin development can be found at www.stanford.edu/class/humbio103/ParaSites2005/Ivermectin/History.htm; https://mectizan.org/what/history/; and in Geary,

2005; and Crump and Omura, 2011. The website for the Mectizan Donation Program is www.mectizan.org.

17. Information about ESPEN can be found at https://espen.afro.who.int/about-espen.

18. Information about progress in treating onchocerciasis in 2019 and progress toward elimination can be found in World Health Organization, 2020a.

19. An important paper on possible emerging ivermectin resistance is Osei-Atweneboana et al., 2007; the accompanying commentary is Hotez, 2007. Details of moxidectin are in Milton et al., 2020. The challenges of eliminating onchocerciasis are highlighted in NTD Modelling Consortium Onchocerciasis Group, 2020.

20. The latest GBD 2019 analysis of trachoma (Global Burden of Disease Collaborative Network, 2020) is at http://www.healthdata.org/results/gbd_summaries/2019/trachoma-level-3-cause. General papers on trachoma and blindness in developing countries include Kasi et al., 2004; and Buchan, 2006.

21. The website of the ITI is www.trachoma.org.

22. The economic burden estimates are in Frick et al., 2003a; and Frick et al., 2003b.

23. Allen and Semba, 2002.

24. Rearwin et al., 1997.

25. A good clinical and epidemiological description of trachoma can be found in McGavin, 2002.

26. An excellent description of the formation of the ITI and the use of azithromycin for the control of trachoma in Morocco and elsewhere can be found in Levine and the What Works Working Group, 2004, p. 83–89; International Trachoma Initiative, 2010; and at www.trachoma.org.

27. The list of countries and estimates of progress can be found in World Health Organization, 2020b. The work of the MORDOR group is reported in Porca et al., 2009; and Keenan et al., 2018.

References

Al-Kubati AS, Mackenzie CD, Boakye D, Al-Qubati Y, Al-Samie AR, Awad IE, Thylefors B, Hopkins A. 2018. Onchocerciasis in Yemen: moving forward towards an elimination program. *Int Health* 10(suppl_1):i89–i96.

Allen SK, Semba RD. 2002. The trachoma menace in the United States, 1897-1960. *Surv Ophthalmol* 47:500–509.

Amazigo U, Noma M, Bump J, Benton B, Liese B, Yameogo L, Zoure H, Seketeli A. 2006. Onchocerciasis, p 215–222. *In* Jamison DT, Feachem RG, Makgoba MW, Bos ER, Bingana FK, Hofman KJ, Rogo KO (ed), *Disease and Mortality in Sub-Saharan Africa*, 2nd ed. World Bank, Washington, DC.

Amazigo UO, Obikeze DS. 1992. *Socio-Cultural Factors Associated with the Prevalence and Intensity of Onchocerciasis and Onchodermatitis among Adolescent Girls in Rural Nigeria. Who/TDR Final Project Report.* World Health Organization, Geneva, Switzerland.

Basáñez MG, Pion SD, Churcher TS, Breitling LP, Little MP, Boussinesq M. 2006. River blindness: a success story under threat? *PLoS Med* 3:e371.

Blackburn BG, Eigege A, Gotau H, Gerlong G, Miri E, Hawley WA, Mathieu E, Richards F. 2006. Successful integration of insecticide-treated bed net distribution with mass drug administration in Central Nigeria. *Am J Trop Med Hyg* 75:650–655.

Bourne RRA, et al, Vision Loss Expert Group. 2017. Magnitude, temporal trends, and projections of the global prevalence of blindness and distance and near vision impairment: a systematic review and meta-analysis. *Lancet Glob Health* 5:e888–e897.

Brattig NW, Cheke RA, Garms R. 2020. Onchocerciasis (river blindness)—more than a century of research and control. *Acta Trop* 218:105677.

Buchan J. 2006. Visual loss in postconflict southern Sudan. *PLoS Med* 3:e450.

Cobb C Jr. 2001. Africa: elimination of river blindness "possible within ten years." *AllAfrica.* December 14, 2001.

Coffeng LE, Stolk WA, Zouré HG, Veerman JL, Agblewonu KB, Murdoch ME, Noma M, Fobi G, Richardus JH, Bundy

DA, Habbema D, de Vlas SJ, Amazigo UV. 2013. African Programme For Onchocerciasis Control 1995-2015: model-estimated health impact and cost. *PLoS Negl Trop Dis* 7:e2032.

Courtright P, Kim SH, Lee HS, Lewallen S. 1997. Excess mortality associated with blindness in leprosy patients in Korea. *Lepr Rev* 68:326–330.

Crump A, Omura S. 2011. Ivermectin, 'Wonder drug' from Japan: the human use perspective. *Proc Jpn Acad Ser B Phys Biol Sci.* 2011 Feb 10; 87(2):13–28. doi: 10.2183/pjab.87.13 PMCID: PMC3043740IF

Cupp EW, Sauerbrey M, Richards F. 2011. Elimination of human onchocerciasis: history of progress and current feasibility using ivermectin (Mectizan®) monotherapy. *Acta Trop* 120(Suppl 1):S100–S108.

Dandona L, Dandona R. 2006. What is the global burden of visual impairment? *BMC Med* 4:6.

Diawara L, Traoré MO, Badji A, Bissan Y, Doumbia K, Goita SF, Konaté L, Mounkoro K, Sarr MD, Seck AF, Toé L, Tourée S, Remme JH. 2009. Feasibility of onchocerciasis elimination with ivermectin treatment in endemic foci in Africa: first evidence from studies in Mali and Senegal. *PLoS Negl Trop Dis* 3:e497.

Frick KD, Basilion EV, Hanson CL, Colchero MA. 2003a. Estimating the burden and economic impact of trachomatous visual loss. *Ophthalmic Epidemiol* 10:121–132.

Frick KD, Hanson CL, Jacobson GA. 2003b. Global burden of trachoma and economics of the disease. *Am J Trop Med Hyg* 69(Suppl):1–10.

Geary TG. 2005. Ivermectin 20 years on: maturation of a wonder drug. *Trends Parasitol* 21:530–532.

Global Burden of Disease Collaborative Network. 2020. Global Burden of Disease Study 2019 (GBD 2019). GBD Results Tool. Institute for Health Metrics and Evaluation, Seattle, WA.

Gumisiriza N, Mubiru F, Siewe Fodjo JN, Mbonye Kayitale M, Hotterbeekx A, Idro R, Makumbi I, Lakwo T, Opar B, Kaducu J, Wamala JF, Colebunders R. 2020. Prevalence and incidence of nodding syndrome and other forms of epilepsy in onchocerciasis-endemic areas in northern Uganda after the implementation of onchocerciasis control measures. *Infect Dis Poverty* 9:12.

Haddad D. 2012. Trachoma: the beginning of the end? *Community Eye Health* 25:18.

Hotez PJ. 2007. Control of onchocerciasis—the next generation. *Lancet* 369:1979–1980.

Hotez P. 2011. Enlarging the "Audacious Goal": elimination of the world's high prevalence neglected tropical diseases. *Vaccine* 29(Suppl 4):D104–D110.

International Trachoma Initiative 2010. *Zithromax® in the Elimination of Blinding Trachoma: a Program Manager's Guide.* Decatur, GA: International Trachoma Initiative. https://www.trachoma.org/sites/default/files/partner-resource/2017-03/itizithromax-managers-guide0.pdf.

Johnson TP, Sejvar J, Nutman TB, Nath A. 2020. The pathogenesis of nodding syndrome. *Annu Rev Pathol* 15:395–417.

Kale OO. 1998. Onchocerciasis: the burden of disease. *Ann Trop Med Parasitol* 92(Suppl 1):S101–S115.

Kasi PM, Gilani AI, Ahmad K, Janjua NZ. 2004. Blinding trachoma: a disease of poverty. *PLoS Med* 1:e44.

Keenan JD, Bailey RL, West SK, Arzika AM, Hart J, Weaver J, Kalua K, Mrango Z, Ray KJ, Cook C, Lebas E, O'Brien KS, Emerson PM, Porco TC, Lietman TM, MORDOR Study Group. 2018. Azithromycin to reduce childhood mortality in sub-Saharan Africa. *N Engl J Med* 378:1583–1592.

Levine R, What Works Working Group. 2004. *Millions Saved: Proven Successes in Global Health.* Center for Global Development, Washington, DC.

Mackenzie CD, Homeida MM, Hopkins AD, Lawrence JC. 2012. Elimination of onchocerciasis from Africa: possible? *Trends Parasitol* 28:16–22.

McGavin DD. 2002. Ophthalmology in the tropics and sub-tropics, p 301–362. *In* Cook GC, Zumla AI (ed), *Manson's Tropical Diseases,* 21st ed. W B Saunders, New York, NY.

Milton P, Hamley JID, Walker M, Basáñez MG. 2020. Moxidectin: an oral treatment for human onchocerciasis. *Expert Rev Anti Infect Ther* 18:1067–1081.

Ngondi J, Ole-Sempele F, Onsarigo A, Matende I, Baba S, Reacher M, Matthews F, Brayne C, Emerson PM. 2006. Prevalence and causes of blindness and low vision in southern Sudan. *PLoS Med* 3:e477.

NTD Modelling Consortium Onchocerciasis Group. 2019. The World Health Organization 2030 goals for onchocerciasis: insights and perspectives from mathematical modelling: NTD Modelling Consortium Onchocerciasis Group. *Gates Open Res* 3:1545.

Osei-Atweneboana MY, Eng JK, Boakye DA, Gyapong JO, Prichard RK. 2007. Prevalence and intensity of *Onchocerca volvulus* infection and efficacy of ivermectin in endemic communities in Ghana: a two-phase epidemiological study. *Lancet* 369:2021–2029.

Porco TC, Gebre T, Ayele B, House J, Keenan J, Zhou Z, Hong KC, Stoller N, Ray KJ, Emerson P, Gaynor BD, Lietman TM. 2009. Effect of mass distribution of azithromycin for trachoma control on overall mortality in Ethiopian children: a randomized trial. *JAMA* 302:962–968.

Rearwin DT, Tang JH, Hughes JW. 1997. Causes of blindness among Navajo Indians: an update. *J Am Optom Assoc* 68:511–517.

Sauerbrey M, Rakers LJ, Richards FO. 2019. Progress toward elimination of onchocerciasis in the Americas. *Int Health* 10(Suppl_1):i71–i78.

Simonsen PE. 2002. Filariases, p 1487–1526. *In* Cook GC, Zumla AI (ed), *Manson's Tropical Diseases,* 21st ed. W B Saunders, New York, NY.

Smith JL, Haddad D, Polack S, Harding-Esch EM, Hooper PJ, Mabey DC, Solomon AW, Brooker S. 2011. Mapping the global distribution of trachoma: why an updated atlas is needed. *PLoS Negl Trop Dis* 5:e973.

Taylor MJ, von Geldern TW, Ford L, Hübner MP, Marsh K, Johnston KL, Sjoberg HT, Specht S, Pionnier N, Tyrer HE, Clare RH, Cook DAN, Murphy E, Steven A, Archer J, Bloemker D, Lenz F, Koschel M, Ehrens A, Metuge HM, Chunda VC, Ndongmo Chounna PW, Njouendou AJ, Fombad FF, Carr R, Morton HE, Aljayyoussi G, Hoerauf A, Wanji S, Kempf DJ, Turner JD, Ward SA. 2019. Preclinical development of an oral anti-*Wolbachia* macrolide drug for the treatment of lymphatic filariasis and onchocerciasis. *Sci Transl Med* **11**:eaau2086.

Vlassoff C, Weiss M, Ovuga EB, Eneanya C, Nwel PT, Babalola SS, Awedoba AK, Theophilus B, Cofie P, Shetabi P. 2000. Gender and the stigma of onchocercal skin disease in Africa. *Soc Sci Med* **50**:1353–1368.

World Health Organization.1995. *Expert Committee on Onchocerciasis Control. Technical Report Series No. 852*. World Health Organization, Geneva, Switzerland.

World Health Organization. 2020a. Elimination of human onchocerciasis: progress report, 2019-20. *Wkly Epidemiol Rec* **95**:545–556.

World Health Organization. 2020b. WHO Alliance for the Global Elimination of Trachoma by 2020: progress report, 2019. *Wkly Epidemiol Rec* **95**:349–360.

6 The Mycobacterial Infections: Buruli Ulcer and Leprosy

Witches are known mainly for spreading mysterious diseases in Ghanaian societies, like tuberculosis, Buruli, and leprosy. In my case, I think it's caused by witchcraft because I have good hygiene. I hope it is only a witch who can bring such a strange sickness or disease to someone because God the Almighty loved us and will not bring us such a sickness.

BURULI PATIENT FROM GHANA[1]

Now the leper on whom the sore is, his clothes shall be torn and his head bare; and he shall cover his mustache, and cry, "Unclean! Unclean!" He shall be unclean. All the days he has the sore he shall be unclean. He is unclean, and he shall dwell alone; his dwelling shall be outside the camp.

LEVITICUS 13[1]

Mycobacteria are slim bacteria with unusual growth requirements and unique structural and biochemical properties that place them in their own category of microbes. Unlike many other types of bacteria, the mycobacteria exhibit the ability to grow inside cells of our immune system known as macrophages. This ability is a remarkable feature because immunologists consider macrophages "professional killer cells," i.e., the cells best armed to ward off invading bacteria and other pathogens. Through evolution, however, the mycobacteria have adapted such that they not only thrive and multiply in macrophages, but in some cases actually use the macrophages as vehicles for transport to specific body tissues.[2]

Tuberculosis (TB), caused by the bacterium *Mycobacterium tuberculosis*, is the best-known mycobacterial infection of humans. As urbanization increased during the 18th and 19th centuries, TB became known as the "white plague" and may have caused up to 30% of all deaths in Europe, including those of

Forgotten People, Forgotten Diseases: The Neglected Tropical Diseases and Their Impact on Global Health and Development, Third Edition. Peter J. Hotez.
© 2022 American Society for Microbiology. DOI: 10.1128/9781683673903.ch06

some of its leading intellectuals, writers, and musicians, such as the Brontë sisters, Chekhov, Chopin, Goethe, Keats, and Rousseau.[2] Today, TB remains one of the great killers of humankind, responsible for an estimated 1.18 million deaths annually, with the largest number of deaths in Africa, followed by Southeast Asia.[3,4] Up until 2020, the year of coronavirus disease 2019 (COVID-19), TB was the single leading infectious disease cause of mortality. In Africa, TB is a particular crisis because of its high mortality rate in individuals coinfected with HIV/AIDS. TB is now also an important emerging problem in the countries of the former USSR, where a multidrug-resistant form of the infection is becoming widespread.

In response to the enormous global burden of disease from TB, the Group of Seven governments are financing large-scale TB control programs through the Global Fund to Fight AIDS, Tuberculosis, and Malaria. In 2019, these Global Fund grants treated an estimated 5.7 million people through approved therapy known as "DOTS" (originally an acronym used for "direct observed treatment for TB" but now a brand for a multicomponent international TB control strategy).[4,5] Similarly, more than 46 million patients have been treated through the World Health Organization's (WHO) Stop TB Partnership, saving an estimated 6.8 million lives.[4,5] The Stop TB Partnership is a network of international organizations, donors, and countries, as well as governmental and nongovernmental organizations, established in 2000 with a goal of eliminating TB as a public health problem; the organization has designated March 24 as the annual World TB Day.[6] Thus, through high-level advocacy, TB has become a major target for global health activities and philanthropy.

Here we consider two other serious and important mycobacterial infections of the poor—Buruli ulcer and leprosy, which have not benefited from the same level of advocacy as TB. Buruli ulcer and leprosy are two highly disfiguring and stigmatizing neglected tropical diseases (NTDs) that occur almost exclusively among the impoverished living in developing countries.

Buruli Ulcer

Buruli ulcer (also known as Buruli disease) is a disfiguring skin infection caused by *Mycobacterium ulcerans*. The name of this condition comes from a region located near the Nile River in Uganda where large numbers of cases were reported in 1962[7]; however, earlier descriptions can be traced as far back as 1897, to those by Sir Albert Cook, a British medical missionary.[8] The identification of a mycobacterium as the cause of ulcers in Buruli disease patients was reported in 1948.[8,9] Today, Buruli ulcer occurs primarily in tropical and humid regions of West and Central Africa, where since 1980 large numbers of cases have been reported from Benin, Ghana, and Gabon (Fig. 6.1).[8,9] Roughly 5,000 to 6,000 cases annually are reported from 15 of the 33 countries where the disease is endemic.[9] Some investigators believe that the number of cases of Buruli disease in West Africa has increased since 1980, possibly because of deforestation and other environmental factors.[9] In other parts of sub-Saharan Africa, too, the disease is endemic, and it has been reported in some countries outside Africa, including Australia, where aboriginal populations are particularly vulnerable. The disease has emerged in French Guiana

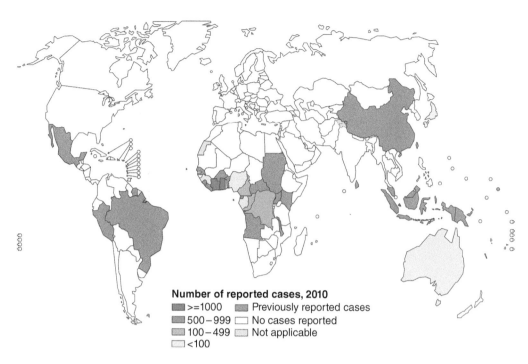

Figure 6.1   Distribution of Buruli ulcer, worldwide, 2010. (© WHO. 2011. Accessed 28 July 2021.)

and neighboring Brazil.[9] After tuberculosis, Buruli ulcer is probably the sec-ond- or third-most-common mycobacterial infection of humans, although because the disease occurs primarily in remote areas of the tropics, we lack good global disease burden estimates.

Figure 6.2 shows the terrible ulceration and destruction that can result from *M. ulcerans* infection. Buruli ulcer strikes primarily school-age chil-dren. Typically, the ulcer begins as a small painless nodule immediately under the skin, but this nodule subsequently breaks down into a large ulcer.[9] The ulcers usually appear on the limbs but can occur almost anywhere, including the face, breast, and genitals.[9] It is believed that the *M. ulcerans* bacteria that cause Buruli create the ulcers by producing a unique chemical toxin that is destructive to the skin and underlying subcutaneous tissue. A fair amount of scientific information is known about the so-called mycolactone toxin of the Buruli bacterium, including its size, its chemical composition, and the biochemical mechanisms by which it destroys human skin.[10] The entire genome of *M. ulcerans* was recently sequenced, with the finding that the genes encoding the enzymes needed to produce the mycolactone exist on genetic elements known as *plasmids*,[9,11] i.e., circular molecules of DNA that exist and replicate separately from chromosomal DNA.

Although it is typically painless and rarely results in death, the ulcer of Buruli disease has a number of catastrophic consequences for the patient. For instance, tissue destruction sometimes results in infection of the underlying bone (known as osteomyelitis) or can necessitate limb amputation. Also, if lesions occur near joints such as the knee or elbow, the subsequent healing can result in contractures that prevent the use of the limb.

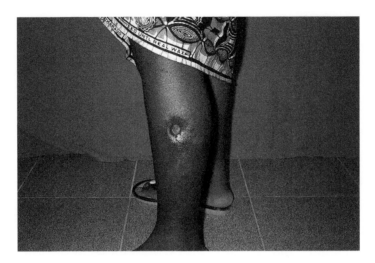

Figure 6.2 Case of Buruli ulcer. (Reproduced from World Health Organization. *Buruli ulcer: recognize act now*. World Health Organization. WHO reference number: WHO/HTM/NTD/IDM/2011 https://www.who.int/publications/i/item/WHO-HTM-NTD-IDM-2015.01.)

The socioeconomic consequences of Buruli ulcer are just as profound. In many cases occurring in West Africa, the cost of treating a case of Buruli ulcer can exceed the per capita government spending on health.[9] Moreover, as suggested by the opening quotation of this chapter, there are widespread beliefs that witchcraft and curses play an important role in transmitting the disease.[1] Such beliefs include the concept of the "evil eye," which holds that the Buruli lesion can be made worse when it is seen by certain individuals. A team from Groningen University Hospital in The Netherlands recently studied beliefs about and attitudes toward Buruli ulcer in Ghana.[1] Through patient interviews, they gained a wealth of information about the stigmatizing and poverty-promoting aspects of the disease. For instance, regarding the concept of the evil eye, the Groningen team obtained this statement about one patient's ulcer[1]:

> If a lot of people see my wound, it might not heal. If all people start to talk about it, stare at it and are surprised by it, it would not heal either. . . . I always want the door closed during changing of dressings, to protect it from evil eyes.

This statement illustrates how someone afflicted with Buruli ulcer could isolate himself from his community and even medical attention. Among the stigmatizing aspects of Buruli ulcer, the Groningen team found that even though the disease is not transmitted from person to person, fear of acquiring the disease was a major reason to shun individuals with the condition. Leading to further isolation was the belief that some people with Buruli ulcer are cursed or were victims of witchcraft.[1]

> If the family has a curse or witchcraft, I would think less of the family. I can hear from rumors in the community if the cause of the disease is a curse or witchcraft.

It was further found that young people with Buruli ulcer have diminished prospects for marriage or leadership positions in their villages.[1]

Although Buruli ulcer is certainly not transmitted through witchcraft, the exact transmission mechanism is still under scientific investigation. It has been observed that Buruli ulcer occurs in tropical areas of sub-Saharan Africa where there are either slow-moving rivers or other bodies of water. One exciting finding is the recovery of *M. ulcerans* or *M. ulcerans* DNA from some of the aquatic insects living in regions of endemicity, usually either from their salivary glands or the biofilm on their legs.[12]

This finding has led to the hypothesis that Buruli is either a waterborne infection (swimming in rivers is a risk factor for infection) or a vector-borne infection or possibly some combination of the two.[9] Possibly, children place themselves at risk for acquiring Buruli ulcer when they first begin swimming and are exposed to either insect bites or the *M. ulcerans* microbes on the insect biofilm.[9] So far, however, information on the possible role of insect vectors in the transmission of *M. ulcerans* has not led to new preventive approaches for Buruli ulcer. Moreover, attempting to discourage children from frequent water contact is not a practical solution for large-scale prevention in regions of endemicity, just as it is not a practical solution for the prevention of schistosomiasis, another waterborne NTD.

The management and treatment of Buruli ulcer are a complex and difficult undertaking, especially in the remote tropical regions of West Africa. Although the *M. ulcerans* bacterium is susceptible to a number of antibiotics when this microorganism is cultured in the laboratory and exposed to different concentrations of antibiotics in the test tube (*in vitro*), it is less clear whether these same antibiotics work well in infected patients. An 8-week course of rifampin and streptomycin, both drugs developed to cure TB, has emerged as a promising treatment regimen.[8] However, streptomycin cannot be given orally, which requires the use of needles and injections in often remote and rural areas. Moreover, when the ulcers become too large and there is massive tissue destruction, it is thought that insufficient amounts of antibiotics can enter the affected area.[8,9] Therefore, the most effective treatment of large ulcers requires not only access to essential antibiotics such as rifampin and streptomycin (i.e., drugs similar to those used for the treatment of other mycobacterial infections, including TB), but also access to some surgical treatments, such as removal of dead tissue (a process known as debridement), excision of the ulcer, and sometimes skin grafting. Even with surgery, however, there is a high relapse rate for the condition,[8] and sadly, there are very few skilled surgeons or other health care professionals who are both adept at these treatments and prepared to establish surgical practices in the remote areas of tropical West Africa where Buruli ulcer is endemic.

In this sense, Buruli ulcer represents a "perfect storm" of neglect. Like so many other NTDs, it only occurs in remote and rural areas of the tropics, so that we do not even know how many people are actually afflicted by this condition. Moreover, the disfigurement and stigma of Buruli ulcer, because of its alleged links to witchcraft, curses, and the evil eye, cause individuals to be shunned by their communities. Such people are not brought to the attention of the medical services even in the few instances when health care is available. Finally, we have no easy solutions for either treating or preventing this condition, and there is an urgent need for research on Buruli disease in order to identify its mode of transmission as a basis for designing innovative prevention strategies. In response to the disease's profoundly neglected status, the WHO launched the Global Buruli Ulcer Initiative in 1998 in order to coordinate international Buruli control efforts, as well as to establish an international research agenda.[8] These efforts led in 2004 to a World Health Assembly resolution that called for increased surveillance and control and for efforts to intensify research on new approaches to treat and prevent Buruli disease.[8] The WHO has also launched a larger integrated neglected skin disease initiative to include addressing the psychosocial aspects of the stigma linked to such illnesses.[13]

There is an urgent need for an effective regimen of Buruli disease treatments that can be administered orally and with ease in resource-poor settings.[8] Alternatively, another promising modality for preventing Buruli ulcer is the development of a safe and effective vaccine for use in high-transmission areas of West Africa. Based on large-scale efforts to develop a vaccine for TB, several modalities of vaccination that could be applied to potential Buruli vaccines have been developed. These approaches include the development of recombinant vaccines and DNA vaccines.[14] Such vaccines could exploit parallel efforts to mine the *M. ulcerans* genome in order to identify new target antigens. These studies would need to also consider the genetic variability of the *M. ulcerans* pathogen.[14] To date, however, no Buruli vaccines are in clinical trials. The hurdles to developing such antipoverty vaccines will be discussed later (in chapter 11) in more detail.

Leprosy

When it comes to stigma, the mycobacterial infection known as leprosy may have it head and shoulders above the other NTDs, at least in its historical record in the "civilized" ancient worlds of Greece and Rome and in medieval Europe.[15,16] Because of its notorious and profoundly disfiguring clinical features, there are several excellent accounts of leprosy in ancient texts, and some scholars believe that leprosy represents the first infectious disease to be accurately described. In ancient Egypt, where some investigators conjecture that leprosy originated, Pharaoh Ramses II (1379–1290? BCE) banished an estimated 80,000 lepers to live on the edge of the Sahara Desert,[15] while Leviticus, the third book of the Pentateuch, records that Moses received specific instructions on how to diagnose lepers and declare them "unclean."[1,15] There are numerous references to "leprosy" in both the Old Testament and New Testament,[1,16] although it is likely that many of the sores described in these writings were probably other infections having cutaneous manifestations. Possibly more than any other ancient text, the Bible probably articulated in the greatest detail the stigmatizing aspects of an NTD.[1] As with so many other NTDs, affected women are especially stigmatized.

While leprosy may have originated in Egypt, some medical historians who have read accounts of diseases resembling leprosy in ancient Chinese texts and in the Hindu writings of the Veda consider it more likely that leprosy originated in Asia.[16] Possibly, it was introduced into Anatolia (Asia Minor) and later ancient Greece and Rome by ancient trade routes or by the armies of Alexander the Great (356–323 BCE).[15,16] Irwin W. Sherman of the University of California states that Hippocrates (470–380 BCE) never described leprosy because he never saw it.[16] This observation partly reinforces the concept that biblical leprosy might have been a constellation of other infectious and even some noninfectious conditions. According to Sherman and others, some of this confusion arises because the Hebrew word *saraath* was used to describe a variety of skin conditions and was translated into Greek about 300 years before Christ as *lepros*, meaning "scaly."

One of the earliest accurate descriptions of leprosy in Europe is dated 150 CE.[16] Analysis of bones recovered from cemeteries in Europe indicates that the disease became widespread in medieval Europe, where it was sometimes known as leontiasis because of the resemblance between the faces of afflicted patients and lions. During this period, specialized homes or hospitals for lepers were established and were known as leprosaria or lazarettos from the biblical, sore-infested beggar, Lazarus.[15,16] In medieval Europe, the lazarettos often functioned more as monasteries than as hospitals as we now know them. Lepers were routinely banished to such institutions because of widespread fear of contracting the disease. The medical writer Berton Roueché found that lepraphobia was extant throughout medieval Europe and reached its pinnacle in the 12th, 13th, and 14th centuries.[15] This period coincided with the height of the leprosy epidemic in Europe. Examples of lepraphobia included the requirement that lepers endure their own mock funeral as a prelude to their banishment either to the lazaretto or to a lifetime of begging, with the added requirement that they announce their presence in the community by a self-imposed ringing of a "Lazarus bell." Other lepers were not so lucky and were burned at the stake or buried alive.[15,16] The origins of the leprosy epidemic in medieval Europe are unknown, although it has been suggested that knights returning from the Crusades were a contributing factor.[16] There was even a spiritual order of leprosy-infected knights known as the Order of Lazarus.[16] Following the medieval period, leprosy apparently became much less common. The reasons for this purported decrease are still a subject of speculation, with one major theory arguing that with the widespread emergence of TB around this time (possibly in part from increasing urbanization), there were large populations of TB-exposed and -infected patients who developed cross-immunity to leprosy.[17] In this way, infection with *M. tuberculosis* functionally may have acted as a partially effective vaccine against *Mycobacterium leprae* infection.

The United States has also had a long and interesting association with leprosy patients. By the early 20th century, it had been noted that considerable numbers of leprosy patients were living in the continental United States, especially in Louisiana. In testimony given before Congress in 1916, witnesses noted that lepers in the United States were living under horrible conditions, with some treated like criminals in solitary confinement.[18] Some 20 years before it was designated as a national leprosarium, the Indian Camp Plantation in Louisiana began housing lepers and treating them with a novel therapy, developed in India, calling for application of chaulmoogra oil. Whether this is an effective treatment remains controversial, although some studies have found antileprosy effects. The treatment and care of the patients were administered by the sisters of an order founded by St. Vincent de Paul known as the Daughters of Charity.[18] The National Hansen's Disease Center, as it ultimately became known, lasted for more than 100 years until it finally closed its doors in 1999 (Fig. 6.3). During that time, it was the only inpatient facility for treating leprosy patients in the continental United States. When patients entered Carville, they frequently left behind their families and friends for life; in some cases they gave up their legal name. For the lepers at Carville, the hospital staff and patients became their families, with the hospital and its grounds functioning as a self-contained village housing its own dental office, cafeteria, cemetery, and even a jail. Well until the middle of the 20th century, patients with leprosy were highly discriminated against in the United States, even to the point where they were refused entry to public restrooms or access to public transportation.[16] Chinese-Americans were particularly victimized on the mistaken belief that they were a frequent source of the disease.[16]

An equally extraordinary story of American leprosy is the formation of the Molokai Colony in Hawaii.[19] During the middle of the 19th century, a Hawaiian king ordered lepers to a quarantine area on the island of Molokai. An area

Figure 6.3   National Hansen's Disease Center, Carville, LA. (Patients' Recreation Building. Photographer unknown, c. 1950. From the permanent collection of the National Hansen's Disease Museum, Carville, LA.)

surrounded by the Pacific Ocean on three sides and by steep cliffs on the fourth was chosen to prevent the inhabitants of the leper colony from escaping. To get there, many of those sentenced to Molokai were pushed off ships several hundred yards from the shore and forced to swim to the island.[16] In 1873, the Honolulu Catholic mission sent Father Damien (Joseph Damien de Veuster), a Belgian Roman Catholic missionary, to Molokai. For more than a decade Father Damien took care of the resident leprosy patients, before he himself contracted leprosy and died 5 years later on the island.[16,19] Known as the "Martyr of Molokai," Father Damien was beatified in 1995 by Pope John Paul II.[16,19] Bronze statues of Father Damien stand in the U.S. Capitol and the Hawaii State Capitol (Fig. 6.4).

Figure 6.4 Bronze statue of Father Damien at the U.S. Capitol. (See https://www.aoc.gov/explore-capitol-campus/art/father-damien. Photo credit: Architect of the Capitol.)

Today, as a result of widespread use of a cocktail of antimycobacterial drugs known as MDT (multidrug therapy), the global registered prevalence of leprosy has been reduced to approximately 200,000 new cases detected annually and almost that many registered cases (Fig. 6.5).[20] India accounts for approximately one-half the number of new cases detected globally, followed by Brazil, Indonesia, Nepal, Bangladesh, Ethiopia, and the Democratic Republic of the Congo.[20] These countries account for almost 90% of the new cases. The Global Burden of Disease Study (GBD) 2019 indicates that the overall prevalence is approximately 0.5 million cases.[20] Previously, the WHO considered leprosy to be eliminated as a public health problem in a country once the prevalence of the disease diminishes to less than 1 leprosy case per 10,000 individuals. As a result, leprosy has been eliminated in most countries, with the Democratic Republic of the Congo, Mozambique, and Timor-Leste being the most recent countries to meet national elimination targets. However, there are about 10 countries, including India and Brazil, that technically have met the elimination criteria but still have pockets of high endemicity. For that reason, the WHO has specified new targets through its Global Leprosy Strategy.[20]

Leprosy is caused by *M. leprae*. The organism was discovered in the late 1800s by the Norwegian scientist G. H. Armauer Hansen, and the bacterium is one of the first to have been implicated as a cause of human disease.[21,22] All mycobacteria, including *M. ulcerans*, *M. tuberculosis*, and *M. leprae*, are known as *acid-fast* organisms because they stain red after specific dyes are used; they are usually seen in clumps or in palisades under the microscope (Fig. 6.6). Leprosy is most likely transmitted from the nasal secretions of infected individuals;

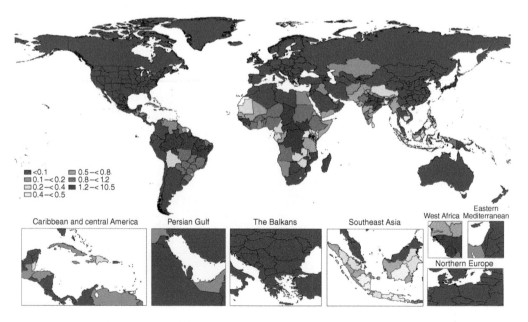

Figure 6.5 Disability caused by leprosy, worldwide. (From Global Burden of Disease Collaborative Network, 2020 [http://www.healthdata.org/results/gbd_summaries/2019/leprosy-level-3-cause].)

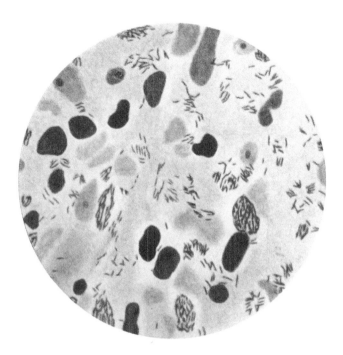

Figure 6.6 Drawing of *M. leprae* under the microscope and following staining. (Image from CDC PHIL [ID#2123]/CDC/1979.)

it has been estimated that a single individual might shed as many as 10 million living leprosy bacilli daily.[21] Exactly how leprosy is subsequently transmitted to another individual is not known, although it is suspected that transmission probably occurs either via the respiratory route or through broken skin.[21] Generally speaking, close contact with a leprosy patient, typically a household member, is required for transmission of leprosy from a patient to a healthy person.[21] Humans are the only major natural host of leprosy, although the infection has also been detected in chimpanzees and other nonhuman primates. Of interest is the finding that the nine-banded armadillo can also support the growth of *M. leprae*.

Not everyone exposed to *M. leprae* develops clinical signs of leprosy. As noted above, the infection is not easily transmitted unless there is very close contact. Even when transmission occurs, a majority of people infected may never show signs or symptoms of the disease. In those who become sick with leprosy, the clinical features usually take years to develop, and frequently the onset is so gradual that it is 3 to 5 years before the patient becomes aware of the changes.[21] By the time clinical features are present, large numbers of *M. leprae* microorganisms are usually replicating in the skin and in the peripheral nerves responsible for motor and sensory functions. In the early stages of the disease, peripheral nerve damage results in tingling or absence of sensation. Sometimes burns occur because the patient does not notice exposure to flames or other sources of heat.[21] When peripheral nerve damage is more extensive, there is weakness of muscles in the hands or feet or sometimes the face. Alternatively, patches of skin become discolored.[21]

Following the initial manifestations of leprosy, the disease usually proceeds along one of two clinical courses.[21] The majority of patients develop the so-called *tuberculoid* form of the disease. Such patients have the ability to mount strong immunological responses against *M. leprae* microbes and, as a result, develop only localized disease such as loss of sensation or weakness (or both) in the face, hand, leg, foot, or some other body part. The localized changes are linked to a thickening and disruption of the associated peripheral nerves in the region.[21] Skin lesions resembling plaques also occur on the face, surfaces of the limbs, the back, and the buttocks. Far more severe is the *lepromatous* form of leprosy. This condition occurs in a much smaller subset of patients who (for reasons that are still not completely elucidated) fail to mount a vigorous immunological response to their *M. leprae* infection.[21] Because replication of the leprosy microorganisms goes unchecked, the bacilli can disseminate widely in the skin and nerves and even invade the eyes, nose, and mouth.[21] Bone invasion can occur, and the testes can also be affected. Massive replication and infiltration of the *M. leprae* bacilli in the skin cause it to become thickened, nodular, and shiny.[21] This condition is accompanied by a deformity of the earlobes and a loss of the eyebrows and eyelashes.[21] The disfigurement is often worsened by burns and other injuries. Blindness and visual impairments result from destruction in the cornea.

The WHO recognizes a grading system of leprosy impairments and deformities. Grade 1 leprosy refers to loss of sensation in the hands, feet, or eyes, but without damage or deformities, while grade 2 includes visible damage to the eyes with loss of visual acuity, as well as damage and deformities to the hands and feet.[21] Even today there is an enormous stigma associated with leprosy in developing countries.[23]

An important breakthrough in the global control of leprosy was the development of antimicrobial drugs to treat *M. leprae* infection and a demonstration of their efficacy. One of the first drugs shown to be effective for leprosy treatment was the sulfone compound dapsone. However, because enormous numbers of *M. leprae* mycobacteria live in a patient with lepromatous leprosy, the rate of mutations in the bacteria's genetic material is such that microbial drug resistance was noted early on as a common problem. Drug resistance is also a common problem in patients being treated for TB. During the 1980s, however, the medical community began to circumvent the problem of drug resistance by designing and administering a combination package of several antimycobacterial drugs. Today, the most widely used drug regimen includes dapsone, which is administered together with the antibacterial drugs rifampin (also known as rifampicin) and clofazimine.[22] The regimen requires 6 to 12 months of treatment, with dapsone and clofazimine administered daily and rifampin once monthly.[21]

Within a decade of widespread use of MDT, the World Health Assembly adopted a resolution to eliminate leprosy as a public health problem by the year 2000.[22] Through the generosity of both the Sasakawa Foundation and the drug company Novartis, millions of people have so far received MDT free of charge, and almost all of the countries where leprosy was once endemic have reached their elimination targets.[2] The barriers to leprosy elimination are those in common with many other NTD control and elimination efforts, including the remoteness of many rural areas where these diseases occur and the lack of adequate health care facilities.[22] However, another problem unique to leprosy as compared to most other NTDs is the relatively complicated three-drug regimen that must be used in order to ensure that a patient is adequately treated.[22]

The WHO and other international health agencies are also facing a larger problem of whether their proposed target of reducing prevalence to less than 1 per 10,000 is appropriate and whether it has an adequate scientific rationale. The suggestion was put forward that the 1-per-10,000 goal is artificial and is more of a political target than a scientific one.[22] In addition, several leprosy researchers believe that there are many more cases of leprosy in the world than the ones currently registered by the WHO and that consequently many more individuals should be on MDT but are not.[22,24] According to some estimates, one-third of newly diagnosed cases present with nerve damage and will progress to disability,[24] a situation that might be prevented with early case detection and treatment. In 1999, the Global Alliance for the Elimination of Leprosy (GAEL) was established in order to make a serious last push for leprosy elimination.[22] Subsequently, an independent evaluation of the GAEL program has recommended changing strategies, with less focus on elimination and more on creating a new approach that focuses on long-term leprosy control activities, together with rehabilitation of patients with nerve damage.[22] A new concern is that while MDT is effective in treating patients, it does not always prevent them from shedding *M. leprae* in their nasal secretions and may not adequately interrupt the transmission of leprosy.[22,24] Therefore, the global community of leprosy scientists and public health experts worries that the concept of an elimination strategy requires serious rethinking.[22]

In response to these concerns, a new Global Leprosy Strategy was endorsed in 2016. Its major goals include the identification of pockets or areas of high endemicity, followed by case detection and treatment. The strategy emphasizes zero new cases in children, overall reductions to fewer than 1 case of grade 2 disability per million population, and bans on laws discriminating against leprosy sufferers.[20] Currently there are approximately 10,000 cases of grade 2 disability leprosy globally. The enhanced global strategy also emphasizes early detection and reduction of disabilities among new cases by ensuring that patients receive a full course of MDT.[20] The complexities of MDT have also led to calls for additional research and development into new and simplified control tools. This strategy includes the possible use of shortened and simplified variations of MDT to prevent the transmission of leprosy, the development of new antileprosy drugs based on the *M. leprae* genome completed in 2001,[25] and opportunities to integrate leprosy control with control of other diseases of poverty,[26] including the integrated package of NTD drugs (chapter 10). Other important research and development activities include examining the potential of administering chemoprophylaxis to household contacts and the development of better diagnostics to detect the infection early in the course of the disease.[27] Efforts to develop leprosy vaccines and explore an expanded role for *Mycobacterium bovis* BCG are also needed.[27]

The Neglected Mycobacterial Infections

Mycobacterial infections

- Mycobacteria are slim bacteria with unusual growth requirements and unique structural and biochemical properties that place them in their own category of microorganisms.
- TB, caused by the bacterium *M. tuberculosis*, is the best-known mycobacterial infection of humans.
- Buruli ulcer and leprosy are two highly disfiguring and stigmatizing mycobacterial NTDs occurring almost exclusively among the impoverished living in developing countries.

Buruli ulcer

- Buruli ulcer is a highly disfiguring skin infection caused by *M. ulcerans*. The name of this condition comes from a region located near the Nile River in Uganda where large numbers of cases were reported in 1962.
- Today, Buruli ulcer occurs primarily in tropical regions of West and Central Africa, although foci also occur in Australia and in French Guiana in the Americas.
- Buruli typically strikes school-age children and manifests as a large ulcer usually appearing on the limbs.
- Although it is typically painless and rarely results in death, the ulcer of Buruli disease has a number of catastrophic consequences for the patient, including a profound socioeconomic impact and the widespread belief that witchcraft and curses play an important role in transmitting the disease.
- Buruli is either a waterborne infection or a vector-borne infection or possibly some combination of the two.
- The management and treatment of Buruli ulcer are a complex and difficult undertaking, especially in remote tropical regions.
- The most effective treatment of large ulcers requires access not only to antibiotics but also to different surgical modalities, including skin grafting.
- Because of its victims' generally remote locations and the associated disfigurement and stigma, as well as inadequate access to health care, Buruli ulcer represents a "perfect storm" of neglect.

Leprosy

- Leprosy (also known as Hansen's disease) is caused by *M. leprae*. Because of its profoundly disfiguring clinical features, there are numerous accounts of leprosy in ancient texts. Lepraphobia was extant throughout medieval Europe. In the United States, the National Hansen's Disease Center lasted for more than 100 years before closing in 1999.
- Today, as a result of widespread availability and use of MDT, the global registered prevalence of leprosy has been reduced to approximately 200,000 cases, with an estimated equal number of new cases detected annually.
- The largest numbers of new cases currently occur in India, Brazil, and Indonesia.
- Leprosy proceeds along either one of two clinical courses. The majority of patients develop the so-called tuberculoid form. Such patients have the ability to mount strong immunological responses against *M. leprae* and, as a result, develop only localized disease. Far more severe is the lepromatous form of leprosy. These patients experience widely disseminated disease in the skin and nerves and even the eyes, nose, mouth, and bones.
- The most widely used MDT drug regimen calls for dapsone to be administered together with the antibacterial drugs rifampin and clofazimine.
- An enhanced global strategy is in place that emphasizes early detection and reduction of severe disabilities among new cases by ensuring completion of a full MDT course.

Notes

1. The first quotation is from Stienstra et al., 2002. The quotation in Leviticus is from https://www.bible.com/bible/114/LEV.13.45-46.NKJV.

2. Plorde, 2004.

3. The latest GBD 2019 mortality estimates for tuberculosis can be found at http://www.healthdata.org/results/gbd_summaries/2019/tuberculosis-level-3-cause (Global Burden of Disease Collaborative Network, 2020).

4. For specific information about the Stop TB Partnership, see http://www.stoptb.org/resources/factsheets/fastfacts.asp; information about the Global Fund is at https://www.theglobalfund.org/en/.

5. The five elements of DOTS include political commitment, case detection, standardized treatment with supervision, an effective drug supply and management system, and a monitoring and evaluation system; see https://www.who.int/tb/dots/en/.

6. See www.stoptb.org/events/world_tb_day.

7. Clancey et al., 1962.

8. Converse et al., 2011; and Johnson et al., 2005.

9. Two disease burden updates for Buruli ulcer are in Simpson et al., 2019; and Omansen et al., 2019. Two excellent clinical reviews on Buruli ulcer were published in 2006: Wansbrough-Jones and Phillips, 2006; and Sizaire et al., 2006.

10. Detailed information about the mycolactone toxin and the *M. ulcerans* genes that encode it can be found in George et al., 1999; and Stinear et al., 2004.

11. Information about the plasmids that express *M. ulcerans* mycolactone toxin-associated genes is found in Stinear et al., 2004; and is summarized in Townsend, 2004.

12. The role of aquatic insects in the transmission of Buruli ulcer was first suggested in 1999 as described in Portaels et al., 1999. More recent summaries of the subsequent literature are found in Silva et al., 2007; and Merritt et al., 2010.

13. Information on the WHO neglected skin disease initiative can be found here: https://www.who.int/activities/promoting-the-integrated-approach-to-skin-related-neglected-tropical-diseases.

14. A review on the prospects of developing a Buruli ulcer vaccine is Einarsdottir and Huygen, 2011. The question of *M. ulcerans* genetic variability is addressed in Walsh et al., 2010.

15. A good historical overview of leprosy can be found in Roueché, 1986, p. 68–86. Berton Roueché wrote about medical mysteries as part of the "Annals of Medicine" feature in the *New Yorker* magazine for more than 40 years. He died in 1994 at the age of 83. An account of his life can be found in Lerner, 2005.

16. Another very good historical account is found in chapter 14 of Sherman, 2006, p. 303–311.

17. This theory was advanced in Donoghue et al., 2005.

18. The history of the National Leprosarium in Carville is told at https://www.hrsa.gov/hansens-disease/history.html and in Gaudet, 2004.

19. A brief description of the history of Father Damien of Molokai can be found at http://www.yourislandroutes.com/articles/leprosy.shtml and https://www.nps.gov/kala/learn/historyculture/damien.htm.

20. World Health Organization, 2020. The GBD 2019 information can be found at http://www.healthdata.org/results/gbd_summaries/2019/leprosy-level-3-cause.

21. An excellent clinical description of leprosy can be found in Leprosy Group, WHO, 2002. The grading system is described in Rathod et al., 2020.

22. Rinaldi, 2005; see also https://www.who.int/health-topics/leprosy.

23. Tsutsumi et al., 2007.

24. Lockwood and Suneetha, 2005; and Rodrigues and Lockwood, 2011.

25. Cole et al., 2001.

26. Lockwood, 2004.

27. Rodrigues and Lockwood, 2011; and Duthie et al., 2011.

References

Clancey J, Dodge R, Lunn HF. 1962. Study of a mycobacterium causing skin ulceration in Uganda. *Ann Soc Belg Med Trop 1920* **42:**585–590.

Cole ST, Eiglmeier K, Parkhill J, James KD, Thomson NR, Wheeler PR, Honoré N, Garnier T, Churcher C, Harris D, Mungall K, Basham D, Brown D, Chillingworth T, Connor R, Davies RM, Devlin K, Duthoy S, Feltwell T, Fraser A, Hamlin N, Holroyd S, Hornsby T, Jagels K, Lacroix C, Maclean J, Moule S, Murphy L, Oliver K, Quail MA, Rajandream MA, Rutherford KM, Rutter S, Seeger K, Simon S, Simmonds M, Skelton J, Squares R, Squares S, Stevens K, Taylor K, Whitehead S, Woodward JR, Barrell BG. 2001. Massive gene decay in the leprosy bacillus. *Nature* **409:**1007–1011.

Converse PJ, Nuermberger EL, Almeida DV, Grosset JH. 2011. Treating *Mycobacterium ulcerans* disease (Buruli ulcer): from surgery to antibiotics, is the pill mightier than the knife? *Future Microbiol* **6:**1185–1198.

Donoghue HD, Marcsik A, Matheson C, Vernon K, Nuorala E, Molto JE, Greenblatt CL, Spigelman M. 2005. Co-infection of *Mycobacterium tuberculosis* and *Mycobacterium leprae* in human archaeological samples: a possible explanation for the historical decline of leprosy. *Proc Biol Sci* **272:**389–394.

Duthie MS, Gillis TP, Reed SG. 2011. Advances and hurdles on the way toward a leprosy vaccine. *Hum Vaccin* **7:**1172–1183.

Einarsdottir T, Huygen K. 2011. Buruli ulcer. *Hum Vaccin* **7:**1198–1203.

Gaudet M. 2004. *Carville: Remembering Leprosy in America.* University Press of Mississippi, Jackson.

George KM, Chatterjee D, Gunawardana G, Welty D, Hayman J, Lee R, Small PL. 1999. Mycolactone: a polyketide toxin from *Mycobacterium ulcerans* required for virulence. *Science* **283:**854–857.

Global Burden of Disease Collaborative Network. 2020. Global Burden of Disease Study 2019 (GBD 2019). GBD Results Tool. Institute for Health Metrics and Evaluation, Seattle, WA.

Johnson PD, Stinear T, Small PL, Pluschke G, Merritt RW, Portaels F, Huygen K, Hayman JA, Asiedu K. 2005. Buruli ulcer (*M. ulcerans* infection): new insights, new hope for disease control. *PLoS Med* **2:**e108.

Leprosy Group. WHO. 2002. Leprosy, p 1065–1084. *In* Cook GC, Zumla AI (ed), *Manson's Tropical Diseases*, 21st ed. W B Saunders, New York, NY.

Lerner BH. 2005. Remembering Berton Roueché—master of medical mysteries. *N Engl J Med* **353:**2428–2431.

Lockwood DN. 2004. Commentary: leprosy and poverty. *Int J Epidemiol* **33:**269–270.

Lockwood DN, Suneetha S. 2005. Leprosy: too complex a disease for a simple elimination paradigm. *Bull World Health Organ* **83:**230–235.

Merritt RW, Walker ED, Small PL, Wallace JR, Johnson PD, Benbow ME, Boakye DA. 2010. Ecology and transmission of Buruli ulcer disease: a systematic review. *PLoS Negl Trop Dis* **4:**e911.

Omansen TF, Erbowor-Becksen A, Yotsu R, van der Werf TS, Tiendrebeogo A, Grout L, Asiedu K. 2019. Global epidemiology of Buruli ulcer, 2010–2017, and analysis of 2014 WHO programmatic targets. *Emerg Infect Dis* **25:**2183–2190.

Plorde JJ. 2004. Mycobacteria, p 439–456. *In* Ryan KJ, Ray CG (ed), *Sherris Medical Microbiology*, 4th ed. McGraw-Hill Medical Publishing Division, New York, NY.

Portaels F, Elsen P, Guimaraes-Peres A, Fonteyne PA, Meyers WM. 1999. Insects in the transmission of *Mycobacterium ulcerans* infection. *Lancet* **353:**986.

Rathod SP, Jagati A, Chowdhary P. 2020. Disabilities in leprosy: an open, retrospective analyses of institutional records. *An Bras Dermatol* **95:**52–56.

Rinaldi A. 2005. The global campaign to eliminate leprosy. *PLoS Med* **2:**e341.

Rodrigues LC, Lockwood DN. 2011. Leprosy now: epidemiology, progress, challenges, and research gaps. *Lancet Infect Dis* **11:**464–470.

Roueché B. 1986. *The Medical Detectives*. Washington Square Press, New York, NY.

Sherman IW. 2006. *The Power of Plagues*. ASM Press, Washington, DC.

Silva MT, Portaels F, Pedrosa J. 2007. Aquatic insects and *Mycobacterium ulcerans*: an association relevant to Buruli ulcer control? *PLoS Med* **4:**e63.

Simpson H, Deribe K, Tabah EN, Peters A, Maman I, Frimpong M, Ampadu E, Phillips R, Saunderson P, Pullan RL, Cano J. 2019. Mapping the global distribution of Buruli ulcer: a systematic review with evidence consensus. *Lancet Glob Health* **7:**e912–e922.

Sizaire V, Nackers F, Comte E, Portaels F. 2006. *Mycobacterium ulcerans* infection: control, diagnosis, and treatment. *Lancet Infect Dis* **6:**288–296.

Stienstra Y, van der Graaf WT, Asamoa K, van der Werf TS. 2002. Beliefs and attitudes toward Buruli ulcer in Ghana. *Am J Trop Med Hyg* **67:**207–213.

Stinear TP, Mve-Obiang A, Small PL, Frigui W, Pryor MJ, Brosch R, Jenkin GA, Johnson PD, Davies JK, Lee RE, Adusumilli S, Garnier T, Haydock SF, Leadlay PF, Cole ST. 2004. Giant plasmid-encoded polyketide synthases produce the macrolide toxin of *Mycobacterium ulcerans*. *Proc Natl Acad Sci USA* **101:**1345–1349.

Townsend CA. 2004. Buruli toxin genes decoded. *Proc Natl Acad Sci USA* **101:**1116–1117.

Tsutsumi A, Izutsu T, Islam AM, Maksuda AN, Kato H, Wakai S. 2007. The quality of life, mental health, and perceived stigma of leprosy patients in Bangladesh. *Soc Sci Med* **64:**2443–2453.

Walsh DS, Portaels F, Meyers WM. 2010. Recent advances in leprosy and Buruli ulcer (*Mycobacterium ulcerans* infection). *Curr Opin Infect Dis* **23:**445–455.

Wansbrough-Jones M, Phillips R. 2006. Buruli ulcer: emerging from obscurity. *Lancet* **367:**1849–1858.

World Health Organization. 2020. Global leprosy (Hansen disease) update, 2019: time to step-up prevention initiatives. *Wkly Epidemiol Rec* **95:**417–440.

7

The Kinetoplastid Infections: Human African Trypanosomiasis (Sleeping Sickness), American Trypanosomiasis (Chagas Disease), and the Leishmaniases

Black shapes crouched, lay, sat between the trees, leaning against the trunk, clinging to the earth, half coming out, half effaced within the dim light, in all the attitudes of pain, abandonment, and despair . . . They were dying slowly . . . nothing but black shadows of disease and starvation, lying confusedly in the greenish gloom.

JOSEPH CONRAD, *HEART OF DARKNESS*

The kinetoplastid infections constitute a group of three major human protozoan infections caused by single-celled parasites with a flagellum and an unusual DNA-containing cell organelle known as the *kinetoplastid*. Like lymphatic filariasis and onchocerciasis, the kinetoplastid infections are transmitted by insect vectors. Together, the three major kinetoplastid infections of humans, human African trypanosomiasis (HAT), Chagas disease, and leishmaniasis, kill approximately 12,000 to 13,000 people annually, making them among the more lethal neglected tropical diseases (NTDs).[1] However, alternative studies find that the number of deaths from these conditions may be 4 or 5 times higher.[1] Equally important are their poverty-promoting effects and their particular propensity to reappear in areas of conflict or post-conflict turmoil, with often devastating consequences for vulnerable and migratory populations. For example, between 1988 and 1994 an estimated 100,000 perished from leishmaniasis in the western Upper Nile region of Sudan, and during the 1990s and 2000s in some regions of war-torn Angola, the Democratic Republic of the Congo (DRC), South Sudan, and Uganda, the death rate from HAT has exceeded that of HIV/AIDS.[1] Many of these deaths could have been avoided if the victims of trypanosomiasis and leishmaniasis had had access to essential medicines. However, even when the drugs are available, they are often highly toxic and can sometimes even kill patients. Many of the drugs used to treat the kinetoplastid infections are older agents that were developed

Forgotten People, Forgotten Diseases: The Neglected Tropical Diseases and Their Impact on Global Health and Development, Third Edition. Peter J. Hotez.
© 2022 American Society for Microbiology. DOI: 10.1128/9781683673903.ch07

in the early or middle parts of the 20th century. Because there is essentially no market in North America, Europe, or Japan for drugs to treat kinetoplastid infections, there has been little economic incentive to develop new and improved versions. However, a nonprofit product development partnership known as the Drugs for Neglected Diseases initiative (DNDi) has made substantial recent progress in developing new kinetoplastid drugs, and vaccine partnerships are also making progress.

HAT

The modern history of sub-Saharan Africa is linked intimately to the transmission of HAT, also known as sleeping sickness. It is a highly lethal NTD caused by protozoan parasites in the bloodstream and central nervous system and currently affects just under 10,000 people, including almost 4,000 with existing infection, with about 2,000 new cases annually. However, the exact number is not known because the disease typically occurs in remote rural areas of Africa living in the "tsetse belt," which extends across the continent from Senegal in the west to Somalia in the east (Fig. 7.1 and 7.2). This belt is the ecological habitat of more than 20 species of the tsetse, a biting fly of the genus *Glossina* that serves as the insect vector of HAT. In modern history, the greatest numbers of HAT infections and the greatest risk of acquiring HAT occurred primarily in belt areas where long-standing conflicts have thwarted public health

Distribution of human African trypanosomiasis (*T.b.gambiense*), worldwide, 2018

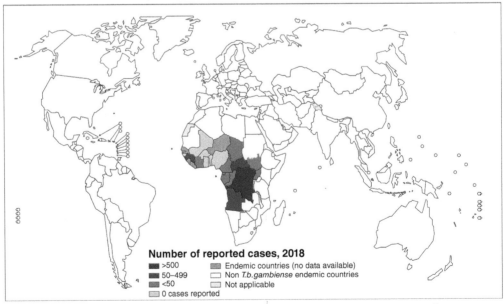

Figure 7.1 Distribution of Gambian HAT (*T. b. gambiense*), worldwide, 2018. (© WHO. 2018. Accessed 5 May 2021.)

Distribution of human African trypanosomiasis (*T.b.rhodesiense*), worldwide, 2018

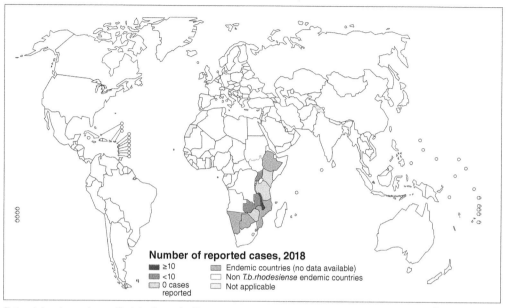

Number of reported cases, 2018
- ≥10
- <10
- 0 cases reported
- Endemic countries (no data available)
- Non *T.b.rhodesiense* endemic countries
- Not applicable

Data Source: World Health Organization
Map Production: Control of Neglected Tropical Diseases (NTD)
World Health Organization

Figure 7.2 Distribution of Rhodesian HAT (*T. b. rhodesiense*), worldwide, 2018. (© WHO. 2018. Accessed 5 May 2021.)

control measures simultaneously aimed at tsetse control and human case detection and treatment.[1] In Angola, Burundi, Central African Republic, Chad, the Democratic Republic of the Congo, Gabon, South Sudan, and northern Uganda, HAT has reemerged in parallel with the past and present human-invented calamities in these countries.[1] The latest estimates indicate that HAT remains prevalent in these nations, as well as elsewhere in sub-Saharan Africa.[1]

HAT is caused by infection with one of two subspecies of *Trypanosoma brucei*, a slender and graceful-appearing protozoan that swims in the human bloodstream with the aid of a flagellum (Fig. 7.3). The graceful motion of *trypanosomes* as seen under a microscope results from the membrane attachment of the flagellum along the length of the parasite. The term "undulating membrane" has been used to describe this structure.

There are three major morphologically indistinguishable subspecies of *T. brucei* causing human and animal disease (Table 7.1). *T. brucei gambiense* (*T. b. gambiense*) is the cause of human sleeping sickness in West Africa (known as Gambian HAT and responsible for 95% of the cases[1]). *T. brucei rhodesiense* (*T. b. rhodesiense*) causes HAT in East Africa (known as Rhodesian HAT).[2] A third subspecies, *T. brucei brucei* (*T. b. brucei*), does not infect humans. Instead, infection with *T. b. brucei* and other animal trypanosomes has a serious impact on the health of the rural poor in Africa because it causes a serious wasting disease in cattle known as nagana (sometimes written as n'gana), which severely

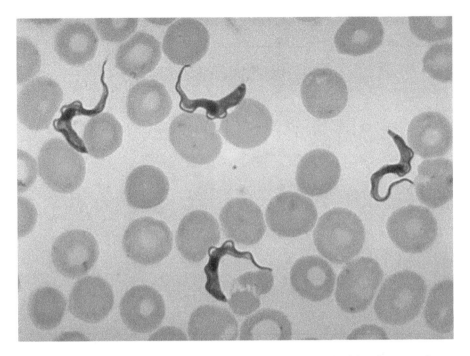

Figure 7.3 Photomicrograph of stained trypanosomes in the bloodstream. (Image from CDC-PHIL [ID#613]/CDC/Dr. Myron G. Schultz, DVM, MD, DCMT, 1970.)

reduces meat and dairy production in the region. By some estimates, 30% of sub-Saharan Africa's 150 million cattle are at risk for nagana, and the losses in milk and meat reach an estimated US$5 billion annually.[3]

Humans and cattle become infected with trypanosomes when the infective form of the parasite (known as a metacyclic trypanosome) living in the salivary glands of the female tsetse enters mammalian tissues as the fly bites (Fig. 7.4). Following inflammation at the site of the bite (a lesion known as a chancre), the parasites enter the lymph nodes before they eventually enter

Table 7.1 The major species of HAT and animal trypanosomiasis

Disease	Parasite	Tsetse vector	Geographic distribution	Estimated no. of human cases	Major animal reservoir of infection?
West African (Gambian) HAT	*Trypano-soma brucei gambiense*	*Glossina* species of the palpalis group	West Africa extending east to Sudan and parts of Uganda	2,000	No
East African (Rhodesian) HAT	*Trypanosoma brucei rhodesiense*	*Glossina* species of the mor-sitans group	East Africa (more limited distribu-tion than *T. b. gambiense*)	<1,000	Yes
Nagana (bovine trypanosomiasis)	*Trypanosoma brucei brucei, Trypano-soma vivax, Trypanosoma congolense*	*Glossina* species	Throughout the tsetse belt	None	Yes

Sleeping Sickness, African (African trypanosomiasis)
(*Trypanosoma brucei gambiense*)
(*Trypanosoma brucei rhodesiense*)

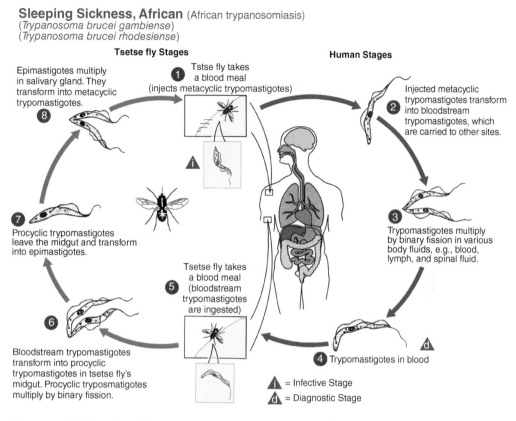

Tsetse fly Stages   **Human Stages**

Epimastigotes multiply in salivary gland. They transform into metacyclic trypomastigotes.
8

① Tstse fly takes a blood meal (injects metacyclic trypomastigotes)

Injected metacyclic trypomastigotes transform into bloodstream trypomastigotes, which are carried to other sites. ②

i

7
Procyclic trypomastigotes leave the midgut and transform into epimastigotes.

③ Trypomastigotes multiply by binary fission in various body fluids, e.g., blood, lymph, and spinal fluid.

Tsetse fly takes a blood meal (bloodstream trypomastigotes are ingested) ⑤

6

Bloodstream trypomastigotes transform into procyclic trypomastigotes in tsetse fly's midgut. Procyclic trypomastigotes multiply by binary fission.

④ Trypomastigotes in blood **d**

i = Infective Stage
d = Diagnostic Stage

Figure 7.4 Life cycle of human trypanosomes and HAT. (Image from CDC-PHIL [ID#3418]/ CDC-Alexander de Silva, PhD; Melanie Moser, 2003.)

the bloodstream and multiply through asexual division as the trypomastigote form. The presence of trypanosomes in the blood is known as *parasitemia*. When another tsetse bites an infected person or animal with parasitemia, the trypanosomes are taken up by the fly, undergo development in the insect gut, and then migrate from the insect gut to the salivary glands. The disease is subsequently transmitted by the bite of another tsetse fly.

Scientific research conducted since the 1970s has shown that both humans and trypanosomes have developed some highly interesting invasion and defense mechanisms as a consequence of the evolution of the host-parasite relationship. For instance, the *T. b. brucei* subspecies causing nagana in cattle cannot survive in human blood because it is vulnerable to the toxic effects of our unique high-density lipoprotein (HDL).[4] In the presence of human HDL, *T. b. brucei* trypanosomes are quickly destroyed. It is therefore possible that HDL (often referred to as "the good cholesterol") evolved in humans as a natural defense mechanism against *T. b. brucei* in addition to its role in facilitating cholesterol metabolism and in countering the effects of low-density lipoprotein in coronary artery disease. In contrast, both *T. b. gambiense* and *T. b. rhodesiense* have evolved biochemical mechanisms to resist the trypanocidal effects of HDL. Another important survival mechanism of trypanosomes is their unique ability to evade host mammalian immune responses. Ordinarily, a human or cow would combat infectious pathogens by producing specific antibodies against them, usually within a few days or weeks after initial infection. However, trypanosomes have evolved a means of changing their surface antigens, a phenomenon known as *antigenic variation*, such that by the time there is a host response to the invading parasites through specific antibody production, a proportion of the trypanosomes have changed the composition of their surface antigen glycoproteins.[5] The newly masked trypanosomes are thereby rendered refractory to antibody-mediated attack. This process goes on for multiple generations of parasites living in the bloodstream until they ultimately gain access to and invade the central nervous system.

There is minimal geographic overlap between the two forms of HAT—currently Uganda is among the only nations with both West and East African forms of the disease.[2] There are also pronounced differences between the clinical features of West African or Gambian HAT (*T. b. gambiense* infection) and those of East African or Rhodesian HAT (*T. b. rhodesiense* infection), as well as pronounced differences in their epidemiology and ecology. However, the ultimate outcome of the two diseases—coma and death—is the same.[2] HAT is usually fatal unless treated.

Gambian HAT, which accounts for most (possibly more than 95%) of the human cases, typically occurs around rivers, especially in areas of dense vegetation where tsetses of the *Glossina palpalis* group are abundant. The largest numbers of infections occur in conflict and post-conflict regions of Angola, the Central African Republic, Chad, South Sudan, and the DRC.[1] This form of HAT can last for years before the classical features of sleeping sickness appear. Early in its course, Gambian HAT presents in a fairly nonspecific way with fever, fatigue, and headache, although some telltale signs also occur, such as enlargement of the lymph nodes in the posterior region of the neck, just behind the ear, as well as swelling of the face and itching.[2] Enlargement of the lymph nodes in the neck is known as Winterbottom's sign, named after a British colonial physician who linked this finding to what he called "the Negro lethargy" while working in Sierra Leone.[6] Both anemia and endocrine disturbances also occur. This first phase of the illness lasts approximately 2 years; unless treated, this phase is inevitably followed by the sleeping sickness stage, which corresponds to invasion by the parasite of the central nervous system. In the brain and in the overlying tissues known as the meninges, the multiplying trypanosomes produce chronic inflammation (referred to as meningoencephalitis), characterized initially by severe and unrelenting headache and disturbances of gait and parkinsonian-like movements. There are also profound changes in the patient's personality, such as extreme paranoia and aggression and, between bouts of such behavior, daytime sleeping. Eventually, the patient becomes comatose and dies. For decades, the classic teaching about HAT was that the disease is invariably fatal, although some have challenged this concept and introduced the idea that some individuals may be "trypanotolerant."[6]

East African or Rhodesian HAT, which occurs predominantly in Kenya, Malawi, Mozambique, Tanzania, Uganda, and Zambia, has a much more rapid course, with death usually occurring within a year following the onset of early symptoms.[2] An epidemic of Rhodesian HAT in British East Africa between 1902 and 1905 may have killed as many as 250,000 people.[6] Rhodesian HAT also exhibits important differences in its epidemiology compared with Gambian HAT. The former is transmitted by the morsitans group of *Glossina* tsetses, which preferentially feed on domestic animals (e.g., pigs, dogs, sheep, cattle, and goats) and wild game animals (such as bushbuck) rather than on humans. *G. morsitans* also transmits HAT primarily in woodland savanna settings rather than near rivers.[2] *T. b. rhodesiense*, unlike *T. b. gambiense*, has the ability to replicate in a wide variety of mammals, so humans are more or less accidental hosts of this parasite. In contrast, *T. b. gambiense* is probably a parasite exclusively of humans. We will see below how the fact that Rhodesian HAT is primarily a zoonosis, i.e., a disease transmitted from animals to humans, has important implications for controlling epidemics of this disease.

Because of its dramatic clinical features, inevitable lethality, and ability to decimate entire African villages, sleeping sickness has made an indelible imprint on the history of sub-Saharan Africa. In terms of its impact on colonial history, sleeping sickness did much to thwart European ambitions on the African continent, and the colonial administrations of the United Kingdom, as well as of Belgium, France, Germany, and Portugal, responded to this threat by investing a considerable component of their medical activities in its study. At least as much as any other single disease, it was HAT that stimulated colonial European governments to commission a first generation of schools of tropical medicine in Antwerp (Belgium), Basel (Switzerland), Hamburg (Germany), Lisbon (Portugal), and Liverpool and London (England) during the last decade of the 19th century and in the first decades of the 20th century. The founding of these schools mostly in European seaports reflected the impact of HAT and other tropical diseases on returning seamen and practitioners of other occupations connected with vital shipping interests.[7] Sir Alfred Lewis Jones, the founder of the Liverpool School of Tropical Medicine, was one of Liverpool's leading businessmen who invested heavily in the shipping trade with West Africa.[7]

It was during this period that the toll of sleeping sickness in sub-Saharan Africa began to be widely appreciated, particularly with colonial expansion into the Belgian Congo, the same location where Joseph Conrad in *Heart of Darkness* placed his abandoned individuals presumably suffering from sleeping sickness.[6,7] Working in Zululand (the site of present-day KwaZulu-Natal Province of South Africa) in 1894, Surgeon-Captain David Bruce, a Scottish bacteriologist working in the British Army, first identified trypanosomes from cattle with nagana and demonstrated that he could reproduce a wasting disease in dogs by injecting them with the same microorganisms; a year later he identified the tsetse as the natural vector for nagana.[6] In 1902, John Everett Dutton of the Liverpool School of Tropical Medicine provided one of the first descriptions of a human trypanosome from a seaman who worked on steamships on the Gambia River.[6] A reproduction of Dutton's original watercolor drawing of what he named *T. gambiense* is shown in Fig. 7.5. Three years

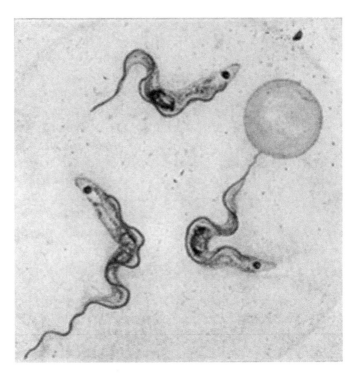

Figure 7.5 Reproduction of J. E. Dutton's original watercolor drawing of *T. b. gambiense* from the blood of an infected patient, Gambia, 1902.

after his discovery, Dutton died from relapsing fever (an unrelated bacterial infection) while working in the Belgian Congo. Today, there stands a church in Dutton's home village of Bunbury, Cheshire (outside Liverpool), with a stained-glass panel commemorating both Dutton and the accomplishments of the Liverpool School.[6,7] Subsequently, while investigating a sleeping sickness outbreak in Uganda as part of a sleeping sickness commission sponsored by the Royal Society of London, David Bruce, together with Aldo Castellani, demonstrated that this human trypanosome was also transmitted by tsetses.[6] Bruce also determined that the organism causing Rhodesian HAT was transmitted from animals to humans.[6]

Concurrent with the activities of the British at the turn of the 20th century, there were also efforts in France, Germany, Switzerland, and England to develop new treatments for HAT. As early as 1858, the explorer David Livingstone was using arsenic to treat tsetse bites, and in 1902 Alphonse Laveran, the French codiscoverer of the malaria parasite, found that sodium arsenite was a potential treatment for HAT.[8] Shortly thereafter, it was discovered that if arsenic was modified by chemically linking it with an organic compound, its toxicity diminished considerably.[8] In 1907, Paul Ehrlich, the father of chemotherapy, synthesized and accurately characterized atoxyl, one of the first such organified arsenic compounds.[8] Ehrlich was a Prussian Jew who worked initially with Robert Koch in Berlin and then moved to Frankfurt, where he headed the Royal Institute for Experimental Therapy. Subsequently, at the Liverpool School, Anton Breinl showed that atoxyl could be developed as a treatment for trypanosomiasis in laboratory animals, and although the compound was extremely toxic, this discovery paved the way for the development of a related arsenic-containing compound, melarsoprol. A second organified arsenic compound, tryparsamide, was also developed around this period at the Rockefeller Institute for Medical Research, the forerunner of today's Rockefeller University.[8] These discoveries laid the foundation for the development and testing of melarsoprol, which became the first successful arsenical drug for sleeping sickness. First developed by Ernst Friedheim (1899–1989), a Swiss physician and chemist, melarsoprol has a better safety profile than does the first generation of arsenicals, atoxyl and tryparsamide. However, because of its arsenic component, melarsoprol is still considered highly toxic, and even today up to 10% of patients undergoing treatment for sleeping sickness with the drug either die or experience a severe toxic reaction known as posttreatment reactive encephalopathy.[9] Sadly, despite its toxicity, melarsoprol still remains one of only a few drugs available for the treatment of central nervous system HAT.

In addition to melarsoprol, there are two other drugs, pentamidine and suramin (neither of which is an arsenical), still in widespread use for the earlier stage of HAT (prior to parasite invasion of the central nervous system). Both drugs were developed even before melarsoprol. Discovered in 1916, suramin is a derivative of the dye trypan red; by noting that many dyes bound selectively to microorganisms, Paul Ehrlich had earlier pioneered the concept that they could also be developed into chemotherapeutic agents. Subsequently, the drug pentamidine was discovered in 1936.[8] Therefore, until recently much of our pharmacopeia for HAT has relied on drugs that are more than 75 years old.

Since 2005, the entire genome of *T. brucei* has been sequenced, and it is theoretically possible to mine this information in order to develop a new generation of antitrypanosomal drugs.[10] You might ask how it can be that we use such old-fashioned drugs with relatively high levels of toxicity when it is possible to develop safer and more effective drugs. There are a number of reasons for this overall absence in research and development activities, but primarily it is because HAT is an NTD and there is no commercial incentive for drug companies to develop and test new compounds. We saw previously in the chapters about onchocerciasis, lymphatic filariasis, and trachoma that the large pharmaceutical companies are prepared to donate drugs, but it is another matter altogether for any publicly held company to convince its shareholders to launch and invest in a discovery program for guaranteed money-losing products. The problem of access to essential medicines for HAT is therefore twofold: first, the existing drugs are highly toxic and often not available; and second, there is no incentive to develop new drugs. Another example of this second problem is illustrated by a newer agent for HAT known as eflornithine. Eflornithine was originally developed and tested as a new anticancer drug, but it was subsequently abandoned for that purpose. Although it might have failed as a breakthrough against cancer, during the 1990s it was discovered that eflornithine was effective against the central nervous system stage of Gambian HAT. This discovery was considered a breakthrough because eflornithine was the first drug for late-stage sleeping sickness that did not contain toxic arsenic. However, without a commercially viable market, there were no incentives for pharmaceutical companies to develop eflornithine for use in sub-Saharan Africa. In 2001, the French nonprofit organization Médecins Sans Frontières (Doctors Without Borders) together with the World Health Organization (WHO) successfully lobbied to encourage the pharmaceutical company Sanofi-Aventis to manufacture eflornithine and make it widely available in Africa. In a landmark agreement, Sanofi Pasteur also agreed to produce pentamidine and melarsoprol. Thus, the WHO advanced public-private partnerships with Sanofi-Aventis to produce and release pentamidine, melarsoprol, and eflornithine and with Bayer AG to produce suramin and nifurtimox and to provide them free of charge to countries where HAT is endemic.[1] Later (in chapter 11), we will also see how new product development partnerships, such as the Geneva-based DNDi, developed new and less toxic antitrypanosomal drugs that are suitable for resource-poor settings.[11] They include combination therapy with the drugs nifurtimox and eflornithine.[13] By combining these two drugs as a package, they can be administered over a shorter period of time and at reduced dosages compared with either drug alone or other combinations.[13] However, in perhaps the most exciting development of all, in 2018, DNDi announced results of studies with fexinidazole, a 5-nitroimidazole derivative that began as a partnership with the Swiss Tropical and Public Health Institute.[13] In clinical trials conducted on patients with Gambian HAT in the DRC and the Central African Republic, fexinidazole was shown to be a safe and effective drug regimen in both adults and children.[13] Equally exciting were findings that the drug could be delivered orally in a 10-day course, a tremendous advantage given that HAT typically occurs in remote and rural areas without access to sophisticated health systems.[13]

By the middle of the 20th century, tremendous progress in the control and prevention of HAT had been made in both West and East Africa. However, based on the different epidemiological patterns of the two forms of HAT, a different approach was required to control each form of the disease.[14] Because Gambian HAT is exclusively a human disease (i.e., no significant animal reservoir is involved in disease transmission), the major approach for control has been to conduct massive screenings for *T. b. gambiense* infection and then treat infected patients. Throughout the first part of the 20th century in French and Belgian colonial West Africa, tens of millions of people were screened by mobile field teams. If they were found to be positive, they were treated with the arsenical compound tryparsamide. Through this activity, the human reservoir of trypanosomes was depleted and transmission was interrupted. The French physician and scientist Eugene Jamot (1879–1937) is largely credited with developing this strategy. While working in Cameroon, for instance, Jamot is believed to have reduced the incidence of Gambian HAT by 300-fold. For this work, Jamot was nominated for the Nobel Prize.[15] In contrast, because East African HAT has a significant animal reservoir for transmission and *T. b.*

rhodesiense trypanosomes can be found in abundance in cattle and bushbuck, the Jamot method would not be expected to work. One draconian practice during the early 20th century in British-held East Africa was forced human resettlement away from regions inhabited by bushbuck and other animals.[6] In eastern Uganda, Ian Maudlin and Sue Welburn have determined that 18% of domestic cattle harbor *T. b. rhodesiense* and represent a significant reservoir of human-infective trypanosomes. By treating infected cattle with either anti-trypanosomal drugs or pour-on insecticides, they are exploring the possibility that this approach could interrupt the human transmission of HAT. Both West and East African approaches benefit from vector control methods, including the use of baited tsetse traps and widespread application of insecticides.[14]

Compared with mass drug administrations for NTDs such as the soil-transmitted helminth infections, schistosomiasis, lymphatic filariasis, onchocerciasis, and trachoma, the control practices for both Gambian HAT and Rhodesian HAT are extremely labor-intensive. Therefore, while it is often feasible to control Gambian HAT through case identification and treatment or Rhodesian HAT by targeting animal reservoirs, when conditions make it difficult to apply these methods, it does not take much to derail public health control measures for HAT and allow a recrudescence of the disease. This is exactly what has happened in many conflict states of sub-Saharan Africa. One of the better-documented examples of a truly horrific humanitarian catastrophe is represented by the nation of Angola.[16] With a total area of almost 500,000 sq. mi. (and 12 million people), Angola is one of the largest countries in Africa. During the years between 1926 and 1952, a series of aggressive campaigns were launched to control Gambian HAT. Based on the classic Jamot model of case identification and treatment, an important component of the Angolan control program was the Brigade for Pentamidinization, in which infected individuals were identified and were treated using the antitrypanosomal drug pentamidine.[15] In the year prior to Angola's independence in 1975, only three cases in the country could be found. Unfortunately, this was the beginning of the end. Starting in 1976, clashes between UNITA (National Union for the Total Independence of Angola) and the Soviet-backed Marxist state of the MPLA (Movement for the Liberation of Angola) ignited almost 40 years of civil conflict. A major hurdle to implementing screening measures was inaccessibility as a result of insecurity and land mines. HAT-infected patients were unable to reach diagnosis and treatment centers, while at the same time mobile health units could not travel. In addition, there was massive looting of technical equipment, including drugs and microscopes. As a result, the number of people screened for HAT went from a high of 12 million (the entire population) during the 1950s to only about 150,000 by 1998. By this time almost 7,000 cases had been detected, but presumably there were at least 10 times that, based on studies in Uganda indicating that an estimated 12 deaths caused by sleeping sickness go undetected for every 1 reported.[17]

During the war years, only the Catholic Church was permitted to conduct cross-border initiatives through ANGOTRIP, an organization created for HAT control in the region.[16] At local health centers established in regions of HAT endemicity in the northwestern part of the country, a diagnosis was established by palpating neck lymph nodes (Winterbottom's sign) or by identifying trypanosomes in the blood by microscopy, as well as using a serological method that detects antitrypanosomal antibodies. Treatments of infected patients were conducted with either pentamidine or melarsoprol, depending on whether there was central nervous system involvement. Between 1996 and 2001, ANGOTRIP screened almost 200,000 patients and achieved a substantial reduction in HAT mortality.[16] Some suggest that the death of UNITA opposition leader Jonas Savimbi in 2002 helped to restore some semblance of public health infrastructure.[16] Sadly, there are similar stories of the resurgence of Gambian HAT because of warfare in the DRC, Sudan, and elsewhere, while in East Africa, there remains a similar vulnerability to epidemic outbreaks of Rhodesian HAT. For instance, around the Lake Victoria basin, epidemics killing tens of thousands of people have occurred periodically, including a major outbreak in Uganda during the 1980s.[17,18]

With diminished hostilities in some parts of the tsetse belt in Africa, there is renewed optimism that HAT may one day be eliminated, although achieving such a target may require new methods of diagnosis and treatment, as well as new approaches to vector and animal reservoir control.[18] While the relationship between "conflict and contagion" remains a major theme of HAT's reemergence in sub-Saharan Africa,[19] redoubled efforts led by the WHO are under way to work toward meeting elimination targets. It has become apparent that there has been a more than 90% reduction in the number of new cases of HAT in the past 20 years, with only an estimated 2,000 cases remaining.[19] For Gambian HAT, the decline has been noted to occur despite an actual increase in the number of health facilities capable of HAT detection and treatment, meaning the decrease goes beyond underreporting.[19] For Rhodesian HAT, similar decreases have been observed, but there is less confidence in the actual numbers due to the remoteness of the areas where cases occur and some weakening of surveillance infrastructures.[19] Mobilizing resources to expand case detection and treatment will be essential to ensure that we reach the last mile and eliminate HAT.

Chagas Disease (American Trypanosomiasis)

American trypanosomiasis is also named Chagas disease in honor of Carlos Chagas, who as a young Brazilian physician in 1909 identified *Trypanosoma cruzi* (named after his mentor, Oswaldo Cruz) as the causative agent and then went on to elucidate the entire life cycle of the infection.[20] Except for being caused by a trypanosome, American trypanosomiasis bears very little resemblance to HAT in either its clinical features or its epidemiology. An estimated 6 million to 7 million suffer from Chagas disease,[21] an insect vector-borne infection occurring almost exclusively among the very poorest inhabitants of Latin America (Fig. 7.6). Accordingly, the highest rates of *T. cruzi* transmission and disease prevalence tend to occur in the most impoverished Latin American countries. Currently, the nation with the highest percentage of people infected with *T. cruzi* is probably Bolivia, where the seroprevalence in many villages

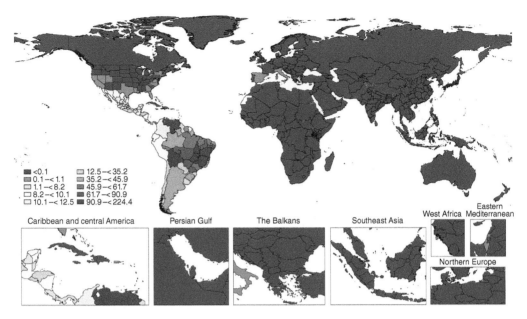

Figure 7.6 Distribution of Chagas disease, worldwide. (From Global Burden of Disease Collaborative Network, 2020 [http://www.healthdata.org/results/gbd_summaries/2019/chagas-disease-level-3-cause].)

and even cities exceeds 25%.[22] However, the prevalence rates are almost as high in many areas of Venezuela and eastern Brazil.[21] Other major foci of endemic Chagas disease occur in southern Mexico and Central America; the Andean region comprising Colombia, Ecuador, and Peru (DNA from *T. cruzi* has been detected in the mummified remains of people from this region, especially Peru and northern Chile, from as far back as 2000 BCE)[6]; and in an area known as the Gran Chaco, a lowland region that incorporates parts of Bolivia, Paraguay, northern Argentina, and southern Brazil. Throughout these regions of endemicity in the Americas, Chagas disease continues to be a significant cause of chronic heart disease. In addition, cases of Chagas disease now occur in the Amazon region, and the disease has been exported to the United States, Europe (especially Spain), and elsewhere through immigration. This phenomenon is sometimes known as the "globalization of Chagas disease."[23] The Chagas disease situation in the United States is described in more detail in chapter 9, including evidence for transmission of the disease in Texas and other southern states.

One major success story in the war against Chagas disease has been an innovative vector control program known as the Southern Cone Initiative or INCOSUR (Incitativa de Salud del Cono Sur), through which there have been dramatic gains in reducing the prevalence of the disease.[24] Therefore, there is some optimism that, through vector control, Chagas disease transmission could be eliminated in the Southern Cone and possibly elsewhere in the coming decades. However, just as there are important differences between the East and West African forms of HAT, there are also some regional differences in the geographical variants of Chagas disease. Of particular importance is an endemic focus in Mesoamerica (i.e., Mexico and Central America), where, because of the unique ecology of the infection and its vector, there are significant concerns that Chagas disease may be difficult to eliminate. Another huge and largely neglected issue has been the high rates of Chagas disease during pregnancy and mother-to-child transmission of the disease; this problem occurs even in areas where vectorial transmission has been eliminated.[25] According to some estimates, approximately 5% of pregnant women in Latin America may be infected with *T. cruzi*.[25]

Imagine a large cockroach-like bug endowed with the ability to suck blood. The vector transmitting *T. cruzi* trypanosomes is the triatomine bug, known locally as the assassin bug, kissing bug, *vinchuca* (*benchuca*), or *barbeiro* (Fig. 7.7). Triatomine bugs operate as rural cousins of the cockroach, typically living in the crevices of walls and thatch of low-quality dwellings in the rural areas of the Americas. At night, they leave the safety of their crevices and thatch, often dropping down on their sleeping victims in order to feed on blood. In his travel journals from Argentina, Charles Darwin—who in his later years suffered from a variety of ailments that some authors have attributed to the chronic consequences of Chagas disease—provided an excellent description of a vinchuca and its voracious appetite for blood[26]:

> At night I experienced an attack (for it deserves no less a name) of benchuca . . . the great black bug of the pampas. It is most disgusting to feel soft wingless insects, about an inch long, crawling over one's body. Before sucking they are quite thin but afterwards they become round and bloated with blood, and in this state are easily crushed. One which I caught . . . was very empty. When placed on a table, and though surrounded by people, if a finger was presented, the bold insect would immediately protrude its sucker, make a charge, and if allowed, draw blood. No pain was caused by the wound. It was curious to watch its body during the act of sucking, as in less than ten minutes it changed from being as flat as a wafer to a globular form.

Figure 7.7 Triatomine bug, *Rhodnius prolixus*. (Courtesy of Erwin Huebner, University of Manitoba, Winnipeg, Canada.)

The possibility that the triatomine bug could transmit American trypano-somiasis was demonstrated by Carlos Chagas. As a young physician and a member of one of Brazil's first biomedical research institutes, the Institute of Experimental Pathology of Manguinhos (later named Instituto Oswaldo Cruz and now a major component of FIOCRUZ [the Oswaldo Cruz Foundation], Brazil's largest and most extensive biomedical and research organization devoted to infectious diseases and other conditions), Chagas had developed a reputation for his ability to organize and lead antimalaria campaigns in the interior of the country.[20] While he was on assignment, a railway company engineer brought to the young Chagas's attention the vinchucas inhabiting the poor dwellings of the region and their nocturnal biting behavior. Chagas began to examine vinchucas and identified trypanosomes, which he named in honor of Oswaldo Cruz, his institute director.[20] He subsequently found a trypano-some in the blood of a 22-month-old girl named Bernice who was suffering from fever with enlargement of the liver, spleen, and lymph nodes, as well as facial edema (swelling of the face). He also identified the trypanosomes in a cat that lived in the same house. Ultimately, Bernice recovered from her acute illness, and 52 years later (27 years after the death of Chagas), Bernice was rediscovered as a grandmother of three living on a farm in Minas Gerais State. Blood tests conducted at a major university teaching hospital at that time sub-sequently confirmed a diagnosis of Chagas disease. Bernice ultimately died in 1981 at the age of 73.[20]

Humans become infected with *T. cruzi* in an interesting way (Fig. 7.8). Unlike metacyclic African trypanosomes, which live in the insect salivary glands and enter through a bite, the metacyclic American trypanosome lives in the hindgut of the triatomine bug. While working in Brazil around the same time as Chagas, the eminent French parasitologist Emile Brumpt dem-onstrated that *T. cruzi* infection first occurs by the process of autoinoculation. This transmission phenomenon occurs when the triatomine bug defecates as it feeds; during sleep, the unsuspecting victim of the assassin bug rubs the trypanosome-infected bug feces into either the puncture wound caused by the bug or the mucous membranes of the eyes or mouth. Once the metacy-clic trypanosomes invade the mucous membranes, they continue to behave quite differently from *T. brucei*. Rather than invade the bloodstream imme-diately and multiply as the trypomastigote form, the metacyclic *T. cruzi* try-pomastigotes exhibit the ability to invade cells and multiply as intracellular pathogens.[27] Initially, this stage of invasion results in inflammation near the site of entry, causing swelling around the eye and face (known as Romana's sign) or some other region of the skin, where it is known as a chagoma. In other cases, more general swelling, referred to as "facial edema," occurs. Once inside cells, the parasite replicates as a rounded, so-called amastigote form, but subsequently the cells burst and release a new round of trypomastigote forms of the parasite, which then invade the bloodstream before besieging multiple organs. Initially, organ invasion results in acute Chagas disease associated with fever, headache, enlargement of the spleen and liver, and facial or generalized edema.[28] However, the acute presentation is often less dramatic and sometimes

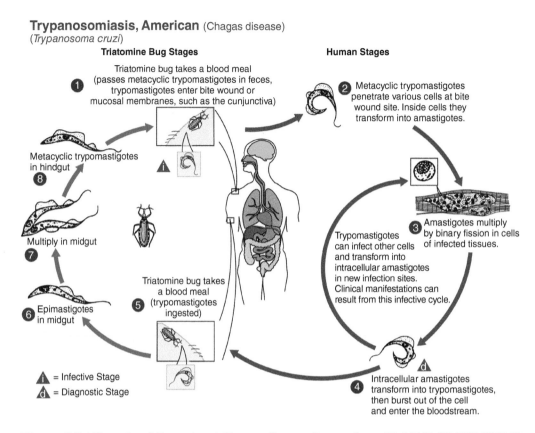

Trypanosomiasis, American (Chagas disease)
(*Trypanosoma cruzi*)

Triatomine Bug Stages　　　　　　　　　　**Human Stages**

① Triatomine bug takes a blood meal (passes metacyclic trypomastigotes in feces, trypomastigotes enter bite wound or mucosal membranes, such as the cunjunctiva)

② Metacyclic trypomastigotes penetrate various cells at bite wound site. Inside cells they transform into amastigotes.

⑧ Metacyclic trypomastigotes in hindgut

⑦ Multiply in midgut

⑥ Epimastigotes in midgut

⑤ Triatomine bug takes a blood meal (trypomastigotes ingested)

③ Amastigotes multiply by binary fission in cells of infected tissues.

Trypomastigotes can infect other cells and transform into intracellular amastigotes in new infection sites. Clinical manifestations can result from this infective cycle.

④ Intracellular amastigotes transform into trypomastigotes, then burst out of the cell and enter the bloodstream.

🔺**i** = Infective Stage
🔺**d** = Diagnostic Stage

Figure 7.8 Life cycle of *T. cruzi* and Chagas disease. (Image from CDC-PHIL [ID#3384]/CDC/ Alexander J. de Silva, PhD; Melanie Moser, 2002.)

without obvious symptoms. During this stage of the illness, it is sometimes possible to identify trypanosomes in the blood. Acute Chagas disease can also cause inflammation of the heart (myocarditis), which can be fatal because it can cause either conduction disturbances in the heart or even heart failure. An individual can also become acutely infected with *T. cruzi* through the accidental ingestion of crushed kissing bugs in juices prepared from sugarcane or other foods. Oral transmission is considered an important route of *T. cruzi* transmission in the Amazon region and possibly elsewhere, and was recently noted to be an important mode of transmission in Venezuela, where it has been linked to breakdowns in the social and economic structures.[29]

Following this acute phase of the illness, many patients recover and remain asymptomatic over a period of years (sometimes known as indeterminate Chagas disease) before entering a new phase of *T. cruzi* infection known as the chronic phase. This syndrome—determinate Chagas disease—occurs in roughly 30% of patients who previously experienced acute infection.[28] In the chronically affected patients, there is a severe disturbance of the electrical system of the heart leading to arrhythmias, palpitations, chest pain, difficulties with breathing on exertion, and fainting. In some of these patients, there are a profound enlargement of the heart and distortion of its architecture, a condition known as chronic chagasic cardiomyopathy (CCC). Over time, CCC can lead to heart failure, aneurysms of the heart wall, and even sudden death. In addition to the heart changes, some patients also develop a curious disorder of their colon and esophagus in which both organs lose their tone and become grossly enlarged. Megacolon and megaesophagus are associated with regurgitation, pain on swallowing, and severe constipation. Discovery of the link between megacolon and megaesophagus (and even CCC) and Chagas disease has been attributed to the work of Fritz Koberle in the 1960s.[6,30] A number of hypotheses have been advanced to explain the mechanisms by which long-standing *T. cruzi* infections lead to cardiomyopathy and megacolon and megaesophagus.[30] These hypotheses are controversial because it is not clear whether these maladies result from the continuous low-grade presence of *T. cruzi* parasites in the heart and other organs or whether host immune responses are responsible. Increasingly, the persistence of parasites in the heart has been recognized as the most important element of determinate Chagas disease.[31] In addition, the persistence of the parasites in the heart over time leads to inflammation and fibrosis, two of the hallmarks of CCC.[31] It was recently estimated that the risk of an indeterminate patient progressing to CCC is approximately 2% annually.[31] Another important component of the chronic phase is thought to be an abnormality of the autonomic nervous system controlling both the heart and gastrointestinal tract, again from either parasite invasion or autoimmunity. A number of clinicians and investigators have noted that the frequency of cardiac versus gastrointestinal complications varies geographically and may reflect genetic differences among the six strains of *T. cruzi*.[32] One highly innovative but also highly controversial hypothesis (and one not widely accepted) to account for the pathology of Chagas is based on the observation that in association with the parasite's invasion of the cells, *T. cruzi* DNA, particularly DNA from the parasite's kinetoplast, can be incorporated into the genome of mammalian host cells.[33] Such acquired genetic changes could ultimately alter the structure of host cell proteins or possibly trigger autoimmunity. Finally, roughly 1 in 20 *T. cruzi*-infected pregnant women transmit the infection vertically to their fetus, resulting in a severe condition known as congenital Chagas disease. As noted above, maternal and congenital infections affect hundreds of thousands and tens of thousands of people in the Americas, respectively.[22,25] Mother-to-child transmission represents the new face of Chagas disease.

The treatment of the chronic complications of Chagas disease requires complex modalities. CCC is treated with antiarrhythmia drugs and pacemakers. However, for severe disease, sometimes only heart transplantation is effective. Not surprisingly, a generation of Latin American cardiac surgeons has become adept at this procedure. Surgery is also frequently required for the gastrointestinal complications. For acute Chagas disease, there are two antitrypanosomal drugs when they are available: nifurtimox (also used for the treatment of HAT) and benznidazole (not to be confused with the benzimidazole anthelmintics [mebendazole and albendazole] used for soil-transmitted helminth infections). Unfortunately, many patients with acute Chagas disease are not identified in time for effective treatment, while the few who are diagnosed appropriately require 1 to 2 months of daily doses of these drugs. There is even considerable controversy about whether these drugs have a significant impact on the disease. For example, in most patients the organisms persist despite treatment, and for many, treatment does not prevent severe heart lesions or other long-term complications. A large multicenter clinical trial known as BENEFIT (Benznidazole Evaluation for Interrupting Trypanosomiasis) found that patients with CCC experience progression of their heart disease, with high mortality rates approaching 18% over 5 years even after completing benznidazole antiparasitic therapy.[34] These sad facts, coupled with the high acute and chronic toxicities of nifurtimox and benznidazole and the fact that they cannot be used in pregnant women (so that we have nothing to offer a pregnant woman with Chagas disease), suggest an urgent need for new anti-*T. cruzi* drugs. The completion of the *T. cruzi* genome[34] should in theory lead to the development of new drugs, but the limited commercial markets have restricted these activities to just a handful of product development partnerships such as DNDi and others as outlined above for HAT. In an article published in *PLoS Neglected Tropical Diseases*, a number of us noted an eerie resemblance between the current situation with Chagas disease in the Americas, i.e., the links with poverty, difficulties in diagnosing the illness, absence of safe and effective drugs, and high rates of maternal-to-child-transmission (and also contamination of the blood supply), and that of HIV/AIDS in the early years of the epidemic in Africa.[35] To help correct this situation, efforts are under way to shape a better infrastructure for developing and testing new Chagas disease drugs, including an international patient registry and consensus target product profiles, among other measures.[35]

There are no simple preventive chemotherapy approaches for the control of Chagas disease, nor is it practical to apply wide-scale case detection and management with antitrypanosomal drugs such as what occurred with the pentamidization campaigns against HAT launched in the 20th century. Nevertheless, the INCOSUR initiative has had great success in controlling Chagas disease in the Southern Cone through widespread insect control.[24] In this region of South America, *Triatoma infestans* is the major assassin bug vector of Chagas disease. Because this species of assassin bug lives exclusively inside the poor dwellings of the region, it is possible to reduce domiciliary infestation by large-scale residual spraying with pyrethroid insecticides.[24] This strategy also includes applications of insecticides in chicken houses and other dwellings located near the home, combined with ongoing surveillance and additional spraying of any reinfested houses.[24] In areas of the Southern Cone where *T. infestans* is responsible for transmission of Chagas disease, there has been a dramatic decrease in transmission in Argentina, Brazil, Chile, Paraguay, and Uruguay. Elimination of disease transmission has now been certified in Chile and Uruguay, as well as in four provinces of Argentina, 10 states in Brazil, and one state in Paraguay.[24] Since Bolivia is the poorest member state of the Southern Cone, its elimination efforts are still in an earlier stage. INCOSUR therefore represents one of the very best examples of cooperative programs between Latin American nations.

In Central America and elsewhere, the elimination of Chagas disease through large-scale residual spraying has not been as straightforward. Efforts to reproduce INCOSUR's successes through a Central American initiative known as IPCA (Iniciativa de los paises de Centroamerica) have been complicated by the fact that *Triatoma dimidiata*, rather than *T. infestans*, is the major vector of Chagas disease in this region. Unlike *T. infestans*, *T. dimidiata* can live in the palm trees surrounding the houses, and there are no effective methods for insecticide spraying of palms.[24] Therefore, control efforts require multiple sprayings each time the vectors enter households. This problem makes community-based surveillance, including case detection and management, all the more important.[24,36]

With exception of Venezuela, efforts in the Andean region have not progressed nearly as far as in the Southern Cone, with baseline surveys only beginning in many regions. As a long-term measure, it is theoretically possible to develop a Chagas disease vaccine, but the development testing of a new generation of antipoverty vaccines remains elusive because of the challenges of taking on product development for the world's poorest people.[37] In chapter 11, I will describe some new efforts we have embarked on to develop a therapeutic vaccine for Chagas disease as an alternative to the medicines currently available.[37] In the meantime, the new Global Chagas Disease Coalition (www.coalicionchagas.org/), based at the Barcelona Institute of Global Health (IS Global) in Spain and partnering with several key organizations committed to Chagas disease, is working to raise awareness and to promote access to essential antiparasitic medicines. The stark reality is that 99% of the world's people infected with *T. cruzi* likely do not have access to diagnosis and treatment for this condition.

When I travel to Central America, it always is amazing for me to realize that from Miami or Houston I can reach within just a few hours many regions of Central America where one of the most important neglected diseases of poverty remains endemic. On most days, I can reach the regions of Honduras where Chagas is endemic faster than I can reach the West Coast of the United States. On several occasions I have awakened in the morning in my home in the United States, and by late in the afternoon am making pediatric rounds in the children's wards of the public hospitals in Central America, where I can see sick kids with Chagas disease and other NTDs as well as malnutrition. While so much attention is now being paid to sub-Saharan Africa, it is a sad state of affairs that some of our closest neighboring countries cannot benefit from the very best that Western biotechnology has to offer.

Leishmaniasis

Leishmaniasis is a serious parasitic disease affecting millions of poor people in developing countries (Fig. 7.9 and 7.10). There are two major forms of the disease—visceral leishmaniasis (VL; also known as kala-azar) is a disseminated infection affecting multiple organs, including the liver, spleen, blood, and bone marrow; and cutaneous leishmaniasis (CL) is an ulcerative and often highly

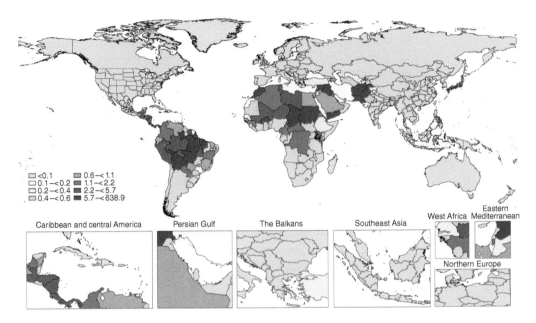

Figure 7.9 Distribution of CL, worldwide, 2009. (From Global Burden of Disease Collaborative Network, 2020 [http://www.healthdata.org/results/gbd_summaries/2019/cutaneous-and-mucocutaneous-leishmaniasis-level-4-cause].)

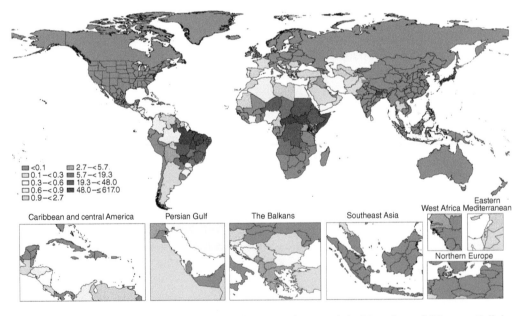

Figure 7.10 Distribution of VL, worldwide, 2009. (From Global Burden of Disease Collaborative Network, 2020 [http://www.healthdata.org/results/gbd_summaries/2019/visceral-leishmaniasis-level-4-cause].)

disfiguring chronic skin infection. A less common third, mucocutaneous form (MCL) is also found in Central and South America. Ancient accounts of leishmaniasis were recorded by the Persian physician and philosopher Avicenna

(Abū Alī al-Ḥusayn ibn Abd Allāh ibn Sīnā) in the 10th century, and there are several descriptions of the illness going as far back as the 7th century BCE.[6] Annually, leishmaniasis affects approximately 4 million to 5 million people, causing approximately 5,000 to 6,000 deaths, in a wide geographic range across the tropics and subtropics.[38] Most VL cases currently occur in conflict and post-conflict areas of sub-Saharan Africa, including South Sudan, Central African Republic, and Somalia, as well as in tropical areas of Brazil and in Bihar State in India.[38] CL cases occur disproportionately in a few African nations such as Chad and Sudan, as well as in nations of the Middle East and Central Asia such as Afghanistan, Iran, Saudi Arabia, and Syria and in Brazil and Bolivia in the Americas.[38] The reported numbers of cases of CL are underestimates since they do not generally consider those living with chronic facial scars and the resulting social stigma, especially for girls and women.[39] Not surprisingly, therefore, there is also an important mental health aspect, including depression, to CL, which is frequently underestimated.[39] Indeed, Professor David Molyneux together with Dr. Freddie Bailey for the last few years have been exploring these important mental health links for NTDs.[39] According to revised disease burden estimates based on this information, the healthy life years lost from premature death or disability (disability-adjusted life years) make leishmaniasis one of the most important protozoan infections of humans in terms of deaths.

Except for a small focus of the disease in southern Europe and in Texas and Oklahoma in the United States, leishmaniasis occurs almost exclusively in low- and middle-income countries. In these regions, poor rural housing conditions such as cracked walls and earthen floors allow the sandfly vectors that transmit both VL and CL to flourish, while in some urban areas with poor sanitation, the uncollected garbage provides additional sandfly niches, because it promotes habitats for stray dogs, which can sometimes serve as animal reservoirs of the infection.[40,41] Another important link between leishmaniasis and poverty is the limited access to essential medicines. Today, our lack of effective medicines and access to populations living in conflict areas represent key obstacles to disease control. This is particularly true for women, who are severely stigmatized as a result of this disfiguring infection and as a result are often denied access to both health care and treatments.[41] Conflict promotes the rise of leishmaniasis just as it does for HAT. Renewed conflict in what is now South Sudan promoted a massive epidemic of VL in the western Upper Nile Province of Sudan, particularly among war refugees who settled along the Sudanese-Ethiopian border.[42] The major reasons for the epidemic of VL were interruption of residual spraying for malaria (which is also effective at controlling sandfly populations); movement of military personnel from areas of Ethiopia where VL was also endemic; and malnutrition among the refugees, which may have increased their susceptibility to infection.[42] Up to 30,000 people died from VL in 1988, and another 19,000 patients were treated in emergency centers set up by Médecins Sans Frontières between 1989 and 1995.[42] Some estimates suggest that between 1984 and 1994, approximately 100,000 deaths occurred among an estimated 280,000 infected people.[42] Thus, the conflict in Sudan has allowed at least three serious NTDs to flourish—HAT, trachoma, and VL—on top of the soil-transmitted helminthiases and schistosomiasis endemic to the region. For similar reasons, CL has also become widespread in the conflict and post-conflict areas of the Middle East, especially in the current and former ISIS-occupied areas of Iraq and Syria, and in Yemen.[42] In some of these regions, CL is considered hyperendemic, with long-term consequences related to its impact on the mental health of girls and women. Both VL and CL are also common diseases in patients with HIV/AIDS. There is a unique HIV-associated focus of VL in southern Europe (where coinfections are significant) and elsewhere, especially Brazil.

There are multiple species of *Leishmania* parasites that cause either VL or CL (Table 7.2). Leishmaniasis is transmitted by the bite of small and delicate-appearing sandflies, which inoculate flagellated forms of the parasite (known as promastigotes) into the skin (Fig. 7.11).[40,43] What then happens is remarkable. Rather than avoid being ingested and killed by macrophages—professional killer cells designed by our bodies to ward off infectious agents—the *Leishmania* parasites actually attract these cells so that they can be ingested by them more efficiently. Once inside macrophages, the parasites multiply as amastigote forms, which actually thrive and multiply in a toxic environment of microbicidal enzymes and chemicals that would ordinarily kill almost any other microorganism. Eventually, the macrophages rupture, releasing new amastigotes that in turn invade new macrophages. In the case of CL, the infected macrophages and amastigotes remain in the skin, while in VL, the macrophages travel to the liver, spleen, and bone marrow, where these infected cells release new rounds of amastigotes and cause significant systemic effects.[40] In humans, therefore, the kinetoplastid trypanosomes causing HAT live an entirely extracellular existence, while American trypanosomes live both intracellularly and extracellularly and *Leishmania* parasites are almost entirely intracellular.

In the early stages, the clinical features of CL resemble those of Buruli ulcer, beginning as a nodule and then over a period of weeks to months degenerating

Table 7.2 Simplified summary of the human leishmaniases[a]

Disease	Major Old World species	Major Old World geographic distribution	Major New World species	Major New World geographic distribution
Visceral leishmaniasis	*Leishmania donovani*	Sudan and Indian subcontinent (Nepal and Bangladesh)	*Leishmania chagasi*	Brazil
	Leishmania infantum	Southern Europe		
Cutaneous leishmaniasis	*Leishmania tropica, Leishmania major*	Afghanistan, Algeria, Iran, Pakistan, Saudi Arabia, Sudan, and Syria	Numerous species	Peru, Brazil, and Central America
Mucocutaneous leishmaniasis			*Leishmania braziliensis*	Bolivia, Brazil, Peru, and Central America

[a] Based on information from Alvar et al., 2006a.

Leishmaniasis
(*Leishmania spp.*)

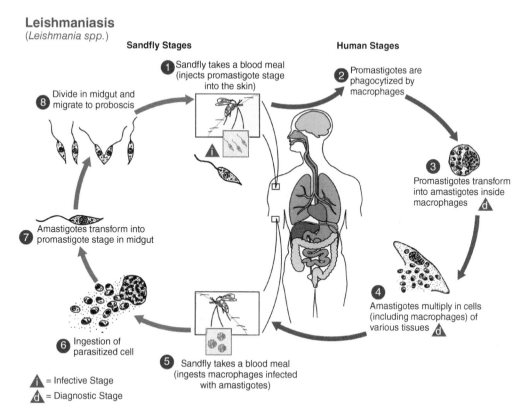

Figure 7.11 Life cycle of human leishmaniasis infection. (Image from CDC-PHIL [ID#3400]/ CDC/Alexander de Silva, PhD, 2002.)

into an ulcer. The ulcer, which is known by a number of different names depending on its geographic location (such as Baghdad boil, Delhi boil, Aleppo evil, Balkh sore, and oriental sore), has a cratered center and raised margins (Fig. 7.12). The term "pizza-like" is frequently used. With time, most CL lesions self-heal, but often not before they produce a disfiguring scar. MCL is a major variant of CL in which ulceration erodes the cartilage of the nose to produce horrific destruction of the face.

In places such as Afghanistan, the Middle East, and India, a CL lesion or scar can have catastrophic social consequences. The stigma of CL, like the stigma of many NTDs, is particularly severe for girls and women of reproductive age. For instance, in Afghanistan, even though CL is not transmitted by contact between people (it requires the sandfly vector), mothers with CL are prevented from touching their children and young women with CL are unable to marry. In Colombia, husbands abandon their infected wives, and in Bangladesh, the husband's family will not pay for medical care.[41,42]

Because it affects multiple internal organs and the blood, VL produces a far more extensive illness, manifesting as high fever, with fever spikes several times a day, night sweats, loss of appetite, and severe weight loss proceeding over a period of months.[40] In many regions VL has a predilection for young

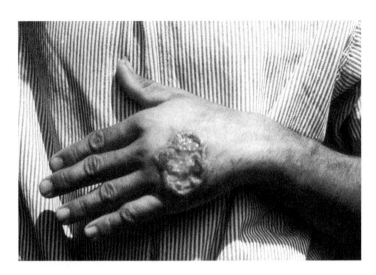

Figure 7.12 The "pizza-like" lesion of CL. (Image from CDC-PHIL [ID#352/CDC/ Dr. D.S. Martin.])

children, who often have a surprising level of energy and activity despite their severe and profound level of disease. The large number of parasites in the bone marrow of these children results in an inability of the patient to produce both white blood cells and red blood cells. As a result, infected children and adults experience pancytopenia, a decrease in their circulating blood cell counts leading to anemia and other blood abnormalities. Their livers and spleens become dramatically enlarged. When cases of VL are imported to the United States, such children are often erroneously diagnosed as having advanced leukemias or lymphomas. In some patients their skin becomes dark; in India and elsewhere, VL is sometimes named kala-azar, meaning black fever.[40] Among patients who died from VL in the Sudanese epidemic during the late 1980s and early 1990s, the most important risk factors for death were age (the very young and the very old), duration of illness greater than 5 months, severe anemia and malnutrition, a large spleen, and a high parasite load in the tissues.[42] Working in India in 1899, the eminent malariologist Sir Ronald Ross thought that VL was a severe form of malaria. It was not until a few years later that William Leishman, a Scottish army doctor, and Charles Donovan, a medical school professor at Madras University, found the cause of kala-azar.[6,43]

Both CL and VL are treatable infections, but many of the drugs used produce severe toxicities, and in many cases they are not available. For the treatment of central nervous system HAT, we learned about continuing reliance on highly toxic arsenical compounds even though it is possible to design a new generation of antitrypanosomal drugs. The situation for leishmaniasis is not much better—rather than arsenic, we instead give drugs containing the heavy metal antimony. In most of the world, compounds containing antimony are still the treatment of choice for leishmaniasis. Usually, a month of daily injections with the antimony-containing compound stibogluconate is required for the treatment of VL. Such treatments are expensive both in terms of actual money and the precious time patients lose from work. Estimates for a course of treatment range from US$30 to US$1,500 just for the medicines, and when the overall costs of lost productivity and impact on affected families are measured, the results are staggering. For instance, in Nepal, the median cost for one VL patient was equivalent to the yearly median per capita income.[41] The treatment of CL also requires multiple toxic injections, and in that regard it is questionable whether there is sufficient therapeutic benefit in administering the drug rather than allowing the cutaneous ulcers to self-heal. Better clinical trials that accurately measure cure rates (in terms of both clinical response rates and reduction of parasite loads) are required for both forms of leishmaniasis.

There are alternative approaches to treat CL and VL, which include the use of a medicine known as liposomal amphotericin B, which was developed primarily for the treatment of opportunistic fungal infections in immunocompromised patients with cancer and other serious disorders. Although in many cases a single dose of liposomal amphotericin B is effective for the treatment of VL, the hundreds of dollars for even a single dose render the drug prohibitively expensive in Bihar State in India (where there is widespread stibogluconate resistance) and other low-income areas. Similarly, there is a new oral drug for VL known as miltefosine, which would be ideal for treating poor people with VL in an outpatient setting, but it too is often too expensive. Therefore, the poorest patients with VL must endure lengthy and toxic treatments with antimony-containing medicines or, in many cases, have no access at all to essential medicine. Recently in Bihar, India, the effectiveness of several therapies was compared for the treatment of VL, including a liposomal amphotericin B formulation known as AmBisome, AmBisome with miltefosine, and miltefosine with paromomycin (developed by one of the first nonprofit product development partnerships known as the Institute for OneWorld Health). Unfortunately, all three groups were shown to have relapsed at follow-up visits several months to years later, indicating that we still urgently need new medications.[44]

Efforts to control leishmaniasis worldwide are hindered by the complex epidemiology of both VL and CL.[39] In some areas of the world, such as in Brazil and southern Europe, sandflies can transmit leishmaniasis from dogs to humans. The control of so-called zoonotic leishmaniasis is typically accomplished by targeting these animals through culling or mass treatments with antimonials or other agents. However, much of the world's serious disease burden results from so-called anthroponotic leishmaniasis, meaning that humans are the sole major reservoir of the disease-causing parasites.[39,40] Unfortunately, the containment of anthroponotic leishmaniasis generally requires fairly labor-intensive practices, such as early diagnosis and treatment and the distribution of insecticide-treated bed nets in order to reduce the likelihood of sandfly bites.[39] Therefore, in conflict and post-conflict settings where both VL and CL occur, the lack of access to essential medicines is a major obstacle to control. Among the factors impairing access are the high cost of the medicines, the requirement for precious hospital beds in order to administer drugs by injection, and the unwillingness of patients to miss time from work in order to receive their treatments.[39] Access to essential medicines has been problematic in areas such as Bihar State, India, where widespread antimony drug resistance has developed and where more-expensive drugs are required.[39] However, there is renewed optimism that elimination may one day be possible in the South Asian nations of India, Bangladesh, and Nepal through case detection and simplified diagnostic testing, treatment with antileishmanial drugs, and aggressive vector control.[45] Later (in chapter 11), we will explore the possibility of developing an antileishmaniasis vaccine.

SUMMARY POINTS · The Kinetoplastid Infections

- The kinetoplastid infections are a group of three major human protozoan infections of the poor caused by single-celled parasites with a flagellum and an unusual DNA-containing cell organelle known as the kinetoplastid.
- They frequently occur in areas of conflict or post-conflict disruption, "conflict and contagion."

HAT

- HAT, also known as sleeping sickness, occurs in an estimated 2,000 people living in sub-Saharan Africa, with the highest rates of infection in conflict and post-conflict regions of Central African Republic, Chad, Democratic Republic of the Congo, South Sudan, and Uganda.
- HAT is caused by two different species of trypanosomes. *T. b. gambiense* is the cause in West Africa, while *T. b. rhodesiense* occurs in East Africa. Both forms are often fatal if untreated. Only Uganda has both types of HAT.
- HAT is a vector-transmitted disease, resulting from the bite of tsetses of the genus *Glossina*.
- West African or Gambian HAT is the more chronic form. Parasites circulate in the bloodstream or lymph nodes for years through a process of antigenic variation before the central nervous system phase begins. There are no major animal reservoirs of the disease.
- East African or Rhodesian HAT is the more acute form, and there is a significant animal reservoir. Patients die within 1 year.

- For decades the major drugs used to treat HAT were highly toxic and were developed in the early or middle part of the 20th century. However, through the stalwart efforts of DNDi, fexinidazole, an oral agent taken for 10 days, has become a new breakthrough therapy.
- Control of West African HAT relies largely on case detection and treatment as well as vector control, while fighting East African HAT relies on control in animal reservoirs and vector control. With continued progress, we may be on the verge of eliminating HAT from the African continent.

Chagas disease

- Chagas disease is also known as American trypanosomiasis and is caused by *T. cruzi*. *T. cruzi* has the ability to invade host cells and replicate as amastigotes.
- Approximately 6 million to 7 million people are infected with Chagas disease, most of whom are the poorest people living in Latin America. Chagas disease has also emerged in Texas and elsewhere in the United States and in southern Europe (especially Spain).
- The infection is transmitted by triatomine kissing bugs. They often live in the cracks or thatch of low-quality dwellings. Chagas victims often autoinoculate themselves by rubbing contaminated bug feces into their eyes or mouth or through an open wound.
- A significant percentage of pregnant women are also infected with *T. cruzi*, and maternal-to-child transmission occurs 5% of the time.

- Acute Chagas disease is associated with fever, edema, and changes in the conducting system of the heart.
- About 30% of those with acute Chagas disease will go on to develop the chronic form, associated with cardiomyopathy and megacolon and megaesophagus.
- Chronic chagasic cardiomyopathy (CCC) is a leading cause of heart disease in Latin America. It is associated with the persistence of parasites in the heart, leading to fibrosis and inflammation that result in conduction disturbances and arrhythmias, heart failure, and even sudden death. CCC is a debilitating and fatal condition.
- The efficacy of the two major anti-Chagas disease drugs is questionable, especially once CCC begins, and they are often toxic.
- Control of Chagas disease in the Southern Cone has been highly effective through indoor spraying that targets the vector, *T. infestans*.
- In Mexico, Central America, and elsewhere, control and the prospects for elimination are more problematic. A therapeutic vaccine (immunotherapy) is under development.

Leishmaniasis

- Approximately 4 million to 5 million people are infected with *Leishmania* parasites. In war-torn areas of Africa, including South

Sudan, Sudan, and elsewhere, highly lethal epidemics of visceral leishmaniasis (VL) have occurred over the last 3 decades. Conflict zones in the Middle East and Central Asia promote the rise of cutaneous leishmaniasis (CL).
- Leishmaniasis is transmitted by the bite of a sandfly.
- CL is a disfiguring, pizza-like lesion, which often self-heals but can leave a scar. The scar is often deeply stigmatizing for women in developing countries. CL represents an example of the mental health impact of NTDs.
- VL, or kala-azar, produces a febrile wasting syndrome with signs and symptoms that resemble leukemia. It can be highly fatal. VL can also be an opportunistic infection in patients with HIV/AIDS.
- Drugs containing antimony are still widely used for the treatment of CL and VL. However, they are toxic and difficult to administer. A liposomal form of amphotericin B is being used increasingly, but it is expensive to produce and purchase.
- Because control of CL and VL relies on successful treatment programs, the disease remains widespread in many low-income countries. However, a concerted program of elimination of VL is under way in India, Nepal, and Bangladesh (South Asia) through case detection and diagnosis, treatment, and vector control.

Notes

1. Estimates of the number of cases of HAT and disease burden estimates are provided by the GBD 2019 at http://www.healthdata.org/results/gbd_summaries/2019/african-trypanosomiasis-level-3-cause (Global Burden of Disease Collaborative Network, 2020). The reemergence of HAT and leishmaniasis as a result of conflict in Africa is described in Anonymous, 1997; Ritmeijer and Davidson, 2003; and Collin et al., 2006. Higher estimates for deaths resulting from Chagas disease and kinetoplastid infections are in Pecoul et al., 2012.

2. Good overviews of the epidemiology and clinical features of HAT can be found in Burri and Brun, 2002; Pepin and Meda, 2001; and Fèvre et al., 2006. Jamonneau et al. reported that *T. b. gambiense* infections are not universally fatal, as previously believed (Jamonneau et al., 2012). See also note 6.

3. Information is from Sherman, 2006, p. 325–330.

4. The lytic properties of HDL against trypanosomes were discovered in the 1970s by Mary ("Miki") Rifkin, then at

the Rockefeller University. A more recent review of this phenomenon is Vanhollebeke and Pays, 2010.

5. The term "variable surface glycoproteins" or "VSGs" is used to describe the antigens responsible for variation. The cellular and molecular mechanisms of antigenic variation in trypanosomes were first described in the 1970s by Keith Vickerman, working at the University of Glasgow, and George Cross, then at Cambridge University and now at the Rockefeller University. Recent reviews of this phenomenon include Pays et al., 2004; and Rudenko, 2011.

6. Historical accounts of HAT and other kinetoplastid infections are in Cox, 2004; Sherman, 2006; and Hoppe, 2003. The concept of human trypanotolerance can be found in Jamonneau et al., 2012. The quotation at the opening of this chapter is from Conrad, 1902, p. 31–32.

7. Both the Liverpool School of Tropical Medicine and the London School of Tropical Medicine were established in 1899. The

London School was renamed the London School of Hygiene in Tropical Medicine coinciding with funding from the Rockefeller Foundation in 1924. An account of the Liverpool School can be found in Power, 1999. An account of the London School can be found in Haynes, 2001.

8. Paul Ehrlich defined chemotherapy as "the use of drugs to injure an invading organism without injury to the host." This information is contained in Riethmiller, 2005. Work on the development of tryparsamide by Louise Pearce at the Rockefeller Institute for Medical Research is described in Pearce, 1921.

9. A description of posttreatment reactive encephalopathy and other neurologic sequelae of HAT in the central nervous system is in Kennedy, 2006.

10. The summary paper on the *T. brucei* genome is Berriman et al., 2005. The activities of DNDi are described in Pecoul, 2004.

11. The website of the Drugs for Neglected Diseases initiative is www.dndi.org.

12. High-throughput screening for antitrypanosomal drugs is described in McKerrow, 2005.

13. Information about treatment regimens for HAT can be found in Bisser et al., 2007; Pepin, 2007; and Simarro et al., 2012. The story of fexinidazole can be found at https://dndi.org/research-development/portfolio/fexinidazole/ and in Pelfrene et al., 2019; and Neau et al., 2020.

14. The two major approaches for the control of HAT are described in Fèvre et al., 2006.

15. A more detailed account of the career of Eugene Jamot can be found in Mbopi-Keou F-X et al., 2014.

16. The story of the reemergence of human trypanosomiasis in Angola is told in Stanghellini and Josenando, 2001; and Abel et al., 2004.

17. Odiit et al., 2004.

18. Welburn et al., 2006; Malvy et al., 2011; Brun and Blum, 2012; and Aksoy, 2011.

19. I credit the phrase "conflict and contagion" to Michael Moody, who heads the Chemical and Biological Arms Control Institute (CBACI), a Washington, DC-based policy think tank. Information on progress toward eliminating HAT is found in Franco et al., 2018.

20. Excellent historical descriptions of Carlos Chagas and his contributions to medical science are in Lewinsohn, 2003; and Sá, 2005.

21. Estimates of Chagas disease prevalence and deaths are from http://www.healthdata.org/results/gbd_summaries/2019/chagas-disease-level-3-cause (Global Burden of Disease Collaborative Network, 2020).

22. Salas Clavijo et al., 2012.

23. Gascon et al., 2010; and Schmunis and Yadon, 2010.

24. The vector control programs for Chagas disease are described in Yamagata and Nakagawa, 2006.

25. Gebrekristos and Buekens, 2014.

26. The account from Darwin is found in Sá, 2005. Darwin's possible affliction with Chagas disease is described in Bernstein, 1984.

27. The molecular and cellular mechanisms by which *T. cruzi* invades cells and then replicates are elegantly described in Andrade and Andrews, 2005.

28. Excellent clinical descriptions of Chagas disease are found in Miles, 2002; Teixeira et al., 2006b; and Rassi et al, 2012.

29. Information regarding oral transmission is from Shikanai-Yasuda and Carvalho, 2012. The situation from Venezuela is reported from Noya et al., 2017.

30. Information on the long-term pathogenesis of cardiomyopathy is found in Marin-Neto et al., 2007. A classical description of Chagas pathology by Fritz Koberle is found in Koberle, 1968.

31. Machado et al., 2012; and Hoffman et al., 2019. The annual risk of indeterminate patients developing CCC is in Chadalawada et al., 2020.

32. Bellini et al., 2012.

33. The horizontal kDNA hypothesis is described in Nitz et al., 2004; and Teixeira et al., 2006a.

34. The results of the BENEFIT study are in Morillo et al., 2015. The publication of the *T. cruzi* genome is in El-Sayed et al., 2005.

35. The comparisons between HIV/AIDS and Chagas disease are in Hotez et al., 2012. The frameworks for patient registries and target product profiles are in Hotez et al., 2020; and Alonso-Padilla et al., 2020.

36. Schofield et al., 2006.

37. Efforts to develop new Chagas disease vaccines have been reviewed recently in Dumonteil et al., 2012; Lee et al., 2012; Jones et al., 2018; and Barry et al., 2019.

38. Information is from www.who.int/leishmaniasis/burden/en. Leishmaniasis disease burden estimates from the GBD 2019 (Global Burden of Disease Collaborative Network, 2020) are found at http://www.healthdata.org/results/gbd_summaries/2019/leishmaniasis-level-3-cause. For VL, estimates are at http://www.healthdata.org/results/gbd_summaries/2019/visceral-leishmaniasis-level-4-cause. For CL, estimates are at http://www.healthdata.org/results/gbd_summaries/2019/cutaneous-and-mucocutaneous-leishmaniasis-level-4-cause.

39. Information is from Alvar et al., 2006a. Information about the scarring and stigmatization resulting from CL is found in Bailey et al., 2017; Bailey et al., 2019a; and Bailey et al., 2019b.

40. Information about the life cycle and clinical features of CL and VL can be found in Murray et al., 2005.

41. The stigmatizing and poverty-promoting aspects of leishmaniasis are described in Alvar et al., 2006b.

42. An account of the Sudanese leishmaniasis epidemic is in Zijlstra and El-Hassan, 2001; with an update in Al-Salem et al., 2016a. The story of CL among refugees and others fleeing conflict areas in the Middle East can be found in Al-Salem et al., 2016b.

43. The role of sandflies in the transmission of leishmaniasis was not conclusively demonstrated until 1941, by Saul Adler and his colleague M. Ber. Adler became a professor at Hebrew University and was considered one of the founding fathers of modern parasitology in Israel.

44. The comparison of drug regimens is found in Goyal et al., 2019. The challenges of coming up with new drugs has been reported by Caridha et al., 2019.

45. Information about eliminating leishmaniasis in South Asia is at https://www.who.int/news/item/05-03-2020-VL-India-takes-decisive-steps-overcome-last-mile-challenges and in Selvapandivan et al., 2019; and Kumar et al., 2020.

References

Abel PM, Kiala G, Lôa V, Behrend M, Musolf J, Fleischmann H, Théophile J, Krishna S, Stich A. 2004. Retaking sleeping sickness control in Angola. *Trop Med Int Health* **9**:141–148.

Aksoy S. 2011. Sleeping sickness elimination in sight: time to celebrate and reflect, but not relax. *PLoS Negl Trop Dis* **5**:e1008.

Alonso-Padilla J, Abril M, Alarcón de Noya B, Almeida IC, Angheben A, Araujo Jorge T, Chatelain E, Esteva M, Gascón J, Grijalva MJ, Guhl F, Hasslocher-Moreno AM, López MC, Luquetti A, Noya O, Pinazo MJ, Ramsey JM, Ribeiro I, Ruiz AM, Schijman AG, Sosa-Estani S, Thomas MC, Torrico F, Zrein M, Picado A. 2020. Target product profile for a test for the early assessment of treatment efficacy in Chagas disease patients: an expert consensus. *PLoS Negl Trop Dis* **14**:e0008035.

Al-Salem W, Herricks JR, Hotez PJ. 2016a. A review of visceral leishmaniasis during the conflict in South Sudan and the consequences for East African countries. *Parasit Vectors* **9**:460.

Al-Salem WS, Pigott DM, Subramaniam K, Haines LR, Kelly-Hope L, Molyneux DH, Hay SI, Acosta-Serrano A. 2016. Cutaneous leishmaniasis and conflict in Syria. *Emerg Infect Dis* **22**:931–933.

Alvar J, Croft S, Olliaro P. 2006a. Chemotherapy in the treatment and control of leishmaniasis. *Adv Parasitol* **61**:223–274.

Alvar J, Yactayo S, Bern C. 2006b. Leishmaniasis and poverty. *Trends Parasitol* **22**:552–557.

Andrade LO, Andrews NW. 2005. The *Trypanosoma cruzi*-host-cell interplay: location, invasion, retention. *Nat Rev Microbiol* **3**:819–823.

Anonymous. 1997. Trypanosomiasis re-emerges under cover of war. *Afr Health* **19**:3.

Bailey F, Eaton J, Jidda M, van Brakel WH, Addiss DG, Molyneux DH. 2019a. Neglected tropical diseases and mental health: progress, partnerships, and integration. *Trends Parasitol* **35**:23–31.

Bailey F, Mondragon-Shem K, Haines LR, Olabi A, Alorfi A, Ruiz-Postigo JA, Alvar J, Hotez P, Adams ER, Vélez ID, Al-Salem W, Eaton J, Acosta-Serrano Á, Molyneux DH. 2019b. Cutaneous leishmaniasis and co-morbid major depressive disorder: a systematic review with burden estimates. *PLoS Negl Trop Dis* **13**:e0007092.

Bailey F, Mondragon-Shem K, Hotez P, Ruiz-Postigo JA, Al-Salem W, Acosta-Serrano Á, Molyneux DH. 2017. A new perspective on cutaneous leishmaniasis—implications for global prevalence and burden of disease estimates. *PLoS Negl Trop Dis* **11**:e0005739.

Bellini MF, Silistino-Souza R, Varella-Garcia M, de Azeredo-Oliveira MT, Silva AE. 2012. Biologic and genetics aspects of Chagas disease at endemic areas. *J Trop Med* **2012**:357948.

Bernstein RE. 1984. Darwin's illness: Chagas' disease resurgens. *J R Soc Med* **77**:608–609.

Berriman M, et al. 2005. The genome of the African trypanosome *Trypanosoma brucei*. *Science* **309**:416–422.

Bisser S, N'Siesi FX, Lejon V, Preux PM, Van Nieuwenhove S, Miaka Mia Bilenge C, Büscher P. 2007. Equivalence trial of melarsoprol and nifurtimox monotherapy and combination therapy for the treatment of second-stage *Trypanosoma brucei gambiense* sleeping sickness. *J Infect Dis* **195**:322–329.

Brun R, Blum J. 2012. Human African trypanosomiasis. *Infect Dis Clin North Am* **26**:261–273.

Burri C, Brun R. 2002. Human African trypanosomiasis, p 1303–1323. *In* Cook GC, Zumla AI (ed), *Manson's Tropical Diseases*, 21st ed. W B Saunders, New York, NY.

Caridha D, Vesely B, van Bocxlaer K, Arana B, Mowbray CE, Rafati S, Uliana S, Reguera R, Kreishman-Deitrick M, Sciotti R, Buffet P, Croft SL. 2019. Route map for the discovery and preclinical development of new drugs and treatments for cutaneous leishmaniasis. *Int J Parasitol Drugs Drug Resist* **11**:106–117.

Chadalawada S, Sillau S, Archuleta S, Mundo W, Bandali M, Parra-Henao G, Rodriguez-Morales AJ, Villamil-Gomez WE, Suárez JA, Shapiro L, Hotez PJ, Woc-Colburn L, DeSanto K, Rassi A Jr, Franco-Paredes C, Henao-Martínez AF. 2020. Risk of chronic cardiomyopathy among patients with the acute phase or indeterminate form of Chagas disease: a systematic review and meta-analysis. *JAMA Netw Open* **3**:e2015072.

Collin SM, Coleman PG, Ritmeijer K, Davidson RN. 2006. Unseen Kala-azar deaths in south Sudan (1999-2002). *Trop Med Int Health* **11**:509–512.

Conrad J. 1996. Murfin RC (ed), *1902. Heart of Darkness*, 2nd ed. Bedford Books of St. Martin's Press, Boston, MA.

Cox FE. 2004. History of sleeping sickness (African trypanosomiasis). *Infect Dis Clin North Am* **18**:231–245.

Croft SL, Olliaro P. 2011. Leishmaniasis chemotherapy—challenges and opportunities. *Clin Microbiol Infect* **17**:1478–1483.

Du R, Hotez PJ, Al-Salem WS, Acosta-Serrano A. 2016. Old World cutaneous leishmaniasis and refugee crises in the Middle East and North Africa. *PLoS Negl Trop Dis* **10**:e0004545.

Dumonteil E, Bottazzi ME, Zhan B, Heffernan MJ, Jones K, Valenzuela JG, Kamhawi S, Ortega J, de Leon Rosales SP, Lee BY, Bacon KM, Fleischer B, Slingsby BT, Cravioto MB, Tapia-Conyer R, Hotez PJ. 2012. Accelerating the development of a therapeutic vaccine for human Chagas disease: rationale and prospects. *Expert Rev Vaccines* **11**:1043–1055.

El-Sayed NM, et al. 2005. The genome sequence of *Trypanosoma cruzi*, etiologic agent of Chagas disease. *Science* **309**:409–415.

Fèvre EM, Picozzi K, Jannin J, Welburn SC, Maudlin I. 2006. Human African trypanosomiasis: epidemiology and control. *Adv Parasitol* **61**:167–221.

Franco JR, Cecchi G, Priotto G, Paone M, Diarra A, Grout L, Simarro PP, Zhao W, Argaw D. 2018. Monitoring the elimination of human African trypanosomiasis: update to 2016. *PLoS Negl Trop Dis* **12**:e0006890.

Gascon J, Bern C, Pinazo MJ. 2010. Chagas disease in Spain, the United States and other non-endemic countries. *Acta Trop* **115:**22–27.

Gebrekristos HT, Buekens P. 2014. Mother-to-child transmission of *Trypanosoma cruzi*. *J Pediatric Infect Dis Soc* **3** (Suppl 1)**:** S36–S40.

Global Burden of Disease Collaborative Network. 2020. Global Burden of Disease Study 2019 (GBD 2019). GBD Results Tool. Institute for Health Metrics and Evaluation, Seattle, WA.

Goyal V, Burza S, Pandey K, Singh SN, Singh RS, Strub-Wourgaft N, Das VNR, Bern C, Hightower A, Rijal S, Sunyoto T, Alves F, Lima N, Das P, Alvar J. 2019. Field effectiveness of new visceral leishmaniasis regimens after 1 year following treatment within public health facilities in Bihar, India. *PLoS Negl Trop Dis* **13:**e0007726.

Haynes DM. 2001. *Imperial Medicine: Patrick Manson and the Conquest of Tropical Disease.* University of Pennsylvania Press, Philadelphia, PA.

Hoppe KA. 2003. *Lords of the Fly: Sleeping Sickness Control in British East Africa, 1900–1960.* Praeger, Westport, CT.

Hoffman KA, Reynolds C, Bottazzi ME, Hotez P, Jones K. 2019. Improved biomarker and imaging analysis for characterizing progressive cardiac fibrosis in a mouse model of chronic chagasic cardiomyopathy. *J Am Heart Assoc* **8:**e013365.

Hotez P, Bottazzi ME, Strub-Wourgaft N, Sosa-Estani S, Torrico F, Pajín L, Abril M, Sancho J. 2020. A new patient registry for Chagas disease. *PLoS Negl Trop Dis* **14:**e0008418.

Hotez PJ, Dumonteil E, Woc-Colburn L, Serpa JA, Bezek S, Edwards MS, Hallmark CJ, Musselwhite LW, Flink BJ, Bottazzi ME. 2012. Chagas disease: "the new HIV/AIDS of the Americas". *PLoS Negl Trop Dis* **6:**e1498.

Jamonneau V, Ilboudo H, Kaboré J, Kaba D, Koffi M, Solano P, Garcia A, Courtin D, Laveissière C, Lingue K, Büscher P, Bucheton B. 2012. Untreated human infections by *Trypanosoma brucei gambiense* are not 100% fatal. *PLoS Negl Trop Dis* **6:**e1691.

Jones K, Versteeg L, Damania A, Keegan B, Kendricks A, Pollet J, Cruz-Chan JV, Gusovsky F, Hotez PJ, Bottazzi ME. 2018. Vaccine-linked chemotherapy improves benznidazole efficacy for acute Chagas disease. *Infect Immun* **86:**e00876–e17.

Kaye P, Scott P. 2011. Leishmaniasis: complexity at the host-pathogen interface. *Nat Rev Microbiol* **9:**604–615.

Kennedy PG. 2006. Human African trypanosomiasis—neurological aspects. *J Neurol* **253:**411–416.

Kirchhoff LV. 2011. Epidemiology of American trypanosomiasis (Chagas disease). *Adv Parasitol* **75:**1–18.

Köberle F. 1968. Chagas' disease and Chagas' syndromes: the pathology of American trypanosomiasis. *Adv Parasitol* **6:**63–116.

Kumar V, Mandal R, Das S, Kesari S, Dinesh DS, Pandey K, Das VR, Topno RK, Sharma MP, Dasgupta RK, Das P. 2020. Kala-azar elimination in a highly-endemic district of Bihar, India: A success story. *PLoS Negl Trop Dis* **14:**e0008254.

Lee BY, Bacon KM, Wateska AR, Bottazzi ME, Dumonteil E, Hotez PJ. 2012. Modeling the economic value of a Chagas' disease therapeutic vaccine. *Hum Vaccin Immunother* **8:**1293–1301.

Lewinsohn R. 2003. Prophet in his own country: Carlos Chagas and the Nobel Prize. *Perspect Biol Med* **46:**532–549.

Machado FS, Tyler KM, Brant F, Esper L, Teixeira MM, Tanowitz HB. 2012. Pathogenesis of Chagas disease: time to move on. *Front Biosci (Elite Ed)* **4:**1743–1758.

Malvy D, Chappuis F. 2011. Sleeping sickness. *Clin Microbiol Infect* **17:**986–995.

Marin-Neto JA, Cunha-Neto E, Maciel BC, Simões MV. 2007. Pathogenesis of chronic Chagas heart disease. *Circulation* **115:**1109–1123.

Mbopi-Keou F-X, Bélec L, Milleliri J-M, Teo C-G. 2014. The legacies of Eugène Jamot and *La Jamotique*. *PLoS Negl Trop Dis* **8:**e2635.

McKerrow JH. 2005. Designing drugs for parasitic diseases of the developing world. *PLoS Med* **2:**e210.

Miles MA. 2002. American trypanosomiasis (Chagas disease), p 1325–1137. *In* Cook GC, Zumla AI (ed), *Manson's Tropical Diseases*, 21st ed. W B Saunders, New York, NY.

Morillo CA, Marin-Neto JA, Avezum A, Sosa-Estani S, Rassi A Jr, Rosas F, Villena E, Quiroz R, Bonilla R, Britto C, Guhl F, Velazquez E, Bonilla L, Meeks B, Rao-Melacini P, Pogue J, Mattos A, Lazdins J, Rassi A, Connolly SJ, Yusuf S, BENEFIT Investigators. 2015. Randomized trial of benznidazole for chronic Chagas' cardiomyopathy. *N Engl J Med* **373:**1295–1306.

Murray HW, Berman JD, Davies CR, Saravia NG. 2005. Advances in leishmaniasis. *Lancet* **366:**1561–1577.

Neau P, Hänel H, Lameyre V, Strub-Wourgaft N, Kuykens L. 2020. Innovative partnerships for the elimination of human African trypanosomiasis and the development of fexinidazole. *Trop Med Infect Dis* **5:**17.

Nitz N, Gomes C, de Cássia Rosa A, D'Souza-Ault MR, Moreno F, Lauria-Pires L, Nascimento RJ, Teixeira AR. 2004. Heritable integration of kDNA minicircle sequences from *Trypanosoma cruzi* into the avian genome: insights into human Chagas disease. *Cell* **118:**175–186.

Noya BA, Pérez-Chacón G, Díaz-Bello Z, Dickson S, Muñoz-Calderón A, Hernández C, Pérez Y, Mauriello L, Moronta E. 2017. Description of an oral Chagas disease outbreak in Venezuela, including a vertically transmitted case. *Mem Inst Oswaldo Cruz* **112:**569–571.

Odiit M, Shaw A, Welburn SC, Fèvre EM, Coleman PG, McDermott JJ. 2004. Assessing the patterns of health-seeking behaviour and awareness among sleeping-sickness patients in eastern Uganda. *Ann Trop Med Parasitol* **98:**339–348.

Pays E, Vanhamme L, Pérez-Morga D. 2004. Antigenic variation in *Trypanosoma brucei*: facts, challenges and mysteries. *Curr Opin Microbiol* **7:**369–374.

Pays E, Vanhollebeke B, Vanhamme L, Paturiaux-Hanocq F, Nolan DP, Pérez-Morga D. 2006. The trypanolytic factor of human serum. *Nat Rev Microbiol* **4:**477–486.

Pearce L. 1921. Studies on the treatment of human trypanosomiasis with tryparsamide (the sodium salt of N-phenylglycineamide-p-arsonic acid). *J Exp Med* **24:**1–104.

Pécoul B. 2004. New drugs for neglected diseases: from pipeline to patients. *PLoS Med* **1:**e6.

Pecoul B, Batista C, Stobbaerts E, Ribeiro I, Vilasanjuan R, Gascon J, Pinazo MJ, Moriana S, Gold S, Pereiro A, Navarro M, Torrico F, Bottazzi ME, Hotez PJ. 2016. The BENEFIT trial: where do we go from here? *PLoS Negl Trop Dis* **10:**e0004343.

Pelfrene E, Harvey Allchurch M, Ntamabyaliro N, Nambasa V, Ventura FV, Nagercoil N, Cavaleri M. 2019. The European Medicines Agency's scientific opinion on oral fexinidazole for human African trypanosomiasis. *PLoS Negl Trop Dis* **13:**e0007381.

Pepin J. 2007. Combination therapy for sleeping sickness: a wake-up call. *J Infect Dis* **195:**311–313.

Pépin J, Méda HA. 2001. The epidemiology and control of human African trypanosomiasis. *Adv Parasitol* **49:**71–132.

Power HJ. 1999. *Tropical Medicine in the Twentieth Century: a History of the Liverpool School of Tropical Medicine 1898–1990.* Kegan Paul International, London, United Kingdom.

Rassi A Jr, Rassi A, Marcondes de Rezende J. 2012. American trypanosomiasis (Chagas disease). *Infect Dis Clin North Am* **26:**275–291.

Riethmiller S. 2005. From Atoxyl to Salvarsan: searching for the magic bullet. *Chemotherapy* **51:**234–242.

Ritmeijer K, Davidson RN. 2003. Royal Society of Tropical Medicine and Hygiene joint meeting with Médecins Sans Frontières at Manson House, London, 20 March 2003: field research in humanitarian medical programmes. Médecins Sans Frontières interventions against kala-azar in the Sudan, 1989-2003. *Trans R Soc Trop Med Hyg* **97:**609–613.

Rudenko G. 2011. African trypanosomes: the genome and adaptations for immune evasion. *Essays Biochem* **51:**47–62.

Sá MR. 2005. The history of tropical medicine in Brazil: the discovery of *Trypanosoma cruzi* by Carlos Chagas and the German School of Protozoology. *Parassitologia* **47:**309–317.

Salas Clavijo NA, Postigo JR, Schneider D, Santalla JA, Brutus L, Chippaux JP. 2012. Prevalence of Chagas disease in pregnant women and incidence of congenital transmission in Santa Cruz de la Sierra, Bolivia. *Acta Trop* **124:**87–91.

Schmunis GA, Yadon ZE. 2010. Chagas disease: a Latin American health problem becoming a world health problem. *Acta Trop* **115:**14–21.

Schofield CJ, Jannin J, Salvatella R. 2006. The future of Chagas disease control. *Trends Parasitol* **22:**583–588.

Selvapandiyan A, Croft SL, Rijal S, Nakhasi HL, Ganguly NK. 2019. Innovations for the elimination and control of visceral leishmaniasis. *PLoS Negl Trop Dis* **13:**e0007616.

Sherman IW. 2006. *The Power of Plagues.* ASM Press, Washington, DC.

Shikanai-Yasuda MA, Carvalho NB. 2012. Oral transmission of Chagas disease. *Clin Infect Dis* **54:**845–852.

Simarro PP, Franco J, Diarra A, Postigo JA, Jannin J. 2012. Update on field use of the available drugs for the chemotherapy of human African trypanosomiasis. *Parasitology* **139:**842–846.

Stanghellini A, Josenando T. 2001. The situation of sleeping sickness in Angola: a calamity. *Trop Med Int Health* **6:**330–334.

Teixeira AR, Nascimento RJ, Sturm NR. 2006a. Evolution and pathology in Chagas disease—a review. *Mem Inst Oswaldo Cruz* **101:**463–491.

Teixeira AR, Nitz N, Guimaro MC, Gomes C, Santos-Buch CA. 2006b. Chagas disease. *Postgrad Med J* **82:**788–798.

Vanhollebeke B, Pays E. 2010. The trypanolytic factor of human serum: many ways to enter the parasite, a single way to kill. *Mol Microbiol* **76:**806–814.

Welburn SC, Coleman PG, Maudlin I, Fèvre EM, Odiit M, Eisler MC. 2006. Crisis, what crisis? Control of Rhodesian sleeping sickness. *Trends Parasitol* **22:**123–128.

World Health Organization. 2002. Expert committee on the control of Chagas disease. Technical report series no. 905. World Health Organization, Geneva, Switzerland.

Yamagata Y, Nakagawa J. 2006. Control of Chagas disease. *Adv Parasitol* **61:**129–165.

Zijlstra EE, El-Hassan AM. 2001. Leishmaniasis in Sudan. 3. Visceral leishmaniasis. *Trans R Soc Trop Med Hyg* **95** (Suppl 1): S1/27–S1/58.

8

The Urban Neglected Tropical Diseases: Leptospirosis, Dengue and Zika, and Rabies

If a dog is mad and the authorities have brought the fact to the knowledge of its owner, if he does not keep it in and it bites a man and causes his death, then the owner shall pay two-thirds of a min (40 shekels) of silver. If it bites a slave and causes his death, he shall pay 15 shekels of silver.

Eshmuna Code of Babylon, 23rd century BCE

The neglected tropical diseases (NTDs) were originally thought of as chronic infections that occur primarily in regions of rural poverty in low- and middle-income countries (LMICs). However, some of the NTDs discussed in the first seven chapters also occur in urban slums and favelas. For instance, ascariasis, one of the unholy trinity of soil-transmitted helminth infections, occurs in both poor urban and rural environments in part because the parasite eggs responsible for transmission are extremely hardy and can sometimes survive in the otherwise unfavorable conditions found in slums. In addition, because some urban-dwelling mosquitoes can transmit *Wuchereria bancrofti* in Africa, India, and elsewhere, lymphatic filariasis (LF) is a problem in some poor urban areas. It has been estimated that almost 30% of people at risk of acquiring LF live in urban settings.[1] Similarly, leishmaniasis, especially the zoonotic form transmitted from dogs, is frequently found in urban slums. Increasingly, we are recognizing an urbanization of NTDs. In part, this might reflect shifting global demographics. Now for the first time a higher percentage of the human population lives in urban settings than in rural areas, and this trend is expected to continue. For example, by 2050 some experts project that two-thirds of the population will live in cities and peri-urban areas.[1] A major consequence of this urban shift will be the formation of megacities—cities with 10 million people or more—especially in LMICs. In all, there may be as many as 40 megacities created over the next decade, including 3 in India—Delhi, Kolkata,

---

Forgotten People, Forgotten Diseases: The Neglected Tropical Diseases and Their Impact on Global Health and Development, Third Edition. Peter J. Hotez.
© 2022 American Society for Microbiology. DOI: 10.1128/9781683673903.ch08

and Mumbai—and at least 4 in Africa—Dar es Salaam, Kinshasa, Lagos, and Luanda.[1] There is further recognition that helminth infections are increasingly found in the lowest-income megacity neighborhoods; this includes the infections highlighted above as well as schistosomiasis.[1] An unknown is whether there is some parasite adaptation to these new urban landscapes, possibly enhanced by warming temperatures and shifting rainfall patterns due to climate change. Therefore, urbanization may work in concert with climate change to produce shifting patterns.[1]

The NTDs leptospirosis, dengue (and related arbovirus infections including Zika virus infection), and rabies have particular importance as urban health problems in LMICs. For both leptospirosis and dengue, urban flooding is a key component of transmission. As global warming produces an increase in flooding in some regions of the developing world, we can also expect urbanization to combine with climate change to accelerate these infections.[2] Also because of climate change, there is a possibility that these NTDs could emerge in urban regions of the United States (discussed in chapter 9).

Leptospirosis

On the edges of Brazil's major urban centers, including Rio de Janeiro, Salvador, and Fortaleza, are crowded slums and shantytowns known as favelas, so named for a plant that thrived in a region where returning Brazilian soldiers battled against rebel forces at the end of the 19th century. Upon their return, the soldiers occupied squatters' settlements on the hillsides of Rio de Janeiro. Today, favelas are made up of poor-quality, self-constructed dwellings, usually crammed together randomly on the hillsides of Brazil's major cities. They are notorious for the heavy presence of gangs, turf battles, and rampant crime and the inadequacy of their police presence and other city services. Of relevance to the NTDs, the poorest favelas do not benefit from regular garbage collection or sewage treatment, thereby creating excellent niches for rats and stray dogs.

It is in this setting of squalor and environmental degradation that the disease leptospirosis finds a welcome home. Leptospirosis is a zoonosis transmitted primarily from rats and dogs. The bacterium that causes leptospirosis (sometimes called a leptospire) is a graceful-looking spiral bacterium (Fig. 8.1) that lives for long periods of time in the kidneys of rats and dogs.[3] Leptospires live in association with cells that line the drainage system of the kidneys of these animals. As a result, the urine of rats and dogs living in the favelas teems with millions of leptospires. During certain months of the year when stagnant rainwater collects in the favelas or urban slums (the infection is even found in some slums of inner cities in the United States[4]), the subsequent contamination with rat or dog urine provides a large environmental pool of leptospires and a ready-made mechanism for transmission to humans. Infection typically results when the leptospires penetrate small cuts or abrasions in human skin, such as when favela residents bathe in the canals and other bodies of water. The leptospires also frequently enter through the mucosa of the eyes, nose, and mouth.

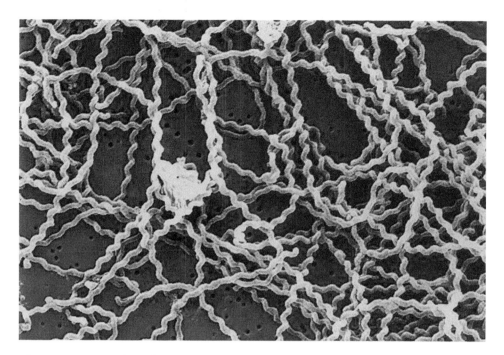

Figure 8.1 Scanning electron micrograph of *Leptospira* sp. bacteria atop a 0.1-μm polycarbonate filter. (Image from CDC-PHIL [ID#138]/Rob Weyant, Janice Haney Carr, CDC, 1998.)

Leptospirosis can also occur in rural areas, where leptospires contained in the urine of domestic animals contaminate the soil as well as water.[3] In some developing regions, such as in the Amazon basin, these microbes are also found in a rich diversity of tropical mammals, including bats, rodents, and marsupials.[3] Not surprisingly, people with occupations that require standing in stagnant bodies of water contaminated with animal urine, such as working in sewers and canals, rice and sugarcane harvesting, fish farming, and food processing, are at extremely high risk for the disease.[5] Some of the greatest numbers of cases are found in Papua New Guinea and Indonesia in the Asia-Pacific region, as well as in East Africa. High rates of infection are also found in Guyana and in some Caribbean islands.[5] Because leptospires can live in the kidneys of so many different types of mammals, rats are so ubiquitous in urban slums, and transmission is relatively facile, leptospirosis is extremely common in developing countries. Some investigators consider leptospirosis to be the most common zoonosis of humans, although there are few data on the actual prevalence and incidence of this disease.[3] One reason why we know so little about the extent of leptospirosis is that many patients with the infection develop very nonspecific symptoms that can resemble those of other infections.

Following their entry through the skin, the highly motile leptospires have the ability to invade and disseminate to a variety of tissues. Many patients with leptospirosis develop only mild symptoms, while others go on to develop fever, headache, nausea, and vomiting.[3,5] Many of these patients, in turn, have liver involvement as evidenced by jaundice, a condition where the skin turns yellow, similar to viral hepatitis. In about one-fourth of the patients with leptospirosis, this phase is often followed by a second one in which the leptospires invade the central nervous system to produce meningitis,[3] characterized by severe headache and pain behind the eyes. Patients with either the first or second phase of the illness seldom die. However, others can develop a severe form of leptospirosis known as Weil's disease, which is characterized by jaundice, renal failure, and pulmonary hemorrhage (bleeding into the lungs).[3,5] Some patients experience heart involvement. Between 5 and 15% of Weil's disease patients die.[3] It has been proposed that Weil's disease brought by rodents aboard English ships to the Plymouth colony helped to decimate the Native American population.[3] Why some patients develop a mild form of the infection rather than Weil's disease is not known—it may be related to strain differences of the leptospires, host differences, or a combination of the two factors. The recent sequencing of *Leptospira* spp. genomes might help to answer this question.[5]

Leptospires are sensitive to many antibiotics in the laboratory, although whether these antibiotics are effective in treating the patient and reducing the severity of the illness is controversial.[5] Many patients have resolution of their illness even if they do not receive antibiotics. Military personnel and travelers to areas where leptospirosis is endemic may benefit from receiving one or more doses of doxycycline as a form of prophylaxis against the illness,[3] but this approach is not practical for widespread use in the favelas. For individuals with high rates of occupational exposure such as sewer workers, wearing protective clothing may help; also, burning sugarcane prior to harvesting diminishes the number of sharp young shoots, which can cut the hands and provide a portal of entry for leptospires.[5] It may be possible to develop a leptospirosis vaccine. However, until such a vaccine is made readily available, the public health control of leptospirosis relies heavily on having a functioning health system with adequate surveillance for the disease. Adequate surveillance, in turn, usually requires a functional laboratory capable of conducting microbiological diagnosis of leptospirosis through culturing of the organism or through more modern PCR-based assays.[6] Pest (rat) control and local water chlorination are also key features of programs targeting local elimination of leptospirosis.[6] All of these measures require a strong political will on the part of the health ministries in countries where leptospirosis is endemic as well as a highly intact and functional health care system. There is also an urgent need for an international leptospirosis database in order to fully realize the true global burden of this important zoonosis.[7]

Dengue fever (break bone fever) and Zika virus infection

When the rains fall and surface water collects in canals and sewers, leptospirosis is not the only NTD that the residents of favelas and other urban slums must fear. As water collects in gutters, cisterns, and other water storage containers and even discarded tires, it creates a breeding ground for mosquitoes. Among the major mosquitoes that breed in this environment are *Anopheles*

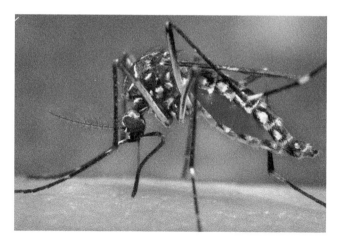

Figure 8.2 *A. aegypti*, the mosquito vector of dengue fever. (Image from CDC-PHIL [ID#9261]/CDC/Prof. Frank Hadley Collins, Dir., Cntr. for Global Health and Infectious Diseases, Univ. of Notre Dame/James Gathany, 2006.)

mosquitoes, which can transmit LF and malaria, as well as *Aedes* mosquitoes, especially *Aedes aegypti* (Fig. 8.2), which transmit the flaviviruses that cause dengue fever, Zika virus infection, and yellow fever.

A. aegypti is an urban-dwelling mosquito that prefers to live in artificial containers that collect rainwater, and is therefore especially common in areas of urban poverty without controlled sewage or piped water. The female mosquito preferentially bites humans. Although this species originated from Africa, it was carried on ships headed for the New World, where it rapidly colonized tropical and subtropical regions. Both dengue and yellow fever epidemics became common beginning in the 1600s, with the first established epidemic on the Yucatan Peninsula in 1647 to 1648, followed by epidemics in both North and South America over the ensuing 300 years.[8,9] Through aggressive vector control efforts led by the Pan American Health Organization of the World Health Organization (PAHO-WHO) in South America during the 1960s, the urban foci of both yellow fever and dengue began to diminish. The PAHO-WHO Eradication Program relied on campaigns described as "military-style" and emphasized removal of indoor and outdoor containers of water, together with spraying with DDT.[9] Often mosquito control efforts were conducted inside households, which meant government workers were allowed access.[9] By the late-1960s yellow fever and dengue had almost been eradicated in the tropical regions of the continent.[8,9] However, with the subsequent suspension of many of these programs during the 1970s, *A. aegypti* mosquito breeding resumed. Potentially, mosquitoes were reintroduced to South America from the U.S. Gulf Coast, where intrusive domestic mosquito control efforts were not permitted. It's worth pointing out that when mosquito trucks unleash a fog of insecticidal spray in southern U.S. states during the summer months, those primarily target *Culex* mosquitoes, whereas American households never welcomed the indoor spraying required for *A. aegypti*.[9]

As a result, by the mid-1990s, maps showing the geographic distribution of this insect vector increasingly resembled vector maps made 30 years earlier.[9] This situation is not too different from the reemergence of human African trypanosomiasis in Angola, the Democratic Republic of the Congo, and Sudan, which resulted from interruption of tsetse control efforts, but in the case of South America it was a voluntary suspension of vector control rather than one that was forced by conflict. Today in Brazil, it has been estimated that every year there are thousands of cases of dengue fever, most of them in the favelas of the hot and tropical northeastern regions.[8] In such Brazilian favelas, epidemics occur from May through July, a period that follows the heaviest rains by approximately 2 to 3 months. This period corresponds to that required for several life cycles of the *Aedes* vector to build up and transmit dengue fever on a community-wide scale.[8] According to the WHO, in 2010, more than 1.6 million cases were reported from the Americas.[10]

Dengue fever is distributed throughout the poor urban areas of the tropics. The latest estimates show 57 million people infected annually; approximately 500,000 are hospitalized, and 2.5% of those hospitalized die from dengue and its complications[10] (Fig. 8.3). Moreover, up to 4 billion people are at risk for acquiring the infection, making this the most common arboviral infection (arthropod-borne viral infection) worldwide.[10] Dengue is also one of the most serious consequences of increasing urbanization in the developing world. As human populations continue to migrate from rural areas to the cities, the ever-expanding slums, shantytowns, and favelas provide ideal ecological niches for *A. aegypti* mosquitoes.[8] Such migrations, together with flooding that can result from global warming, largely account for the estimated 30-fold increase in dengue outbreaks over the last 50 years.[11] In the Americas, the first dengue outbreaks occurred in Cuba in 1977 to 1978, followed by subsequent outbreaks there in 1981 and 1997, as well as outbreaks in Venezuela.[7] In all, more than 30 Latin American countries have reported dengue outbreaks, fueled by the forces of poverty and rapid urbanization, lack of sanitation, and possibly climate change, together with an absence of political will to enforce strict mosquito control and other public health practices.[11] Dengue has also become common in the Northern Triangle area of Central America, possibly in part because of the destabilization and breakdowns in health system infrastructures resulting from the illegal drug trade, together with marked shifts due to climate alterations that create a so-called "dry corridor."[11]

Because *A. aegypti* mosquitoes also thrive in Texas, Louisiana, Florida, and other parts of the American South, it would not be too surprising if outbreaks appeared among the poor living in the slums of urban centers along the Gulf Coast, including Houston and New Orleans.[11] So far, we know that dengue has already emerged in South Texas along the border with Mexico and in Florida; in the South Texas outbreaks, impoverished conditions were shown to be a strong risk factor for exposure to *A. aegypti* mosquitoes and contraction of dengue.[11] A new predictive model from Professor Simon Hay and his colleagues indicates that dengue cases will continue to accumulate on the U.S. Gulf Coast in association with climate change.[11] While the Latin American

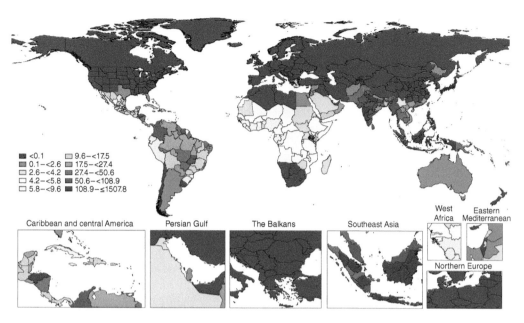

Figure 8.3 Distribution of dengue, worldwide. (From Global Burden of Disease Collaborative Network, 2020 [http://www.healthdata.org/results/gbd_summaries/2019/dengue-level-3-cause].)

dengue outbreak is exploding, the largest number of cases of dengue fever still occur in Asia, particularly Southeast Asia (Indonesia and Philippines), India and Sri Lanka, southern China (especially Hainan Province), and the South Pacific islands (especially Fiji).[11] Dengue fever is also common in coastal regions of sub-Saharan Africa, especially West Africa, and is emerging on the Arabian Peninsula.

The virus that causes dengue fever belongs to a family of flaviviruses, which also comprise other tropical pathogens, including the viruses that cause yellow fever, West Nile fever, Zika virus infection, and Japanese encephalitis. A characteristic feature of the flaviviruses is that their genome is composed of a single strand of RNA, as opposed to our human double-stranded DNA genome.[11] Another important feature is that there are four antigenically distinct types (called serotypes) of dengue viruses.[11] This observation will become relevant when we discuss some of the severe complications that can result from dengue infection. Dengue virus replicates in the salivary glands of *Aedes* mosquitoes. *A. aegypti* is the major but not the only species of the *Aedes* genus. *A. aegypti* usually bites during the day. After humans become infected, the virus migrates to the lymph nodes and then to the bloodstream, where it disseminates widely throughout the body.[12] Infants and young children, when they become infected for the first time, often develop an illness characterized by high fever, which is difficult to distinguish from other common childhood illnesses. However, older children and adults often develop classical dengue fever, often known as break bone fever, so named because of the high fever accompanied by headache and severe muscle, joint, and bone pain.[12] Flushing is also commonly observed on the face, neck, and chest.

The most dreaded complication is dengue hemorrhagic fever (DHF), which occurs in roughly 0.5% of all dengue patients (about 500,000 cases annually).[11] DHF typically occurs in children (less than 16 years of age) who experience two sequential dengue infections of two different serotypes within an interval of several years.[12] The mechanism by which DHF occurs in children with two different serotypes of dengue is not known, although it has been hypothesized by Scott Halstead (formerly of the Rockefeller Foundation) and others that the host immune response is somehow involved.[13] The major clinical features of DHF include high fever for several days, with hemorrhagic complications, especially bleeding from the gut, gums, and vagina, and a drop in the number of platelets circulating in the bloodstream.[11] Some patients with DHF progress to liver failure and dengue shock syndrome, which is associated with circulatory collapse and profound shock. The mortality rate of shock syndrome can approach 50% and accounts for a large percentage of the estimated several thousand deaths from dengue that occur every year.[11] Epidemics of DHF are dramatic because they primarily target large numbers of children and produce profoundly disturbing hemorrhagic complications. During outbreaks, it is not uncommon to have entire pediatric hospital wards filled with crying and bleeding children. The number of DHF outbreaks is on the rise in Asia; the first epidemic was reported from Manila, the Philippines, in 1954, and was succeeded by a severe and noteworthy rise in epidemics beginning in 1980, including those in Indonesia, Vietnam, Sri Lanka, Hainan Province (China), and Fiji.[11] In these regions, overall dengue transmission is high and there are at least two different dengue serotypes appearing in an interval of a few years. In the Americas, DHF outbreaks have occurred recently in more than 20 countries.[11] The Cuban DHF epidemic in 1981 resulted from a type 2 dengue virus imported from Southeast Asia and occurred 4 years following a type 1 dengue epidemic in 1977.[11,12] In 2001, the World Health Assembly passed a resolution advocating increased global surveillance efforts. In response, the WHO established DengueNet in order to increase the efficiency of global monitoring efforts.[14] Yet another emerging aspect is the finding of severe dengue in adult patients with underlying comorbidities such as diabetes and hypertension.[12] It's curious that a similar observation has been made with the coronavirus disease 2019 (COVID-19), although it's not clear whether the underlying mechanisms are similar. However, with the rise of diabetes and hypertension in South Asia and Africa, we might expect the incidence of severe dengue to continue rising because of this phenomenon.

Ultimately, the long-term control of dengue fever and DHF will require the development of a safe and effective vaccine.[14] Later (in chapter 11), we will discuss ongoing international efforts to do precisely that. In the meantime, dengue control relies on a rather labor-intensive, aggressive, and multidisciplinary approach comprising environmental control methods to reduce the number of available vector breeding sites (through rigorous drainage or destruction of open vessels containing rainwater), insecticidal spraying, and improvements in house design.[11] For instance, during periods of dengue outbreaks in Singapore, the National Environmental Agency employs 500 inspectors who conduct house-to-house searches in any area where two cases of dengue occur within 14 days of each other and within a 150-m radius.[15] In such areas, the inspectors check for standing bodies of water in gutters, in potted plants, or between the leaves of palm trees.[15] Such methods were partly responsible for the reduction of malaria, yellow fever, and other vector-borne diseases that allowed construction of the Panama Canal during the early 20th century. It has been estimated that the costs of dengue and dengue control in Singapore average about US$1 billion annually![15] Unfortunately, many developing countries cannot afford a sophisticated public health infrastructure like the one currently in place in Singapore. Until a vaccine becomes available, however, there are not many alternatives to house-to-house mosquito "search and destroy" missions. In contrast to malaria and LF control efforts in which insecticide-treated bed nets have a role in interrupting transmission by night-biting *Anopheles* mosquitoes, this approach has not been of much value for the day-biting *A. aegypti* mosquito.[11] In some dengue-plagued regions, such as in Vietnam, biological control with copepod crustaceans and other agents[16] has also met with some success.

Most recently, several groups are working to develop innovative methods for genetically modifying mosquitoes to render them resistant to viruses and other pathogens or to cause the mosquitoes to die prematurely. For example, a group from Monash University in Australia led by Professor Scott O'Neill has provided proof of concept that it might be possible to genetically modify mosquitoes through modified *Wolbachia* microorganisms. It has been noted that mosquitoes, like some filarial worms, also depend on *Wolbachia* bacterial endosymbionts. The release of such genetically modified mosquitoes was recently reported in Townsville, Northern Australia, although this is not the only approach. The British-based company Oxitec has genetically modified male mosquitoes, known as OX5034, to create offspring that fail to survive. Such mosquitoes have been released in Florida and elsewhere.[16]

Major drivers for studying genetically modified mosquitoes include the urgency of controlling dengue, but equally important is the stunning emergence of Zika virus infection in the Western Hemisphere. Although it was discovered in Africa during the 1940s, Zika virus emerged as an important global pathogen in the South Pacific during the early 2010s before it entered Brazil and the Americas in 2015.[16] The clinical infection in adults was generally considered mild except for clusters of a peripheral neuropathy known as Guillain-Barré syndrome. Then, starting in 2016, an epidemic of congenital defects was noted in newborns that included severe microcephaly and ultimately profound intellectual deficits in later life. Northeastern Brazil became the epicenter of Zika-induced congenital birth defects syndrome, but then the illness spread across the Caribbean, causing widespread and serious illness in Puerto Rico, and from there into South Florida. Similarly, the virus infection affected Central America and Mexico before it spread northwards into South Texas. By 2017 Zika virus infection had largely disappeared as rapidly as it had emerged the year before. However, small epidemics still affected Cuba at that time. The circumstances by which Zika rapidly exited the Western Hemisphere remain unknown, as does when and where it might appear next.

Rabies

Of all the infections plaguing humankind, rabies is among the most terrifying. Once contracted and once clinical signs become evident, rabies is almost invariably fatal. Most patients die a truly horrible death by first developing a condition known as furious rabies, the hallmark of which is hydrophobia (fear of water).[17] In hydrophobia, an effort to drink water, or in some cases even the sound of water or the thought of water, provokes a reflex contraction of the throat and diaphragm (as well as of other muscles of inspiratory respiration) accompanied by expressions of abject fear and terror. This condition lasts for about 1 week before coma and paralysis take over prior to death.

The dramatic endgame of rabies and the fear that it generated are no doubt responsible for why we can find unambiguous descriptions of rabies in texts from ancient Egypt, Greece, Rome, and China.[18] There is even a reference to rabies in the Babylon Codex, 23 centuries BCE, a quotation from which is provided at the beginning of this chapter.[18] Today, human rabies is mostly an urban disease found in developing countries. While sporadic cases of human rabies transmitted from bats (and, rarely, from skunks and raccoons) occur in the United States and elsewhere in the industrialized world, most of the 14,000 people who contract rabies and die annually live in Asia and Africa, where the infection is endemic in dogs.[19] In India (and likely neighboring Pakistan, Nepal, Bhutan, and Myanmar), where more than half of the annual rabies deaths may occur, approximately 96% of the cases are contracted through dog bites.[20] Most of the deaths in India occur among very poor adult males.[20] In the vast majority of the cases both in India and elsewhere, urban rabies is transmitted by the bite of a stray dog. It is estimated that India has a dog population of about 25 million, most of which are ownerless.[20] The enormous impact of stray dogs and rabies in New Delhi, Mumbai, and other urban centers in India has been profiled in *The New York Times*.[20]

Almost as serious as its enormous toll on human life is the economic impact of rabies.[21] In the United States, pet owners spend an estimated $300 million annually to vaccinate their pet dogs and cats, while in developing countries, there are enormous costs associated with administering rabies vaccine in individuals bitten by dogs.[21] After someone is bitten by a rabid dog, one of their few opportunities to reduce the odds of developing clinical rabies (and dying) is to receive a rabies vaccine through what is known as postexposure prophylaxis (PEP). The fear and anxiety of developing rabies, and the potentially horrific outcome if no vaccine is given, mean that for every fatal case of rabies, thousands more receive PEP annually. For instance, in Thailand more than 100,000 people receive PEP every year at a cost of US$10 million,[21] while throughout Asia, an estimated 7 million people receive PEP. Globally, an estimated 15 million people receive PEP annually, preventing more than 300,000 deaths.[19] The highest rate of PEP administration occurs in Vietnam.[22]

Following a bite by a rabid animal, saliva containing the rabies virus is introduced into the skin and muscle, where the virus replicates before it attaches to neuronal cell membranes by specific receptors.[23] If a bite victim does not receive PEP, the rabies virus then travels retrogradely along peripheral nerves at a rate of approximately 5 to 10 cm (2 to 5 in.) per day, until it reaches the central nervous system.[23] Once the rabies virus reaches the central nervous system, the patient's fate is sealed. Exactly how the rabies virus causes such severe disease in the brain, especially in a part of the brain known as the limbic system, as suggested by the clinical picture of furious rabies, is not well known. Among the hypotheses under investigation are that the virus somehow disrupts neuronal metabolism or causes formation of interneuronal connections.[23] Conceivably, a better understanding of this process could lead to new and innovative interventions for people who already show clinical signs of the disease.

The first rabies vaccine was developed by Louis Pasteur in the 19th century. This development led to the establishment of Pasteur Institutes all over the Francophone world, which made available first-generation rabies vaccine to many developing countries. Since then, the process of developing rabies vaccine has been refined such that modern rabies vaccine production now uses *in vitro* cell culture methods for growing the vaccine strain of the virus rather than Pasteur's original method of using neural tissues from animals. The toxicities and side effects of the older, neural tissue-derived vaccines make them highly undesirable, and there are even concerns that some of the vaccines are not very effective. However, even after a century, high-quality, cell culture-derived rabies vaccine is still not widely available in developing countries. For instance, a sheep brain-derived vaccine known as Semple is produced in the Indian subcontinent, and suckling mouse brain or nervous tissue vaccine is made elsewhere but is gradually being phased out.[22] In their place, global efforts to transfer the technology for cell culture-derived rabies vaccine to developing countries are in progress. The WHO has called on vaccine manufacturers to produce vaccines with improvements in their thermostability and ease of delivery, as well as broadened immunologic profiles to prevent other related lyssaviruses.[22] In addition to developing human vaccines for PEP, there is also a need to increase rates of vaccination of stray dogs in the urban centers of developing countries. For instance, mass vaccination campaigns have resulted in the elimination of rabies from Japan and Taiwan and many urban areas in South America.[17] There is an urgent need to vaccinate stray dogs throughout the urban centers of sub-Saharan Africa, India, Bangladesh, and wherever the number of annual rabies deaths is high. A new World Rabies Day initiative is working to elevate awareness of this ancient scourge and to provide important health educational materials to empower people in the world's poorest countries.[24]

Summary

According to the United Nations Population Fund, now, for the first time, more than half of the world's population is living in cities and towns, and by 2050, two-thirds will live in urban areas, mostly megacities.[1,25] Unregulated urbanization brings with it serious increases in poverty, and the slums, favelas, and shantytowns are only expected to get worse. Unless significant steps are taken to reduce squalor in the urban areas of developing countries, we can anticipate increases in dengue, leptospirosis, rabies, and other urban NTDs. The UN Population Fund asserts that as our world's cities expand, we need to embrace the concepts of respecting the rights of the poor and promoting access to clean water, waste disposal, and power. Otherwise, we risk condemning the next urban generation to a life of poverty and misery.

SUMMARY POINTS   The Urban NTDs

- Leptospirosis, dengue fever, and rabies are important NTDs found in the urban slums and favelas of developing countries.
- Today, approximately one-half of the world's population lives in urban areas, and urbanization of the NTDs will become an important new framework. Increasingly, NTDs will become common in the megacities of South Asia, Africa, and Latin America.

Leptospirosis

- Leptospirosis is one of the world's most common zoonoses, transmitted through contact with bodies of water, e.g., canals and sewers, contaminated with leptospires. Some of the greatest numbers of cases are found in Papua New Guinea and Indonesia in the Asia-Pacific region, as well as in East Africa. High rates of infection are also found in Guyana and some Caribbean islands.
- In urban areas, contamination occurs through exposure to the urine of rats and stray dogs, which teems with leptospires.
- Many patients with leptospirosis develop a nonspecific febrile illness or aseptic meningitis. The most severe complication is Weil's disease, which has a high mortality rate.
- Adequate prevention requires having an intact system of surveillance and case detection as well as environmental control.

Dengue fever and Zika virus infection

- Dengue fever is one of the most common NTDs of urban areas, with 57 million new cases occurring annually worldwide. The largest numbers of cases are in India and Southeast Asia (Indonesia and Philippines), tropical regions of the Americas, and on the West African coast. The number of

dengue cases, possibly more than that of any other NTD, is expected to continue to rise as a consequence of increasing urbanization and global warming.

- Dengue is caused by a single-stranded RNA flavivirus and transmitted by the bite of a female *Aedes* mosquito. *A. aegypti* is the most common insect vector species.
- Like leptospirosis, most dengue cases present as a nonspecific febrile illness. However, classical dengue is sometimes known as break bone fever because of the severe myalgias and headache that can accompany the illness.
- The most feared complication of dengue is DHF, which is associated with severe morbidity. DHF occurs as a result of infection with two different serotypes of the dengue virus separated by a period of years. Some patients with DHF will die from dengue shock syndrome. Still another population at great risk is adults with underlying diabetes and hypertension, similar to patients at risk for severe COVID-19.
- Dengue has emerged in Latin America and the Gulf Coast of the United States, fueled by the forces of poverty and rapid urbanization, lack of sanitation, and possibly climate change together with an absence of political will to enforce strict public health practices.
- No vaccine is yet available for dengue, although at least one is expected to be available soon. Currently, public health control relies on labor-intensive practices of environmental vector control. Such practices are extremely costly; for instance, the costs in Singapore for dengue control average around US$1 billion annually. Recently, new "gene drive"

technologies are producing genetically modified mosquitoes as an innovative approach to interrupting dengue transmission.

- A WHO-based global surveillance network, DengueNet, is in place.
- Zika virus infection emerged in the Western Hemisphere in 2016, causing significant levels of birth defects and congenital microcephaly in northeastern Brazil, Puerto Rico, and elsewhere.

Rabies

- Rabies has been one of the most feared infections since ancient times.
- Approximately 14,000 people die from rabies annually. Most of these deaths occur as the result of stray-dog bites in developing countries. More than half of the cases occur in India, Pakistan, and nearby nations in South Asia.

- There is also an enormous economic toll from rabies. This is because of the need to administer PEP with rabies vaccine to almost anyone bitten by a dog in developing countries. Approximately 15 million people receive PEP annually, preventing more than 300,000 deaths.
- Now, many developing-country manufacturers are pivoting away from rabies vaccines derived from neural tissues. Such vaccines may have associated toxicities and are often not as effective as newer-generation vaccines that use *in vitro* cell culture methods.
- The public health control of rabies in many countries has been achieved through aggressive campaigns to vaccinate dogs in urban centers.
- A new World Rabies Day initiative is educating and empowering people in developing countries about this ancient scourge.

Notes

1. Information is contained in Erlanger et al., 2005; and Utzinger and Keiser, 2006. Reflections on the urbanization of NTDs can be found in Hotez, 2017; and Hotez, 2018.

2. Information is found in Gubler et al., 2001; and Blum and Hotez, 2018.

3. Good overviews of leptospirosis are found in Bharti et al., 2003; and McBride et al., 2005. Information on Weil's disease and Native American populations is found in Marr and Cathey, 2010.

4. A description of urban leptospirosis in the United States and elsewhere is found in Vinetz et al., 1996.

5. Information is found in Scott and Coleman, 2002; Brett-Major and Coldren, 2012; Ko et al., 2009; and Victoriano et al., 2009. Updated information on disease burden and geographic distribution is in Torgerson et al., 2015.

6. John, 2005.

7. Cachay and Vinetz, 2005.

8. A description of dengue fever in the favelas can be found in Heukelbach et al., 2001.

9. The history of *A. aegypti* entering the New World is described in Powell and Tabachnick, 2013. The history of PAHO eradication efforts is in Hotez, 2016. The reemergence of dengue in South America and elsewhere is described in Gubler and Clark, 1995.

10. Information is found in World Health Organization, 2012a; and Brady et al., 2012. The most recent disease burden estimates and geographic distribution data for dengue are found at http://www.healthdata.org/results/gbd_summaries/2019/dengue-level-3-cause (Global Burden of Disease Collaborative Network, 2020).

11. Information about the emergence of dengue in the Americas can be found in Malavige et al., 2004; Tapia-Conyer et al., 2012;

Bouri et al., 2012; and Brunkard et al., 2007. Specific details about dengue in the Northern Triangle of Central America are in Hotez et al., 2020. Evidence for an earlier outbreak of dengue in Houston, Texas, is in Murray et al., 2013; while the predictions on dengue emerging on the U.S. Gulf Coast due to climate change are in Messina et al., 2019.

12. Nimmannitya, 2002. Information about dengue in patients with underlying diabetes and hypertension is found in Mehta and Hotez, 2016.

13. Scott Halstead, formerly of the Rockefeller Foundation and founding director of the Pediatric Dengue Vaccine Initiative, is largely credited with the sequential infection hypothesis of DHF. It is summarized in Halstead, 1981.

14. The WHO's DengueNet activities are described at https://apps.who.int/iris/handle/10665/163773 . An update on the development of dengue vaccines now in clinical trials can be found in Pinheiro-Michelsen et al., 2020.

15. Singapore's efforts to control dengue were reported in Arnold, 2007. The cost of dengue in Singapore can be found in Carrasco et al., 2011.

16. Copepod control in Vietnam is described in Kay et al., 2002. The *Wolbachia*-modified *Aedes* mosquitoes are reported in O'Neill et al., 2019; and the Oxitec modified male mosquitoes are at https://www.bbc.com/news/world-us-canada-53856776. Details about Zika virus infection can be found in Musso et al., 2019.

17. An excellent description of hydrophobia and other features of clinical rabies is found in Warrell, 2002.

18. The historical accounts are summarized in Théodoridès, 1986. Several different versions of the statement on rabies in the Babylon Codex can be found www.stanford.edu/group/virus/1999/sohoni/rhabdovirus.html.

19. The global burden of human rabies is described in Wyatt, 2007; Coleman et al., 2004; and World Health Organization, 2012b. The latest GBD 2019 estimates are at http://www.health-data.org/results/gbd_summaries/2019/rabies-level-3-cause (Global Burden of Disease Collaborative Network, 2020).

20. An assessment of India's disease burden can be found online in Sudarshan et al., 2007. The *New York Times* article referred to is Harris, 2012.

21. The economic impact of rabies is summarized in Meltzer and Rupprecht, 1998.

22. Dinh, 2001. A summary of efforts to phase out Semple and related nervous tissue vaccines is in https://rabiesalliance.org/resource/when-we-will-say-good-bye-rabies-ntv, and discussion of needs for improved vaccines and PEP is in World Health Organization, 2018.

23. An excellent description of the sequence of pathologic events leading to clinical rabies is in Warrell and Warrell, 2004.

24. Wunner and Briggs, 2010. The website for the World Rabies Day initiative is https://www.who.int/rabies/WRD_landing_page/en/.

References

Arnold W. 2007. Mosquitoes have the edge in Singapore's dengue war. *The New York Times* **2007**(June 12).

Bharti AR, Nally JE, Ricaldi JN, Matthias MA, Diaz MM, Lovett MA, Levett PN, Gilman RH, Willig MR, Gotuzzo E, Vinetz JM, Peru-United States Leptospirosis Consortium. 2003. Leptospirosis: a zoonotic disease of global importance. *Lancet Infect Dis* **3:**757–771.

Blum AJ, Hotez PJ. 2018. Global "worming": climate change and its projected general impact on human helminth infections. *PLoS Negl Trop Dis* **12:**e0006370. Global "worming": climate change and its projected general impact on human helminth infections.

Bouri N, Sell TK, Franco C, Adalja AA, Henderson DA, Hynes NA. 2012. Return of epidemic dengue in the United States: implications for the public health practitioner. *Public Health Rep* **127:**259–266.

Brady OJ, Gething PW, Bhatt S, Messina JP, Brownstein JS, Hoen AG, Moyes CL, Farlow AW, Scott TW, Hay SI. 2012. Refining the global spatial limits of dengue virus transmission by evidence-based consensus. *PLoS Negl Trop Dis* **6:**e1760.

Brett-Major DM, Coldren R. 2012. Antibiotics for leptospirosis. *Cochrane Database Syst Rev* **2:**CD008264.

Brunkard JM, Robles López JL, Ramirez J, Cifuentes E, Rothenberg SJ, Hunsperger EA, Moore CG, Brussolo RM, Villarreal NA, Haddad BM. 2007. Dengue fever seroprevalence and risk factors, Texas-Mexico border, 2004. *Emerg Infect Dis* **13:**1477–1483.

Cachay ER, Vinetz JM. 2005. A global research agenda for leptospirosis. *J Postgrad Med* **51:**174–178.

Carrasco LR, Lee LK, Lee VJ, Ooi EE, Shepard DS, Thein TL, Gan V, Cook AR, Lye D, Ng LC, Leo YS. 2011. Economic impact of dengue illness and the cost-effectiveness of future vaccination programs in Singapore. *PLoS Negl Trop Dis* **5:**e1426.

Coleman PG, Fèvre EM, Cleaveland S. 2004. Estimating the public health impact of rabies. *Emerg Infect Dis* **10:**140–142.

Dinh KX. 2001. Rabies in humans in Viet Nam, p 255–256. *In* Dodet B, Meslin FX (ed), *Rabies Control in Asia*. John Libbey Eurotext, Paris, France.

Erlanger TE, Keiser J, Caldas De Castro M, Bos R, Singer BH, Tanner M, Utzinger J. 2005. Effect of water resource development and management on lymphatic filariasis, and estimates of populations at risk. *Am J Trop Med Hyg* **73:**523–533.

Global Burden of Disease Collaborative Network. 2020. Global Burden of Disease Study 2019 (GBD 2019). GBD Results Tool. Institute for Health Metrics and Evaluation, Seattle, WA.

Gubler DJ, Clark GG. 1995. Dengue/dengue hemorrhagic fever: the emergence of a global health problem. *Emerg Infect Dis* **1:**55–57.

Gubler DJ, Reiter P, Ebi KL, Yap W, Nasci R, Patz JA. 2001. Climate variability and change in the United States: potential impacts on vector- and rodent-borne diseases. *Environ Health Perspect* **109**(Suppl 2)**:**223–233.

Halstead SB. 1981. The pathogenesis of dengue: the Alexander D. Langmuir Lecture. *Am J Trop Med Hyg* **114:**632–648.

Harris G. 2012. Where streets are thronged with strays baring fangs. *The New York Times* **2012**(August 6).

Heukelbach J, de Oliveira FA, Kerr-Pontes LR, Feldmeier H. 2001. Risk factors associated with an outbreak of dengue fever in a favela in Fortaleza, north-east Brazil. *Trop Med Int Health* **6:**635–642.

Hotez PJ. 2016. Zika in the United States of America and a fateful 1969 decision. *PLoS Negl Trop Dis* **10:**e0004765.

Hotez PJ. 2017. Global urbanization and the neglected tropical diseases. *PLoS Negl Trop Dis* **11:**e0005308.

Hotez PJ. 2018. Human parasitology and parasitic diseases: heading towards 2050. *Adv Parasitol* **100:**29–38.

Hotez PJ, Damania A, Bottazzi ME. 2020. Central Latin America: two decades of challenges in neglected tropical disease control. *PLoS Negl Trop Dis* **14:**e0007962.

John TJ. 2005. The prevention and control of human leptospirosis. *J Postgrad Med* **51:**205–209.

Kay BH, Nam VS, Tien TV, Yen NT, Phong TV, Diep VT, Ninh TU, Bektas A, Aaskov JG. 2002. Control of *Aedes* vectors of dengue in three provinces of Vietnam by use of *Mesocyclops* (copepoda) and community-based methods validated by entomologic, clinical, and serological surveillance. *Am J Trop Med Hyg* **66:**40–48.

Ko AI, Goarant C, Picardeau M. 2009. *Leptospira*: the dawn of the molecular genetics era for an emerging zoonotic pathogen. *Nat Rev Microbiol* **7**:736–747.

Kroeger A, Nathan M, Hombach J, World Health Organization TDR Reference Group on Dengue. 2004. Dengue. *Nat Rev Microbiol* **2**:360–361.

Malavige GN, Fernando S, Fernando DJ, Seneviratne SL. 2004. Dengue viral infections. *Postgrad Med J* **80**:588–601.

Marr JS, Cathey JT. 2010. New hypothesis for cause of epidemic among native Americans, New England, 1616-1619. *Emerg Infect Dis* **16**:281–286.

McBride AJ, Athanazio DA, Reis MG, Ko AI. 2005. Leptospirosis. *Curr Opin Infect Dis* **18**:376–386.

Messina JP, Brady OJ, Golding N, Kraemer MUG, Wint GRW, Ray SE, Pigott DM, Shearer FM, Johnson K, Earl L, Marczak LB, Shirude S, Davis Weaver N, Gilbert M, Velayudhan R, Jones P, Jaenisch T, Scott TW, Reiner RC Jr, Hay SI. 2019. The current and future global distribution and population at risk of dengue. *Nat Microbiol* **4**:1508–1515.

Mehta P, Hotez PJ. 2016. NTD and NCD co-morbidities: the example of dengue fever. *PLoS Negl Trop Dis* **10**:e0004619.

Meltzer MI, Rupprecht CE. 1998. A review of the economics of the prevention and control of rabies. Part 1: global impact and rabies in humans. *Pharmacoeconomics* **14**:365–383.

Murray KO, Rodriguez LF, Herrington E, Kharat V, Vasilakis N, Walker C, Turner C, Khuwaja S, Arafat R, Weaver SC, Martinez D, Kilborn C, Bueno R, Reyna M. 2013. Identification of dengue fever cases in Houston, Texas, with evidence of autochthonous transmission between 2003 and 2005. *Vector Borne Zoonotic Dis* **13**:835–845.

Musso D, Ko AI, Baud D. 2019. Zika virus infection—after the pandemic. *N Engl J Med* **381**:1444–1457.

Nimmannitya S. 2002. Dengue and dengue haemorrhagic fever, p 765–772. *In* Cook GC, Zumla AI (ed), *Manson's Tropical Diseases*, 21st ed. W B Saunders, New York, NY.

O'Neill SL, Ryan PA, Turley AP, Wilson G, Retzki K, Iturbe-Ormaetxe I, Dong Y, Kenny N, Paton CJ, Ritchie SA, Brown-Kenyon J, Stanford D, Wittmeier N, Jewell NP, Tanamas SK, Anders KL, Simmons CP. 2019. Scaled deployment of *Wolbachia* to protect the community from dengue and other *Aedes* transmitted arboviruses. *Gates Open Res* **2**:36.

Pinheiro-Michelsen JR, Souza RDSO, Santana IVR, da Silva PS, Mendez EC, Luiz WB, Amorim JH. 2020. Anti-dengue vaccines: from development to clinical trials. *Front Immunol* **11**:1252.

Powell JR, Tabachnick WJ. 2013. History of domestication and spread of *Aedes aegypti*—a review. *Mem Inst Oswaldo Cruz* **108**(Suppl 1):11–17.

Scott G, Coleman TJ. 2002. Leptospirosis, p 1165–1171. *In* Cook GC, Zumla AI (ed), *Manson's Tropical Diseases*, 21st ed. W B Saunders, New York, NY.

Sudarshan MK, Madhusudana SN, Mahendra BJ, Rao NS, Ashwath Narayana DH, Abdul Rahman S, Meslin F, Lobo D, Ravikumar K, Gangaboraiah. 2007. Assessing the burden of human rabies in India: results of a national multi-center epidemiological survey. *Int J Infect Dis* **11**:29–35.

Tapia-Conyer R, Betancourt-Cravioto M, Méndez-Galván J. 2012. Dengue: an escalating public health problem in Latin America. *Paediatr Int Child Health* **32**(Suppl 1):14–17.

Théodoridès J. 1986. *Historie de la Rage: Cave Canem*. Masson, Paris, France.

Torgerson PR, Hagan JE, Costa F, Calcagno J, Kane M, Martinez-Silveira MS, Goris MG, Stein C, Ko AI, Abela-Ridder B. 2015. Global burden of leptospirosis: estimated in terms of disability adjusted life years. *PLoS Negl Trop Dis* **9**:e0004122.

Utzinger J, Keiser J. 2006. Urbanization and tropical health—then and now. *Ann Trop Med Parasitol* **100**:517–533.

Victoriano AF, Smythe LD, Gloriani-Barzaga N, Cavinta LL, Kasai T, Limpakarnjanarat K, Ong BL, Gongal G, Hall J, Coulombe CA, Yanagihara Y, Yoshida S, Adler B. 2009. Leptospirosis in the Asia Pacific region. *BMC Infect Dis* **9**:147.

Vinetz JM, Glass GE, Flexner CE, Mueller P, Kaslow DC. 1996. Sporadic urban leptospirosis. *Ann Intern Med* **125**: 794–798.

Warrell MJ. 2002. Rabies, p 808–821. *In* Cook GC, Zumla AI (ed), *Manson's Tropical Diseases*, 21st ed. W B Saunders, New York, NY.

Warrell MJ, Warrell DA. 2004. Rabies and other lyssavirus diseases. *Lancet* **363**:959–969.

World Health Organization. 2012a. Dengue and severe dengue. Fact sheet no. 117. January 2012. World Health Organization, Geneva, Switzerland. www.who.int/mediacentre/factsheets/fs117/en/.

World Health Organization. 2012b. Rabies. Fact sheet no. 99. September 2012. World Health Organization, Geneva, Switzerland. www.who.int/mediacentre/factsheets/fs099/en/.

World Health Organization. 2018. Rabies vaccines: WHO position paper. *Wkly Epidemiol Rec* **93**:201–220.

Wunner WH, Briggs DJ. 2010. Rabies in the 21 century. *PLoS Negl Trop Dis* **4**:e591.

Wyatt J. 2007. Rabies-update on a global disease. *Pediatr Infect Dis J* **26**:351–352.

The Neglected Tropical Diseases of North America

A civilization is judged by the treatment of its minorities.

MAHATMA GANDHI

Although many of the neglected tropical diseases (NTDs) are not as prominent in the United States, Canada, Mexico, and the Caribbean as they are in the low-income countries of Africa, Asia, and Central and South America, these infections are nonetheless significant among the poorest people living in North America. Some of the North American NTDs are also zoonoses, meaning that they are transmitted from animals to people. The term "One Health" has been used to describe public health control for these conditions. Populations at particular risk for acquiring these NTDs include minority populations living in low-income neighborhoods such as African American and Hispanic groups, as well as many of North America's Native Americans. A serious NTD problem resulting from endemic hookworm, lymphatic filariasis (LF), and schistosomiasis also occurs in the Caribbean region, where it has been argued that these diseases represent a glaring legacy of the transatlantic slave trade. Today, the persistence of the NTDs in North America, considered one of the world's wealthiest continents, represents a health disparity of astonishing proportions.

NTDs in the United States

Historically, the most impoverished populations living in the United States have suffered greatly from NTDs. In many respects the rural American South at the beginning of the 20th century resembled a developing country, with high rates of hookworm infection and other soil-transmitted helminthiases, as well as endemic malaria, pellagra (a nutritional deficiency of nicotinic acid),

Forgotten People, Forgotten Diseases: The Neglected Tropical Diseases and Their Impact on Global Health and Development, Third Edition. Peter J. Hotez.
© 2022 American Society for Microbiology. DOI: 10.1128/9781683673903.ch09

typhoid fever, and seasonal epidemics of yellow fever.[1] Indeed, the pejorative concept of the "lazy Southerner" was partly a consequence of the toxic combination of chronic parasitism and nutritional deficiencies that plagued this region, as the NTDs kept the southern population mired in poverty just as they do today in Africa, Asia, and elsewhere. The medical historian Margaret Humphreys has argued effectively that New Deal programs (including the Agricultural Adjustment Act) during the 1930s did more than anything else to reduce the burden of tropical diseases in the United States by relocating agricultural workers to urban areas and transforming the American South from an agrarian economy into an urbanized, industrial one.[1] It is still true that urbanization and parallel economic development are the most potent forces in melting away the NTDs, just as they were in Japan and Korea after World War II and in China beginning in the 1990s.

Today, as a result of dramatic economic gains, tropical afflictions such as malaria, pellagra, and typhoid fever are no longer endemic in the United States. However, several NTDs still remain among our nation's poorest people. According to the U.S. census, in 2019 our nation's official poverty rate was 10.5%, with approximately 34 million Americans living in poverty.[2] Among African American and Hispanic populations, the poverty rate exceeds 18% and 15%, respectively.[2] Moreover, there are even more troubling statistics. For example, in fall of 2017, the United Nations Special Rapporteur on extreme poverty and human rights visited the United States. In his report, he pointed out that approximately one-half of the Americans who live below the poverty line live in "deep poverty," meaning a family income "below one-half of the poverty threshold," while a recent book finds that approximately 1.5 million households live on less than $2 per day, roughly the global poverty level.[2] Poverty is expected to increase even further in 2020 due to COVID-19. Americans who live in poverty are at risk for several NTDs discussed previously in this book, including the soil-transmitted helminth infections, Chagas disease, leishmaniasis, and the major urban NTDs, Zika virus infection and dengue fever. They are also at risk for two zoonotic NTDs not discussed previously, namely, cysticercosis and toxoplasmosis. In all, I estimate that approximately 12 million Americans currently live with one or more of these neglected parasitic infections.[2]

The soil-transmitted helminth infections

Prominent among the American NTDs are the soil-transmitted helminth infections. When last studied during the 1960s and 1970s, pockets of hookworm infection, as well as ascariasis and trichuriasis, still occurred among the rural poor living in regions where those diseases had been highly endemic, especially southeastern Georgia, the Gulf Coast regions of southern Mississippi, the coastal Carolinas, and East Texas near the Sabine River.[3] At one time, the Eastern Cherokees stood out as a population with particularly high rates of infection.[4] While many assert that hookworm is probably no longer a serious public health threat in the rural American South, we do not know this for sure, since, as pointed out in chapter 2, large-scale epidemiological studies of hookworm infection and other soil-transmitted helminth infections in the

United States have not been conducted in the 21st century.[4] Recently, however, our group at the National School of Tropical Medicine, led by Drs. Rojelio Mejia and Megan McKenna, conducted studies with the environmental justice activist Catherine Coleman Flowers to find that hookworm infection may still be endemic in a poor rural area of Alabama where open sewage was present and there was other evidence of absent sanitation.[4] A big unknown is whether this is true throughout the U.S. Gulf Coast states, where extreme poverty is the most prevalent.[4] Therefore, it is important to revisit some of the vulnerable populations living in this part of the United States and reassess the prevalence of the soil-transmitted helminth infections in their communities.

Toxocariasis

Today, we know that at least two soil-transmitted helminth infections are still significant diseases in the United States: toxocariasis and strongyloidiasis. Toxocariasis is a zoonosis that results when humans accidentally ingest the eggs of a dog ascarid parasite known as *Toxocara canis* or the cat ascarid, *Toxocara cati*.[5] A high percentage of puppies and kittens are infected with *Toxocara* worms, which in many respects are equivalent to the canine counterpart of the human soil-transmitted helminth *Ascaris lumbricoides*. Because puppies and kittens indiscriminately defecate on the ground, particularly in urban areas of the United States as well as in some rural areas, *Toxocara* and other ascarid eggs are extremely common in the environment. For instance, in Wallingford, CT, a small city located halfway between Hartford and New Haven, 27.5% of public playground areas were found to be contaminated with *Toxocara* eggs.[5] A more recent study from New York City found that *Toxocara* eggs are widespread in the Bronx and other poor metro areas.[5] Probably more than any other group, children from poor families come into contact with these eggs as a result of exposure to the feces of *Toxocara*-infected stray dogs defecating on dirt and sandy areas of inner-city playgrounds. When children accidentally ingest the eggs, the eggs hatch and the released *T. canis* larvae migrate through the major organs of the body (Fig. 9.1). By so doing, *T. canis* larvae cause a serious syndrome in toddlers known as visceral larva migrans, which is characterized by fever and inflammation in the liver (hepatitis) and lungs (pneumonitis). Visceral larva migrans is also often accompanied by allergic symptoms, including wheezing (Fig. 9.1). A second syndrome resulting from *Toxocara* infection, known as ocular larva migrans, also occurs in older children, in which *T. canis* larvae migrate to the eye, where they cause retinitis and strabismus. Many children also acquire a third form of *Toxocara* infection known as covert toxocariasis. Covert toxocariasis is characterized by eosinophilia, i.e., an increase in a type of white blood cell known as the eosinophil, as well as wheezing and other symptoms that resemble asthma.[6] Because *T. canis* can sometimes migrate through the brain, it is believed that toxocariasis may be linked to epilepsy and neuropsychiatric disturbances.[6]

Unfortunately, there are very few studies to document the extent of any of these three forms of toxocariasis in the United States. During the 1970s, it was determined that between 4.6 and 7.3% of children in different geographic

Toxocariasis
(Toxocara canis, Toxocara cati)

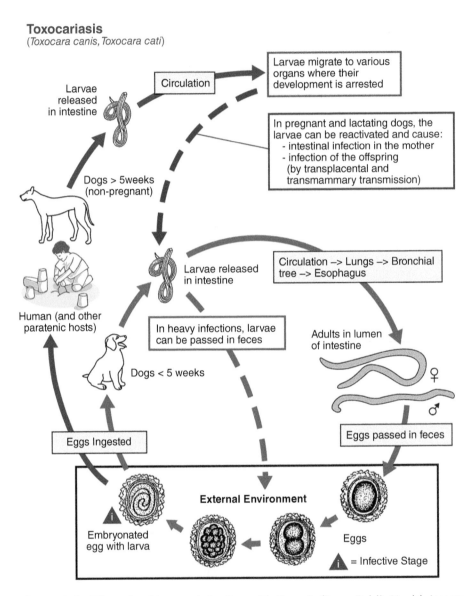

Figure 9.1   Life cycle of human infection with *T. canis*. (From Public Health Image Library, CDC [http://phil.cdc.gov].)

regions of the United States were found to have some form of *T. canis* infection. However, among African American children with lower socioeconomic status, up to 30% were infected.[7] As part of the Third National Health and Nutrition Examination Survey (1988–1994), sera from more than 30,000 subjects were tested for antibodies to *Toxocara*, and it was found that more than 20% of non-Hispanic blacks were seropositive, a much higher prevalence than for any other group. Moreover, *Toxocara* seropositivity was associated with poverty and low education levels.[7] More recent studies published in 2018 found overall lower seroprevalence rates, at around 5%, but the racial, ethnic, and poverty disparity component remains.[7] From these studies I conclude that toxocariasis is a major health disparity affecting African American populations living in the United States, especially in the American South. During the 1990s, a team led by myself and Neda Sharghi, then a Yale University graduate public health student, conducted studies among children in some of Connecticut's poor urban areas (New Haven and Bridgeport) and found that 10% of these inner-city children showed evidence of infection at some point in their lives, as evidenced by having measurable antibody to *T. canis*.[8] Among poor Hispanic children living in Bridgeport, up to 50% showed evidence of previous infection.[8] An earlier study in New York City found that 5% of blood samples from children that were submitted for lead testing were also positive for previous *Toxocara* infection.[9] The relationship between toxocariasis and lead ingestion may be more than coincidental, since the same behavior responsible for ingesting lead paint chips might also promote ingestion of substances contaminated with egg-containing dirt.

Children living in inner cities have a higher frequency and severity of asthma attacks than do other pediatric populations living in the United States.[10] Could covert toxocariasis account for a portion of our nation's asthma epidemic? We were not able to show this relationship conclusively in Connecticut, although studies in Europe and elsewhere have demonstrated an important link between *Toxocara* infection and asthma.[11] Similarly, could toxocariasis and its impact on childhood intellectual development account for the achievement gap among socioeconomically disadvantaged students?[6] Given that 5% or more of such children may have had *Toxocara* infection at some point during their lives, we are likely looking at a large number of U.S. children at risk for covert toxocariasis. I previously estimated that up to 2.8 million African Americans living in poverty are exposed to and infected with *Toxocara* larvae.[12] Because this infection may place these children at risk for asthma, developmental delays, and other conditions, such information suggests an urgent need to conduct larger and better surveillance studies for this illness among U.S. inner-city children.

Strongyloidiasis

Still another potentially important soil-transmitted helminth infection in the United States is strongyloidiasis, a severe cause of enteritis (inflammation of the intestine) with diarrhea. Strongyloidiasis is also a potentially fatal disseminated infection in patients who are immunocompromised from receiving steroids. Worldwide, approximately 30 to 100 million people are infected with *Strongyloides stercoralis*, primarily in rural, developing regions of Asia, Africa, and the Americas. However, strongyloidiasis has also been shown to be endemic in the rural Appalachian region of the United States. Although our existing data are severely limited, studies conducted over the past few decades have shown that the prevalence of endemic strongyloidiasis in the rural counties of eastern Kentucky and Tennessee ranges from 1 to 4%.[13] Overall, the population of rural Appalachia is approximately 6.8 million.[14] If we estimate that 1% of this population is infected with strongyloidiasis, then there are approximately 68,000 cases of strongyloidiasis in the region.[12] A more recent analysis found that strongyloidiasis may extend to areas outside of the southeastern United States.[14] Again, it is important to fully explore the extent of this disease in the United States.

Cysticercosis and neurocysticercosis

Cysticercosis is a very serious parasitic worm infection and NTD that results in seizures and other long-term neurological manifestations. It results from transmission from person to person of the eggs of the pork tapeworm, *Taenia solium*. The life cycle of *T. solium* infection is shown in Fig. 9.2. Humans acquire the pork tapeworm by ingesting the cyst stages of the parasite, known as cysticerci, which are found in the muscles and brains of infected pigs. Following ingestion of these cysts as a result of eating uncooked or improperly cooked pork, an immature larva will fasten onto the intestinal wall and begin growing into an adult tapeworm. The adult pork tapeworm can reach huge lengths in the human gut, often 2 to 3 m, but having a huge tapeworm surprisingly causes very little in the way of symptoms. When a person harbors a pork tapeworm, however, segments of the parasite containing eggs can break off in the intestine. These egg-sac-like segments exit the body in feces. When pigs ingest fecal material containing eggs (in some very poor regions, it is a common practice to feed human feces to pigs), the eggs hatch and give rise to cysticerci in the muscles or brains of the animals.

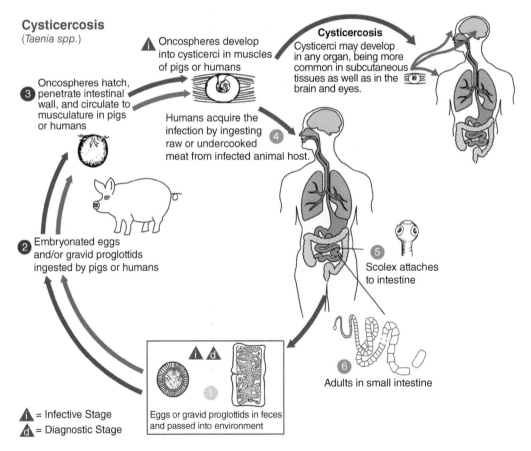

Figure 9.2 Life cycle of *T. solium* and cysticercosis. (Image from CDC-PHIL [ID#3387]/CDC/ Alexander de Silva, PhD; Melanie Moser, 2002.)

Cysticercosis is the more serious condition resulting from *T. solium* infection and occurs when humans accidentally ingest *T. solium* eggs excreted by a household member (often a person from a country such as Mexico or El Salvador where cysticercosis is endemic) who is infected with the adult tapeworm. An infected person can sometimes shed thousands of eggs daily, and the transmission of the eggs from one person to another (typically through accidental fecal-oral contamination) gives rise to this condition. After the eggs are swallowed, larvae known as oncospheres are liberated; these invade the gut wall and enter the human circulatory system. The larvae travel to the muscles and brain, where they form cysts. In the muscles, *T. solium* infection usually does not cause much in the way of symptoms, but the presence of *T. solium* cysts in the brain can trigger severe and unremitting seizures unless the patient is treated immediately with anticonvulsant medication. In some cases, multiple cysts can cause more extensive disease such as encephalitis, while cysts at the base of the brain can cause obstruction of the flow of cerebrospinal fluid, leading to a condition known as hydrocephalus. The diagnosis of cysticercosis in the brain (sometimes known as neurocysticercosis) usually requires special radiographic imaging with either computed tomography or magnetic resonance imaging, while treatment requires use of anticonvulsants and sometimes anthelmintic drugs such as albendazole and praziquantel.

In part because of the large influx of Hispanic migrants from Central America (where cysticercosis is endemic), neurocysticercosis has become an important infection in Texas, New Mexico, Arizona, and California, as well as in Chicago and New York.[15] Across the country, children and adults are today being brought into emergency rooms with seizures and being diagnosed with neurocysticercosis. Today, an estimated 1,000 to 2,000 new cases of neurocysticercosis are diagnosed annually,[16] making it one of the leading causes of epilepsy in the United States. However, at an incidence of 8 to 10 per 100,000 per year among Hispanic populations,[17] and considering that there are approximately 35 million Hispanics living in the United States, as many as 2,800 and 3,500 new cases of cysticercosis may occur annually. In a study conducted in rural Ventura County, CA, it was determined that 1.8% of that population had antibodies to the *Taenia* parasites that cause cysticercosis.[18] If we use this value and extrapolate for the entire rural Hispanic population of the United States (estimated at 2.3 million in 2000), then it is conceivable that the number of people living with cysticercosis in the United States may be much higher, possibly as many as 41,400. However, I have also estimated an upper limit of 169,000 people living with cysticercosis.[12] Another estimate of the disease's prevalence among Hispanics in Oregon is 5.8 cases per 100,000.[18] Not surprisingly, neurocysticercosis was shown previously to account for 10% of neurology and neurosurgery admissions and 10% of all seizures presenting to some emergency rooms in Los Angeles.[18] Presumably, this is the experience of other areas with large Hispanic populations.[18] The large number of possible cases of cysticercosis suggests that there is an urgent need for better surveillance studies on the presence of this condition in the southwestern United States and for treatment of patients who harbor the pork tapeworm as a means of reducing the prevalence of this serious disease of poverty. Globally, the role of preventive treatments with albendazole or praziquantel is under investigation.[18]

Neglected protozoan infections

Toxoplasmosis

One of the most serious protozoan infections that occurs in the United States is toxoplasmosis. Caused by *Toxoplasma gondii*, it has two major routes of infection, either ingestion of a form of the parasite found in cat feces known as the oocyst (which may occur when changing kitty litter) or ingestion of uncooked meat (primarily pork from unconfined pigs, goats, and chickens) contaminated with the bradyzoite stages of the parasite (Fig. 9.3).[19] Roughly

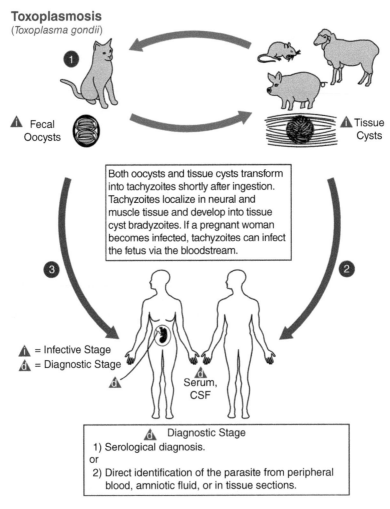

Toxoplasmosis
(*Toxoplasma gondii*)

Fecal Oocysts

Tissue Cysts

Both oocysts and tissue cysts transform into tachyzoites shortly after ingestion. Tachyzoites localize in neural and muscle tissue and develop into tissue cyst bradyzoites. If a pregnant woman becomes infected, tachyzoites can infect the fetus via the bloodstream.

= Infective Stage
= Diagnostic Stage

Serum, CSF

Diagnostic Stage
1) Serological diagnosis.
or
2) Direct identification of the parasite from peripheral blood, amniotic fluid, or in tissue sections.

Figure 9.3 Life cycle of *T. gondii*. (Image from CDC-PHIL [ID#3421]/CDC/Alexander de Silva, PhD; Melanie Moser, 2002.)

one-half of the cases occur by either route. Toxoplasmosis is very common in the United States—approximately 15% of U.S. residents have antibodies to *T. gondii*, meaning that they have been infected with these parasites. However, Mexican Americans and African Americans have a higher rate of infection.[19]

Most people who become infected with toxoplasmosis do not become very ill. Often, this infection is a nonspecific illness accompanied by fever and swelling of the lymph nodes. It is therefore easy to confuse toxoplasmosis with other infectious illnesses, especially many viral infections. However, one population is particularly vulnerable to severe effects of *T. gondii* infection, namely, the unborn fetus. If a pregnant mother becomes infected with *T. gondii* during her pregnancy, especially early in the pregnancy, the newborn infant can be born with congenital toxoplasmosis, a syndrome that includes vision impairment and blindness, hearing loss, seizures, and mental retardation. The Centers for Disease Control and Prevention (CDC) estimates that of the 4 million live infants born each year in the United States, approximately 400 to 4,000 have congenital toxoplasmosis.[20] The National Collaborative Chicago-Based, Congenital Toxoplasmosis Study conducted a 20-year study of 120 infants with congenital toxoplasmosis and found that early diagnosis and treatment for a year with antiparasitic drugs can make a huge difference in the long-term clinical outcomes of these children.[21] However, many infants do not show obvious evidence of congenital toxoplasmosis at birth. Instead, the disease is not overtly manifested until later in the first year, and by then treatment may be too late to prevent the most serious consequences of the infection. Beginning in 1986, the Commonwealth of Massachusetts began conducting universal screening of newborns, a program similar to the screening for phenylketonuria and other genetic disorders. Sadly, only New Hampshire has joined Massachusetts in conducting newborn screening through a New England Newborn Screening Program.[22] Therefore, there is an urgent need to conduct nationwide newborn screening in order to identify the thousands of infants born every year with congenital toxoplasmosis. We will see below that toxoplasmosis is also an important public health threat among the Inuit of the Canadian Arctic, making them a population also worth screening.

Chagas disease

Details about Chagas disease are provided in chapter 7. Of interest is the observation that Chagas disease has emerged in the southern United States, with Texas reporting some of the highest rates of infection. According to the CDC, an estimated 238,000 people are living with Chagas disease in the United States.[23] However, this estimate is based primarily on imported cases from Latin America, whereas some investigators based in Texas have suggested that there may be almost that many infected with *Trypanosoma cruzi* in Texas alone.[23] In South Texas near the Mexican border and possibly elsewhere in the American South, the prevalence of *T. cruzi* infection among dogs is high, suggesting the possibility that transmission to people is also possible. There is also evidence that transmission goes beyond South Texas, including other regions in the state of Texas, and possibly elsewhere in the American Southwest. This is an important area of emphasis for our National School of Tropical Medicine.[23] From my viewpoint, the big question about Chagas disease is whether almost all of the cases of *T. cruzi* infection are being imported from Latin America, as suggested by investigators from the CDC, or there is a significant rate of autochthonous transmission, especially in the extremely poor *colonias* communities in South Texas. One problem is the minimal amount of active surveillance that is being conducted in South Texas or elsewhere in the United States. Another is the problem that very few physicians or other health care providers in the United States have detailed knowledge about Chagas disease or its cardiac manifestations, so that the disease is greatly underdiagnosed. Many obstetricians are also unaware of the significant number of pregnant women with Chagas disease in the United States or the extent of vertical transmission from mother to unborn fetus.[23] Finally, *T. cruzi* has contaminated the U.S. blood supply, and this in itself has emerged as an important transfusion safety issue.

Leishmaniasis

Earlier (in chapter 7), information on the persistence of cutaneous leishmaniasis (CL) in the Americas was presented. Dozens of cases of CL have now been reported from south-central Texas, suggesting that this condition may be becoming an important NTD in the region.[24] In 2007, nine cases of CL were also detected in North Texas, while cases have also been reported in Oklahoma and North Dakota.[24] Indigenous CL in Texas is caused by *Leishmania mexicana*, and it is believed that wood-burrowing rats may serve as reservoir hosts. CL has also been identified as an important NTD among the U.S. troops returning from Operation Iraqi Freedom.[24] Visceral leishmaniasis has been documented from dogs in the United States, especially foxhounds. There is a potential for transmission of leishmaniasis to humans to become a serious public health problem in the United States.[24]

Neglected bacterial and viral infections

Earlier (in chapter 5), we discussed trachoma as an important infectious cause of blindness in developing countries, particularly in dry and dusty areas of the Sahel. Over the last few decades in the American Southwest, trachoma has been a well-known public health problem among the Navajo Indians living on the Navajo Reservation.[25] Other important but seldom discussed neglected bacterial infections include relapsing fever from ticks and murine typhus from fleas, which have emerged in Texas and elsewhere in the United States.[26]

Flavivirus infections caused by dengue, Zika virus, and West Nile virus (WNV) have emerged in the United States, and we can expect that the number of cases will increase in the coming decade, especially among the poor. Details are discussed in chapter 8. Outbreaks of dengue and Zika have emerged in South Texas and in Florida over the last few years,[26] and given the high rates

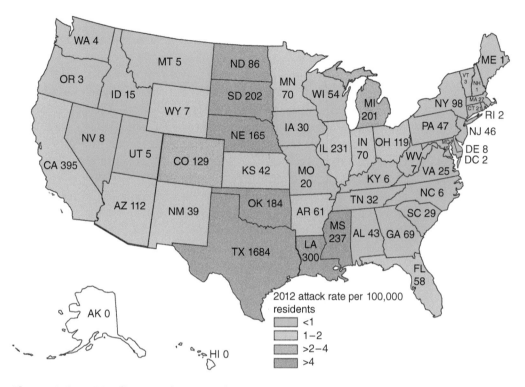

Figure 9.4 WNV, diagnosed cases and attack rate, 2012.

of poverty in the major urban areas of the Gulf Coast, I believe we can expect additional outbreaks in cities such as Houston, New Orleans, and Mobile.[26] The poverty link is based on findings in South Texas and neighboring Mexico regarding the increased risks from the absence of air-conditioning, window screens, and garbage collection and related factors.[26] Dengue transmission also occurs commonly in Puerto Rico, where Zika virus infection was also widespread in 2016.[26] WNV emerged in New York in 1999 and now is associated with seasonal epidemics in the United States, especially in Texas (where one of the nation's worst epidemics occurred in 2012), where urban homeless populations and other people without means living with intense exposure to *Culex* mosquitoes are especially vulnerable (Fig. 9.4).[26] There is evidence that WNV infection can cause long-term neurologic impairments including neuropsychiatric disturbances related to cognition or depression.[26]

NTDs in Texas

Today, the state of Texas is emerging as an epicenter of NTDs in the United States, with some of the highest incidences of diseases such as Chagas disease, cysticercosis, helminth infections, dengue, WNV and Zika virus infection, murine typhus, relapsing fever, and others. The major reasons include the fact that Texas has some of the greatest numbers of people living in poverty among U.S. states (4 to 5 million), together with the warm and often humid climate in South Texas and elsewhere in the state. Climate change may make Texas even more suitable to disease vectors. Still other factors include unchecked and rapid urbanization, and the fact that Texas urban areas attract people from all over the world due to low costs of living and a robust economy.[26] This confluence of disease and poverty is a major reason why in 2011 I launched the National School of Tropical Medicine at Baylor College of Medicine. Located in the renowned Texas Medical Center in Houston—the world's largest medical center, with almost 100,000 employees—the school has a multifold mission, which includes (i) research and development related to NTDs endemic in the United States and globally (including our Texas Children's Center for Vaccine Development); (ii) clinical treatment of NTDs, including a tropical medicine clinic; (iii) education through a unique Diploma in Tropical Medicine and other opportunities and undergraduate offerings with Baylor University in Waco; and (iv) public policy for NTDs in association with the James A. Baker III Institute of Public Policy at the adjacent Rice University and Texas A&M University, where I'm a fellow of the Hagler Institute for Advanced Study. The National School of Tropical Medicine represents an aggressive and innovative effort to focus national and global attention on the NTDs and related diseases of poverty. It is also leading vaccine development and outbreak investigations for COVID-19, Zika virus infection, and other pandemic threats.

NTDs in Arctic Canada and Alaska

The Inuit are a group of approximately 150,000 indigenous people inhabiting the Arctic regions of Canada (including Nunavut, northern Quebec and Labrador, and the Northwest Territories), Alaska, and Greenland, where, for most of their history, they have pursued a largely nomadic lifestyle, with a heavy reliance on hunting and fishing. Over the last few decades, most of the Inuit have relocated to permanent settlements. There they face enormous challenges, including overcrowding, high rates of poverty and unemployment, alcoholism and other substance abuse, depression, and suicide. In the absence of fresh vegetables and fruit in the Arctic tundra and taiga areas, the diet of the Inuit relies heavily on meat, especially from sea mammals and polar bear. Such a diet makes them vulnerable to at least two major food-borne parasitic diseases: trichinellosis (also known as trichinosis) and toxoplasmosis. In addition, the close association between the Inuit and sled dogs, reindeer, and elk makes them susceptible to another NTD known as echinococcosis.

Trichinellosis (trichinosis)

Like toxoplasmosis and pork tapeworm infection, trichinellosis is commonly acquired through the ingestion of uncooked pork. The newborn larval stages of *Trichinella spiralis spiralis*, a nematode parasite, typically live in the flesh of pigs, and outbreaks of the disease occurred throughout the continental United States during the 20th century. Such outbreaks were caused partly by the appalling conditions in U.S. slaughterhouses, which included the practice of allowing pigs to feed on infected rats and uncooked garbage. Such practices were described by Upton Sinclair in his depictions of meat-packing plants in his early-20th-century novel *The Jungle*. Trichinellosis begins as a disease that resembles an attack of food poisoning with nausea and vomiting, usually occurring within a week after ingestion of *Trichinella*-contaminated meat. This so-called enteric phase is followed by the muscle or invasion phase, characterized by severe muscle pain as well as edema around the eyelids (periorbital edema) and eosinophilia. Death can result from large infestations in the muscle or when *Trichinella* larvae invade the heart muscle.

Trichinellosis caused by *T. s. spiralis* is no longer a significant public health problem in the United States, but a variant of the disease caused by *T. s. nativa* is now an important problem among the Inuit. For instance, multiple outbreaks have occurred in Nunavik (the Inuit region in northern Quebec) over the past few decades.[27] Trichinellosis caused by *T. s. nativa* produces a syndrome similar to *T. s. spiralis* infection, but repeated infections cause primarily intestinal symptoms, including prolonged diarrhea, rather than severe muscle pain.[28] Most of the Canadian Arctic trichinellosis outbreaks result from raw or insufficiently cooked walrus meat, although other outbreaks have resulted from the consumption of foxes and polar bears. An estimated 60% of polar bears are infected with *Trichinella* larvae, while only 2 to 4% of walruses are infected.[27,28] Interestingly, *T. s. spiralis*, the major species of *Trichinella* parasites that caused disease in the 20th-century United States, is susceptible to destruction both through heat (cooking) and prolonged freezing. However, *T. s. nativa*, the major Arctic *Trichinella* species found in polar bears and walruses, has specifically adapted through evolution to resist freezing. In Nunavik, an innovative prevention program (developed by the late J. Dick MacLean and his colleagues at the McGill University Centre for Tropical Diseases at the Montreal General Hospital together with Jean-François Proulx from the Nunavik Regional Board of Health and Social Services) has been established in which preconsumption testing of tagged samples from walrus is conducted at a regional laboratory.[27] Any urgent information is communicated immediately to the local community by radio in order to ensure that no meat is consumed. Those who had infected meat are treated prophylactically with the anthelmintic drug mebendazole in order to abort the trichinellosis infection and prevent larvae from entering the muscles. As a result of this prevention program, it is estimated that only 1% of the Inuit population of Nunavik is infected with trichinellosis.[27] Toxoplasmosis and congenital toxoplasmosis are also extremely common among the Inuit, with a seroprevalence of approximately 60%.[29,30] Among the important risk factors are the ingestion of seal and caribou.[29]

Echinococcosis

There are two major forms of the helminth infection known as echinococcosis that affect the Inuit. Cystic echinococcosis, caused by the tapeworm *Echinococcus granulosus*, occurs throughout the North American Arctic, where the natural mammalian hosts of the *E. granulosus* tapeworms are the wolves and sled dogs fed the offal (organ meats) of moose, reindeer, and elk.[30] Humans become infected by a fecal-oral route in which they accidentally ingest the eggs excreted in the feces of sled dogs. Upon ingestion, the eggs liberate larval stages that travel to the lungs and liver, where they produce large fluid-filled cysts. In some patients, the cysts can grow to more than 10 cm in diameter, enabling them to cause mechanical destruction of the affected organs. Alternatively, the cysts can rupture and the leaked fluid can produce severe allergic reactions. The treatment often requires surgical intervention, although anthelmintic drugs such as albendazole appear either to reduce the risks or in some cases to cure the disease entirely. During the first half of the 20th century, when the Inuit led a largely nomadic existence with heavy dependence on sled dogs for travel, cystic echinococcosis was a serious problem. However, as permanent settlements began to develop and as dogs were replaced by snowmobiles and other mechanized vehicles, the life cycle of *E. granulosus* was largely interrupted and today the number of cases is small.[31] According to a seroprevalence study conducted in Nunavik, approximately 12,000 Inuit have been infected with *E. granulosus*.[30]

Alveolar echinococcosis is a second and more severe form. Caused by *Echinococcus multilocularis*, this form of echinococcosis produces cysts that are both highly infiltrative and metastatic and therefore sometimes not treatable by surgical modalities. Alveolar echinococcosis has a high mortality rate. The disease is believed to have been first introduced by arctic foxes and now is endemic in the prairie provinces of Canada as well as in the Dakotas, Montana, and Wyoming in the United States.[32] The number of cases that occur annually is not known.

Summary of NTDs in the United States and the Arctic Region

Table 9.1 shows a summary of the number of possible cases of NTDs in the United States and the Arctic region. For the most part, these numbers were extrapolated on the basis of very limited data, but I present them here because they point out the potential seriousness of the problem and the urgent need for

Table 9.1 Summary of the NTDs of the United States and the Arctic region[a]

Disease	Estimated no. of cases	Population at greatest risk
Toxocariasis	1.3 million–2.8 million	African Americans living in poverty
Strongyloidiasis	68,000–100,000	Residents of Appalachia living in poverty, African refugees
Cysticercosis	41,400–169,000	Hispanic Americans
Congenital toxoplasmosis	<4,000 cases annually	Not determined
Chagas disease	200,000–300,000 cases	Hispanic Americans
Cutaneous leishmaniasis	Not determined	Not determined
Dengue fever	110,000–200,000 new infections annually	Hispanic Americans, African Americans
Leptospirosis	Not determined	Not determined
Trachoma	Not determined	Not determined
Cystic echinococcosis	12,000	Arctic Native American
Trichinellosis	1,500	Arctic Native American
Alveolar echinococcosis	<100	

[a] Compiled from Hotez, 2008b; Hotez, 2010; Proulx et al., 2002; and Siddiqui and Berk, 2001.

meaningful surveillance studies. It is also important to highlight the potential role of climate change in further promoting NTDs, especially as we continue to see unprecedented warming temperatures with each passing year.[27] This might be expected to expand the range of vector-borne diseases, including dengue and other flavivirus infections, to more northerly latitudes.

NTDs in Mexico and the Caribbean

High rates of NTDs occur in many regions of rural Mexico, especially in southern Mexico near the borders of the Central American countries of Guatemala and Belize. For example, in the southern states of Chiapas and Oaxaca, onchocerciasis was eliminated recently (in 2015) through preventive treatment and other measures.[33] In addition, hookworm infection and other soil-transmitted helminth infections are endemic to Chiapas (as well as elsewhere in rural Mexico),[34] trachoma is still focally endemic, and the number of cases of CL has increased because of interruptions in public health control measures forced by guerilla movements in the region.[35] Chagas disease still occurs in focal regions of the country, and between 1990 and 2005, 1,532 new acute infections were recorded.[36] However, under a system of active surveillance that used 1.5 million blood samples collected from patients with fever in more than 16,000 rural areas, fewer than 10 acute cases were identified annually.[36] Many believe that these numbers are gross underestimates, with the possibility that there are presently 2 million to 6 million chronic cases of Chagas disease in Mexico.[36] Additional estimates suggest that approximately 40,000 pregnant women and 2,000 newborns may be infected with *T. cruzi* in North America.[36] Dengue, leptospirosis, and the protozoan infection amebiasis are important infections in Mexico's urban and rural areas.[35] The devastating impact of dengue on Latin America was recently summarized.[37] The Carlos Slim Foundation is helping to support efforts for neglected disease control and innovation in Mexico and the area known as "Mesoamerica," which includes Central America.

Although we cannot consider most of the Caribbean a part of the United States, according to the Caribbean Tourist Organization more than 10 million Americans travel to the Caribbean every year.[38] Today, the Caribbean is considered the most tourism-dependent region in the world, with tourism dollars responsible for an estimated 25% of the region's gross domestic product.[38] Beyond the hotels and resorts, however, there is a hidden side to the Caribbean. High rates of selected NTDs, especially LF, schistosomiasis, hookworm infection, and dengue, exist in the region. The highest rates of LF occur in Haiti, and the highest rates of schistosomiasis occur in Guadeloupe.[39] Hookworm infection and other soil-transmitted helminth infections are also endemic to the Caribbean, particularly Haiti and Belize, but overall there are very few present-day data on their prevalence.[40]

Given that tourism generates more than US$30 billion annually in the Caribbean, it is truly shocking that such obvious diseases of poverty, including thousands of cases of hookworm infection, LF, and schistosomiasis, could be allowed to remain endemic. In the beginning of this book (in chapter 1),

I pointed out how hookworm infection, LF, and schistosomiasis represent a historical legacy of the transatlantic slave trade from Africa (a link first proposed by Pat Lammie and his colleagues at the CDC and the Pan American Health Organization).[41] It is therefore an especially sad fact that we allow these diseases to remain near where Americans go to play. It is especially remarkable that if every American who travels to the Caribbean in the next year were to contribute $1 toward NTD control, we would likely have sufficient funds to eliminate LF, schistosomiasis, and possibly other NTDs from the region.

Conclusions

The frequent unavailability of reliable numbers on the prevalence of the NTDs is reflective of their neglected status and of their disproportionate impact on the poorest of the poor. Many of these NTDs could be easily controlled or eliminated and at relatively modest cost. The most immediate need is to support studies in order to better assess the true prevalence of these infections and then to identify simple and cost-effective public health solutions. There are really no excuses for allowing such glaring health disparities to continue in the backyards of some of the world's wealthiest countries.

SUMMARY POINTS   **NTDs in the United States and Canada**

- NTDs are common among the minority populations and indigenous people of North America, although their true prevalence has not been well studied.

- The major U.S. NTDs include soil-transmitted helminth infections, especially toxocariasis, hookworm infection, and strongyloidiasis; cysticercosis; toxoplasmosis; Chagas disease; and dengue. Cysticercosis may be a leading cause of epilepsy in the southwestern United States, and toxoplasmosis is a major cause of birth defects.

- The Inuit of the Canadian Arctic and Alaska also suffer from several NTDs, including toxoplasmosis, trichinellosis, and echinococcosis.

- Several NTDs, including soil-transmitted helminth infections, trachoma, and CL, occur frequently in rural southern Mexico, especially in Chiapas, and Chagas disease is still widely endemic. Onchocerciasis was eliminated in 2015.

- Schistosomiasis and LF remain serious endemic NTDs in the Caribbean.

- The NTDs represent some of North America's greatest health disparities.

Notes

1. The socioeconomic impact of tropical diseases in the American South is a topic of works by the medical historian Margaret Humphreys, including Humphreys, 2001; and Martin and Humphreys, 2006. Also, the economist Hoyt Bleakley has written about the impact of hookworm in Bleakley, 2007.

2. The poverty figures quoted can be found at https://www.census.gov/library/publications/2020/demo/p60-270.html, https://www.ohchr.org/EN/NewsEvents/Pages/DisplayNews.aspx?NewsID=22533, and http://www.twodollarsaday.com/. My estimate that 12 million Americans living in poverty suffer from at least one neglected parasitic infection is in Hotez et al., 2014a.

3. Some papers on the prevalence of hookworm in the United States during the last half of the 20th century include Henderson, 1957; Farmer, 1983; Arnold, 1949; Disalvo and Melonas, 1970; Sargent et al., 1972; and Martin, 1972.

4. The high prevalence of ascariasis and trichuriasis among Cherokee Indian schoolchildren was reported in Healy et al., 1969. A summary of soil-transmitted helminth studies in the United States since 1940 was presented in Starr and Montgomery, 2011. Studies on hookworm in Alabama are in McKenna et al., 2017. A discussion of disease and poverty on the U.S. Gulf Coast is in Hotez and Jackson Lee, 2017.

5. Data from Connecticut are in Chorazy and Richardson, 2005. The New York City study is in Tyungu et al., 2020. The global prevalence of toxocariasis in humans, dogs, and cats is covered in Rostami et al., 2019; Rostami et al., 2020a; and Rostami et al., 2020b.

6. Covert toxocariasis is described in Sharghi et al., 2000; and Hotez and Wilkins, 2009. Specific information about toxocariasis and neurological illness, including developmental delays, epilepsy, and neurodegenerative disease, is discussed in Hotez, 2014b; Fan et al., 2015; and Luna et al., 2018.

7. Toxocariasis infection rates during the 1970s are found in Hermann et al., 1985. The results of the toxocariasis studies from the Third National Health and Nutrition Examination Survey are found in Won et al., 2008. A summary of these studies is provided in Hotez and Wilkins, 2009. A more recent update is in Liu et al., 2018.

8. The Connecticut study is reported in Sharghi et al., 2001.

9. The New York study is reported in Marmor et al., 1987.

10. Busse and Mitchell, 2007.

11. Information on the relationship between toxocariasis and asthma is found in Buijs et al., 1997; and Kuk et al., 2006.

12. Hotez, 2008b.

13. Data are found in Siddiqui and Berk, 2001.

14. Centers for Disease Control and Prevention, 2002. A more recent modeling exercise by Singer and Sarkar (2020) indicates that strongyloidiasis may be even more widespread in the United States.

15. del la Garza et al., 2005.

16. White and Atmar, 2002.

17. Wallin and Kurtzke, 2004.

18. DeGiorgio et al., 2005a; and DeGiorgio et al., 2005b. Cysticercosis is also an important cause of death, responsible for 221 deaths between 1990 and 2002 (Sorvillo et al., 2007). A recent estimate for the incidence of cysticercosis among Hispanics in Oregon is found in O'Neal et al., 2011. Hospital expenses in the United States incurred as a result of cysticercosis are reported in Neal and Flecker, 2015. Information about mass or preventive treatment for taeniasis and cysticercosis is in Haby et al., 2020.

19. Information about animal sources of toxoplasmosis in the United States is in Guo et al., 2016. The health disparity aspects are found in Jones et al., 2007; and Stebbins et al., 2019.

20. Data on the U.S. disease burden of congenital toxoplasmosis are found in Lopez et al., 2000.

21. McLeod et al., 2006.

22. Kim, 2006. Information about toxoplasmosis newborn screening is in Peyron et al., 2017.

23. Information is found in Bern et al., 2011; Sarkar et al., 2010; Hanford et al., 2007; and Manne-Goehler et al., 2016. Information about Chagas disease transmission in Texas and elsewhere in the United States is in Gunter et al., 2017; and Gunter et al., 2020. Information about mother-to-child transmission in the United States is in Edwards et al., 2019.

24. Information about leishmaniasis in Texas and elsewhere can be found in Maloney et al., 2002; Jacobson, 2007; Willard et al., 2005; Weina et al., 2004; Enserink, 2000; Douvoyiannis et al., 2014; and Kipp et al., 2020.

25. Information about trachoma on the Navajo Reservation is in Rearwin et al., 1997; and Ludlam, 1978.

26. The impacts of dengue, WNV, Zika virus, relapsing fever, and murine typhus in the United States are summarized in Bouri et al., 2012; Brunkard et al., 2007; Gubler et al., 2001; Meyer et al., 2007; Reeves et al., 2008; Murray et al., 2013; Murray et al., 2017; Murray et al., 2018; and Bissett et al., 2018. The confluence of factors making Texas vulnerable to NTDs are summarized in Hotez, 2018.

27. Proulx et al., 2002. Information about NTDs in the Arctic is summarized in Hotez, 2010. Information about the impact of climate change is at https://www.wired.com/story/by-2080-tropical-diseases-could-be-headed-to-alaska/.

28. The clinical syndrome caused by *T. s. nativa* was described first by J. D. MacLean and his associates at the Centre for Tropical Diseases of McGill University; see MacLean et al., 1992; and MacLean et al., 1989. A tribute to the life of J. D. MacLean can be found in Hotez, 2010.

29. McDonald et al., 1990.

30. A study of the seroprevalence of the major zoonoses of Nunavik was conducted by Valerie Messier; see www.arcticnet.ulaval.ca/pdf/posters_2005/messier_et_%20al.pdf.

31. Rausch, 2003.

32. Wilson et al., 1995.

33. Information about onchocerciasis in Mexico is in Fernandez et al., 2020.

34. Information about hookworm and other soil-transmitted helminth infections in Mexico is from Brentlinger et al., 2003; and Quihui-Cota, et al., 2004.

35. Information about CL in Chiapas State as well as in the Yucatan is found in Flisser et al., 2002; Andrade-Narváez et al., 2001; and Rebollar-Tellez et al., 2005.

36. An interesting debate about the extent of Chagas disease in Mexico is in the following *Lancet* correspondence: Attaran, 2006; and Tapia Conyer, 2006. Information about the figure of 2 million to 6 million Chagas disease cases is provided in Hotez et al., 2012. Information about Chagas disease in pregnant women and newborns is provided in Buekens et al., 2008.

37. Tapia-Conyer et al., 2012.

38. Information is from http://www.onecaribbean.org/content/files/2004uSmarketdata.pdf.

39. Information on the prevalence of NTDs in the Caribbean is from the following sources. For schistosomiasis, see Chitsulo et al., 2000. This reference cites 15,000 cases of schistosomiasis in Puerto Rico; however, a more recent reference (Hillyer, 2005) suggests that schistosomiasis is close to being eliminated on the island. For LF, two sources are available: Global Programme to Eliminate Lymphatic Filariasis, 2006; and Pan American Health Organization/World Health Organization, 2000. For dengue, see https://www.paho.org/en/news/23-11-2020-caribbean-told-unite-control-dengue-during-covid-19-pandemic. Much of this information is summarized in Hotez, 2008a.

40. Information on the prevalence of soil-transmitted helminth infections in the Caribbean is summarized in Ehrenberg, 2002.

41. Information about the link between NTDs and slavery from Africa is found in Lammie et al., 2007.

References

Andrade-Narváez FJ, Vargas-González A, Canto-Lara SB, Damián-Centeno AG. 2001. Clinical picture of cutaneous leishmaniases due to *Leishmania (Leishmania) mexicana* in the Yucatan peninsula, Mexico. *Mem Inst Oswaldo Cruz* **96:**163–167.

Arnold JH. 1949. The public health problems of hookworm disease in South Carolina. *J S C Med Assoc* **45:**367–369.

Attaran A. 2006. Chagas' disease in Mexico. *Lancet* **368:**1768, discussion 1768–1769.

Bern C, Kjos S, Yabsley MJ, Montgomery SP. 2011. *Trypanosoma cruzi* and Chagas' disease in the United States. *Clin Microbiol Rev* **24:**655–681.

Bissett JD, Ledet S, Krishnavajhala A, Armstrong BA, Klioueva A, Sexton C, Replogle A, Schriefer ME, Lopez JE. 2018. Detection of tickborne relapsing fever spirochete, Austin, Texas, USA. *Emerg Infect Dis* **24:**2003–2009.

Bleakley H. 2007. Disease and development: evidence from hookworm eradication in the American South. *Q J Econ* **122:**73–117.

Bouri N, Sell TK, Franco C, Adalja AA, Henderson DA, Hynes NA. 2012. Return of epidemic dengue in the United States: implications for the public health practitioner. *Public Health Rep* **127:**259–266.

Brentlinger PE, Capps L, Denson M. 2003. Hookworm infection and anemia in adult women in rural Chiapas, Mexico. *Salud Publica Mex* **45:**117–119.

Brunkard JM, Robles López JL, Ramirez J, Cifuentes E, Rothenberg SJ, Hunsperger EA, Moore CG, Brussolo RM, Villarreal NA, Haddad BM. 2007. Dengue fever seroprevalence and risk factors, Texas-Mexico border, 2004. *Emerg Infect Dis* **13:**1477–1483.

Buekens P, Almendares O, Carlier Y, Dumonteil E, Eberhard M, Gamboa-Leon R, James M, Padilla N, Wesson D, Xiong X. 2008. Mother-to-child transmission of Chagas' disease in North America: why don't we do more? *Matern Child Health J* **12:**283–286.

Buijs J, Borsboom G, Renting M, Hilgersom WJ, van Wieringen JC, Jansen G, Neijens J. 1997. Relationship between allergic manifestations and *Toxocara* seropositivity: a cross-sectional study among elementary school children. *Eur Respir J* **10:**1467–1475.

Busse WW, Mitchell H. 2007. Addressing issues of asthma in inner-city children. *J Allergy Clin Immunol* **119:**43–49.

Centers for Disease Control and Prevention (CDC). 2002. Cancer death rates—Appalachia, 1994-1998. *MMWR Morb Mortal Wkly Rep* **51:**527–529.

Chitsulo L, Engels D, Montresor A, Savioli L. 2000. The global status of schistosomiasis and its control. *Acta Trop* **77:**41–51.

Chorazy ML, Richardson DJ. 2005. A survey of environmental contamination with ascarid ova, Wallingford, Connecticut. *Vector Borne Zoonotic Dis* **5:**33–39.

DeGiorgio C, Pietsch-Escueta S, Tsang V, Corral-Leyva G, Ng L, Medina MT, Astudillo S, Padilla N, Leyva P, Martinez L, Noh J, Levine M, del Villasenor R, Sorvillo F. 2005a. Seroprevalence of *Taenia solium* cysticercosis and *Taenia solium* taeniasis in California, USA. *Acta Neurol Scand* **111:**84–88.

DeGiorgio CM, Sorvillo F, Escueta SP, Wallin MT, Kurtzke JF. 2005b. Neurocysticercosis in the United States: review of an important emerging infection. *Neurology* **64:**1486, author reply 1486.

del la Garza Y, Graviss EA, Daver NG, Gambarin KJ, Shandera WX, Schantz PM, White AC Jr. 2005. Epidemiology of neurocysticercosis in Houston, Texas. *Am J Trop Med Hyg* **73:**766–770.

DiSalvo AF, Melonas J. 1970. Intestinal parasites in South Carolina, 1969. *J S C Med Assoc* **66:**355–358.

Douvoyiannis M, Khromachou T, Byers N, Hargreaves J, Murray HW. 2014. Cutaneous leishmaniasis in North Dakota. *Clin Infect Dis* **59:**e73–e75.

Edwards MS, Stimpert KK, Bialek SR, Montgomery SP. 2019. Evaluation and management of congenital Chagas disease in the United States. *J Pediatric Infect Dis Soc* **8:**461–469.

Ehrenberg JP. 2002. *An Epidemiological Overview of Geohelminth and Schistosomiasis in the Caribbean.* Pan American Health Organization, Washington, DC.

Enserink M. 2000. Infectious diseases. Has leishmaniasis become endemic in the U.S.? *Science* **290:**1881–1883.

Fan CK, Holland CV, Loxton K, Barghouth U. 2015. Cerebral toxocariasis: silent progression to neurodegenerative disorders? *Clin Microbiol Rev* **28:**663–686.

Farmer HF Jr. 1983. The germ of laziness: a Florida historical perspective. *J Fla Med Assoc* **70:**659–662.

Fernández-Santos NA, Unnasch TR, Rodríguez-Luna IC, Prado-Velasco FG, Adeniran AA, Martínez-Montoya H, Rodríguez-Pérez MA. 2020. Post-elimination surveillance in formerly onchocerciasis endemic focus in Southern Mexico. *PLoS Negl Trop Dis* **14:**e0008008.

Flisser A, Velasco-Villa A, Martínez-Campos C, González-Domínguez F, Briseño-García B, García-Suárez R, Caballero-Servín A, Hernández-Monroy I, García-Lozano H, Gutiérrez-Cogco L, Rodríguez-Angeles G, López-Martínez I, Galindo-Virgen S, Vázquez-Campuzano R, Balandrano-Campos S, Guzmán-Bracho C, Olivo-Díaz A, de la Rosa J, Magos C, Escobar-Gutiérrez A, Correa D. 2002. Infectious diseases in Mexico. A survey from 1995-2000. *Arch Med Res* **33:**343–350.

Global Programme to Eliminate Lymphatic Filariasis. 2006. Global Programme to Eliminate Lymphatic Filariasis. *Wkly Epidemiol Rec* **81:**221–232.

Gubler DJ, Reiter P, Ebi KL, Yap W, Nasci R, Patz JA. 2001. Climate variability and change in the United States: potential impacts on vector- and rodent-borne diseases. *Environ Health Perspect* **109**(Suppl 2):223–233.

Gunter SM, Murray KO, Gorchakov R, Beddard R, Rossmann SN, Montgomery SP, Rivera H, Brown EL, Aguilar D, Widman LE, Garcia MN. 2017. Likely autochthonous transmission of *Trypanosoma cruzi* to humans, south central Texas, USA. *Emerg Infect Dis* **23**:500–503.

Gunter SM, Ronca SE, Sandoval M, Coffman K, Leining L, Gorchakov R, Murray KO, Nolan MS. 2020. Chagas disease infection prevalence and vector exposure in a high-risk population of Texas hunters. *Am J Trop Med Hyg* **102**:294–297.

Guo M, Mishra A, Buchanan RL, Dubey JP, Hill DE, Gamble HR, Jones JL, Pradhan AK. 2016. A systematic meta-analysis of *Toxoplasma gondii* prevalence in food animals in the United States. *Foodborne Pathog Dis* **13**:109–118.

Haby MM, Sosa Leon LA, Luciañez A, Nicholls RS, Reveiz L, Donadeu M. 2020. Systematic review of the effectiveness of selected drugs for preventive chemotherapy for *Taenia solium* taeniasis. *PLoS Negl Trop Dis* **14**:e0007873.

Hanford EJ, Zhan FB, Lu Y, Giordano A. 2007. Chagas disease in Texas: recognizing the significance and implications of evidence in the literature. *Soc Sci Med* **65**:60–79.

Healy GR, Gleason NN, Bokat R, Pond H, Roper M. 1969. Prevalence of ascariasis and amebiasis in Cherokee Indian school children. *Public Health Rep* **84**:907–914.

Henderson HE. 1957. Incidence and intensity of hookworm infestation in certain East Texas counties with comparison of technics. *Tex Rep Biol Med* **15**:283–291.

Herrmann N, Glickman LT, Schantz PM, Weston MG, Domanski LM. 1985. Seroprevalence of zoonotic toxocariasis in the United States: 1971-1973. *Am J Epidemiol* **122**:890–896.

Hillyer GV. 2005. The rise and fall of Bilharzia in Puerto Rico: its centennial 1904-2004. *P R Health Sci J* **24**:225–235.

Hotez PJ. 2008a. Holidays in the sun and the Caribbean's forgotten burden of neglected tropical diseases. *PLoS Negl Trop Dis* **2**:e239.

Hotez PJ. 2008b. Neglected infections of poverty in the United States of America. *PLoS Negl Trop Dis* **2**:e256.

Hotez PJ. 2010. Neglected infections of poverty among the indigenous peoples of the Arctic. *PLoS Negl Trop Dis* **4**:e606.

Hotez PJ. 2014a. Neglected parasitic infections and poverty in the United States. *PLoS Negl Trop Dis* **8**:e3012.

Hotez PJ. 2014b. Neglected infections of poverty in the United States and their effects on the brain. *JAMA Psychiatry* **71**:1099–1100.

Hotez PJ. 2018. The rise of neglected tropical diseases in the "new Texas". *PLoS Negl Trop Dis* **12**:e0005581.

Hotez PJ, Bottazzi ME, Dumonteil E, Valenzuela JG, Kamhawi S, Ortega J, Rosales SP, Cravioto MB, Tapia-Conyer R. 2012. Texas and Mexico: sharing a legacy of poverty and neglected tropical diseases. *PLoS Negl Trop Dis* **6**:e1497.

Hotez PJ, Gurwith M. 2011. Europe's neglected infections of poverty. *Int J Infect Dis* **15**:e611–e619.

Hotez PJ, Jackson Lee S. 2017. US Gulf Coast states: the rise of neglected tropical diseases in "flyover nation". *PLoS Negl Trop Dis* **11**:e0005744.

Hotez PJ, Wilkins PP. 2009. Toxocariasis: America's most common neglected infection of poverty and a helminthiasis of global importance? *PLoS Negl Trop Dis* **3**:e400.

Humphreys M. 2001. *Malaria: Poverty, Race, and Public Health in the United States.* The Johns Hopkins University Press, Baltimore, MD.

Jacobson S. 2007. Rare, non-fatal skin disease found in N. Texans. *The Dallas Morning News* 2007(September 14).

Jones JL, Kruzson-Moran D, Sanders-Lewis K, Wilson M. 2007. *Toxoplasma gondii* infection in the United States, 1999 2004, decline from the prior decade. *Am J Trop Med Hyg* **77**:405–410.

Kim K. 2006. Time to screen for congenital toxoplasmosis? *Clin Infect Dis* **42**:1395–1397.

Kipp EJ, de Almeida M, Marcet PL, Bradbury RS, Benedict TK, Lin W, Dotson EM, Hergert M. 2020. An atypical case of autochthonous cutaneous leishmaniasis associated with naturally infected phlebotomine sand flies in Texas, United States. *Am J Trop Med Hyg* **103**:1496–1501.

Kuk S, Özel E, Oğuztürk H, Kirkil G, Kaplan M. 2006. Seroprevalence of *Toxocara* antibodies in patients with adult asthma. *South Med J* **99**:719–722.

Lammie PJ, Lindo JF, Secor WE, Vasquez J, Ault SK, Eberhard ML. 2007. Eliminating lymphatic filariasis, onchocerciasis, and schistosomiasis from the Americas: breaking a historical legacy of slavery. *PLoS Negl Trop Dis* **1**:e71.

Liu EW, Chastain HM, Shin SH, Wiegand RE, Kruzson-Moran D, Handali S, Jones JL. 2018. Seroprevalence of antibodies to *Toxocara* species in the United States and associated risk factors 2011–2014. *Clin Infect Dis* **66**:206–212.

Lopez A, Dietz VJ, Wilson M, Navin TR, Jones JL. 2000. Preventing congenital toxoplasmosis. *MMWR Recomm Rep* **49**(RR02):59–68.

Ludlam JA. 1978. Prevalence of trachoma among Navajo Indian children. *Am J Optom Physiol Opt* **55**:116–118.

Luna J, Cicero CE, Rateau G, Quattrocchi G, Marin B, Bruno E, Dalmay F, Druet-Cabanac M, Nicoletti A, Preux PM. 2018. Updated evidence of the association between toxocariasis and epilepsy: systematic review and meta-analysis. *PLoS Negl Trop Dis* **12**:e0006665.

MacLean JD, Poirier L, Gyorkos TW, Proulx JF, Bourgeault J, Corriveau A, Illisituk S, Staudt M. 1992. Epidemiologic and serologic definition of primary and secondary trichinosis in the Arctic. *J Infect Dis* **165**:908–912.

MacLean JD, Viallet J, Law C, Staudt M. 1989. Trichinosis in the Canadian Arctic: report of five outbreaks and a new clinical syndrome. *J Infect Dis* **160**:513–520.

Maloney DM, Maloney JE, Dotson D, Popov VL, Sanchez RL. 2002. Cutaneous leishmaniasis: Texas case diagnosed by electron microscopy. *J Am Acad Dermatol* **47**:614–616.

Manne-Goehler J, Umeh CA, Montgomery SP, Wirtz VJ. 2016. Estimating the burden of Chagas disease in the United States. *PLoS Negl Trop Dis* **10**:e0005033.

Marmor M, Glickman L, Shofer F, Faich LA, Rosenberg C, Cornblatt B, Friedman S. 1987. *Toxocara canis* infection of children: epidemiologic and neuropsychologic findings. *Am J Public Health* **77**:554–559.

Martin LK. 1972. Hookworm in Georgia. I. Survey of intestinal helminth infections and anemia in rural school children. *Am J Trop Med Hyg* **21**:919–929.

Martin MG, Humphreys ME. 2006. Social consequence of disease in the American South, 1900-World War II. *South Med J* **99**:862–864.

McDonald JC, Gyorkos TW, Alberton B, MacLean JD, Richer G, Juranek D. 1990. An outbreak of toxoplasmosis in pregnant women in northern Québec. *J Infect Dis* **161**:769–774.

McKenna ML, McAtee S, Bryan PE, Jeun R, Ward T, Kraus J, Bottazzi ME, Hotez PJ, Flowers CC, Mejia R. 2017. Human intestinal parasite burden and poor sanitation in rural Alabama. *Am J Trop Med Hyg* **97**:1623–1628.

McLeod R, Boyer K, Karrison T, Kasza K, Swisher C, Roizen N, Jalbrzikowski J, Remington J, Heydemann P, Noble AG, Mets M, Holfels E, Withers S, Latkany P, Meier P, Meier P, Toxoplasmosis Study Group. 2006. Outcome of treatment for congenital toxoplasmosis, 1981-2004: the National Collaborative Chicago-Based, Congenital Toxoplasmosis Study. *Clin Infect Dis* **42**:1383–1394.

Meyer TE, Bull LM, Cain Holmes K, Pascua RF, Travassos da Rosa A, Gutierrez CR, Corbin T, Woodward JL, Taylor JP, Tesh RB, Murray KO. 2007. West Nile virus infection among the homeless, Houston, Texas. *Emerg Infect Dis* **13**:1500–1503.

Murray KO, Evert N, Mayes B, Fonken E, Erickson T, Garcia MN, Sidwa T. 2017. Typhus group rickettsiosis, Texas, USA, 2003-2013. *Emerg Infect Dis* **23**:645–648.

Murray KO, Nolan MS, Ronca SE, Datta S, Govindarajan K, Narayana PA, Salazar L, Woods SP, Hasbun R. 2018. The neurocognitive and MRI outcomes of West Nile virus infection: preliminary analysis using an external control group. *Front Neurol* **9**:111.

Murray KO, Rodriguez LF, Herrington E, Kharat V, Vasilakis N, Walker C, Turner C, Khuwaja S, Arafat R, Weaver SC, Martinez D, Kilborn C, Bueno R, Reyna M. 2013. Identification of dengue fever cases in Houston, Texas, with evidence of autochthonous transmission between 2003 and 2005. *Vector Borne Zoonotic Dis* **13**:835–845.

Nair D. 2001. Screening for *Strongyloides* infection among the institutionalized mentally disabled. *J Am Board Fam Pract* **14**:51–53.

O'Neal SE, Flecker RH. 2015. Hospitalization frequency and charges for neurocysticercosis, United States, 2003-2012. *Emerg Infect Dis* **21**:969–976.

O'Neal S, Noh J, Wilkins P, Keene W, Lambert W, Anderson J, Compton Luman J, Townes J. 2011. *Taenia solium* tapeworm infection, Oregon, 2006-2009. *Emerg Infect Dis* **17**:1030–1036.

Ong S, Talan DA, Moran GJ, Mower W, Newdow M, Tsang VC, Pinner RW, EMERGEncy ID NET Study Group. 2002. Neurocysticercosis in radiographically imaged seizure patients in U.S. emergency departments. *Emerg Infect Dis* **8**:608–613.

Pan American Health Organization/World Health Organization. 2000. *Lymphatic Filariasis Elimination in the Americas: First Regional Program Managers Meeting, Dominican Republic, 9–11 August 2000.* Pan American Health Organization, Washington, DC.

Peyron F, Mc Leod R, Ajzenberg D, Contopoulos-Ioannidis D, Kieffer F, Mandelbrot L, Sibley LD, Pelloux H, Villena I, Wallon M, Montoya JG. 2017. Congenital toxoplasmosis in France and the United States: one parasite, two diverging approaches. *PLoS Negl Trop Dis* **11**:e0005222.

Proulx JF, MacLean JD, Gyorkos TW, Leclair D, Richter AK, Serhir B, Forbes L, Gajadhar AA. 2002. Novel prevention program for trichinellosis in Inuit communities. *Clin Infect Dis* **34**:1508–1514.

Quihui-Cota L, Valencia ME, Crompton DW, Phillips S, Hagan P, Diaz-Camacho SP, Triana Tejas A. 2004. Prevalence and intensity of intestinal parasitic infections in relation to nutritional status in Mexican schoolchildren. *Trans R Soc Trop Med Hyg* **98**:653–659.

Rausch RL. 2003. Cystic echinococcosis in the Arctic and Sub-Arctic. *Parasitology* **127**(Suppl):S73–S85.

Rearwin DT, Tang JH, Hughes JW. 1997. Causes of blindness among Navajo Indians: an update. *J Am Optom Assoc* **68**:511–517.

Rebollar-Téllez EA, Tun-Ku E, Manrique-Saide PC, Andrade-Narvaez FJ. 2005. Relative abundances of sandfly species (Diptera: Phlebotominae) in two villages in the same area of Campeche, in southern Mexico. *Ann Trop Med Parasitol* **99**:193–201.

Reeves WK, Murray KO, Meyer TE, Bull LM, Pascua RF, Holmes KC, Loftis AD. 2008. Serological evidence of typhus group rickettsia in a homeless population in Houston, Texas. *J Vector Ecol* **33**:205–207.

Robinson P, Garza A, Weinstock J, Serpa JA, Goodman JC, Eckols KT, Firozgary B, Tweardy DJ. 2012. Substance P causes seizures in neurocysticercosis. *PLoS Pathog* **8**:e1002489.

Rostami A, Riahi SM, Hofmann A, Ma G, Wang T, Behniafar H, Taghipour A, Fakhri Y, Spotin A, Chang BCH, Macpherson CNL, Hotez PJ, Gasser RB. 2020a. Global prevalence of *Toxocara* infection in dogs. *Adv Parasitol* **109**:561–583.

Rostami A, Riahi SM, Holland CV, Taghipour A, Khalili-Fomeshi M, Fakhri Y, Omrani VF, Hotez PJ, Gasser RB. 2019. Seroprevalence estimates for toxocariasis in people worldwide: A systematic review and meta-analysis. *PLoS Negl Trop Dis* **13**:e0007809.

Rostami A, Sepidarkish M, Ma G, Wang T, Ebrahimi M, Fakhri Y, Mirjalali H, Hofmann A, Macpherson CNL, Hotez PJ, Gasser RB. 2020b. Global prevalence of *Toxocara* infection in cats. *Adv Parasitol* **109**:615–639.

Sargent RG, Dudley BW, Fox AS, Lease EJ. 1972. Intestinal helminths in children in coastal South Carolina: a problem in southeastern United States. *South Med J* **65**:294–298.

Sarkar S, Strutz SE, Frank DM, Rivaldi CL, Sissel B, Sánchez-Cordero V. 2010. Chagas disease risk in Texas. *PLoS Negl Trop Dis* **4**:e836.

Schantz PM, Tsang VC. 2003. The US Centers for Disease Control and Prevention (CDC) and research and control of cysticercosis. *Acta Trop* **87**:161–163.

Serpa JA, Graviss EA, Kass JS, White AC Jr. 2011. Neurocysticercosis in Houston, Texas: an update. *Medicine (Baltimore)* **90**:81–86.

Sharghi N, Schantz PM, Caramico L, Ballas K, Teague BA, Hotez PJ. 2001. Environmental exposure to *Toxocara* as a possible risk factor for asthma: a clinic-based case-control study. *Clin Infect Dis* **32**:E111–E116.

Sharghi N, Schantz P, Hotez PJ. 2000. Toxocariasis: an occult cause of childhood neuropsychological deficits and asthma? *Semin Pediatr Infect Dis* **11**:257–260.

Siddiqui AA, Berk SL. 2001. Diagnosis of *Strongyloides stercoralis* infection. *Clin Infect Dis* **33**:1040–1047.

Singer R, Sarkar S. 2020. Modeling strongyloidiasis risk in the United States. *Int J Infect Dis* **100**:366–372.

Sorvillo FJ, DeGiorgio C, Waterman SH. 2007. Deaths from cysticercosis, United States. *Emerg Infect Dis* **13**:230–235.

Starr MC, Montgomery SP. 2011. Soil-transmitted helminthiasis in the United States: a systematic review—1940-2010. *Am J Trop Med Hyg* **85**:680–684.

Stebbins RC, Noppert GA, Aiello AE, Cordoba E, Ward JB, Feinstein L. 2019. Persistent socioeconomic and racial and ethnic disparities in pathogen burden in the United States, 1999-2014. *Epidemiol Infect* **147**:e301.

Tapia Conyer R. 2006. Chagas' disease in Mexico—response from the Mexican Ministry of Health. *Lancet* **368**:1768–1769.

Tapia-Conyer R, Betancourt-Cravioto M, Méndez-Galván J. 2012. Dengue: an escalating public health problem in Latin America. *Paediatr Int Child Health* **32**(Suppl 1):14–17.

Tyungu DL, McCormick D, Lau CL, Chang M, Murphy JR, Hotez PJ, Mejia R, Pollack H. 2020. *Toxocara* species environmental contamination of public spaces in New York City. *PLoS Negl Trop Dis* **14**:e0008249.

Wallin MT, Kurtzke JF. 2004. Neurocysticercosis in the United States: review of an important emerging infection. *Neurology* **63**:1559–1564.

Weina PJ, Neafie RC, Wortmann G, Polhemus M, Aronson NE. 2004. Old world leishmaniasis: an emerging infection among deployed US military and civilian workers. *Clin Infect Dis* **39**:1674–1680.

White AC Jr, Atmar RL. 2002. Infections in Hispanic immigrants. *Clin Infect Dis* **34**:1627–1632.

Willard RJ, Jeffcoat AM, Benson PM, Walsh DS. 2005. Cutaneous leishmaniasis in soldiers from Fort Campbell, Kentucky returning from Operation Iraqi Freedom highlights diagnostic and therapeutic options. *J Am Acad Dermatol* **52**:977–987.

Wilson JF, Rausch RL, Wilson FR. 1995. Alveolar hydatid disease. Review of the surgical experience in 42 cases of active disease among Alaskan Eskimos. *Ann Surg* **221**:315–323.

Won KY, Kruzson-Moran D, Schantz PM, Jones JL. 2008. National seroprevalence and risk factors for zoonotic *Toxocara* spp. infection. *Am J Trop Med Hyg* **79**:552–557.

10

Uniting to Combat Neglected Tropical Diseases, and a New WHO Roadmap (2021–2030)

And it's not just the story about one billion people who are afflicted with disabling, oftentimes stigmatizing, neglected tropical diseases, such as human hookworm infection and elephantiasis . . . it's all about the faces of dying children and sick mothers who haunt those who have seen them.

FORMER PRESIDENT BILL CLINTON

Comprehensive, Africa-wide control of malaria and NTDs together would probably cost no more than $3 billion a year, or just two days of Pentagon spending. If each of the billion people in the rich world devoted the equivalent of one $3 coffee a year to the cause, several million children every year would be spared death and debility, and the world would be spared the grave risks when disease and despair run unchecked. A new Global Network for Neglected Tropical Disease Control is helping make this opportunity a reality.

JEFFREY SACHS

My experience has taught me that no movement ever stops or languishes for want of funds. This does not mean that any movement can go on without money, but it does mean that wherever it has good men and true at its helm, it is bound to attract to itself the requisite funds.

MAHATMA GANDHI

The neglected tropical diseases (NTDs) are among the leading disabling conditions of humankind. Because of their disfiguring clinical manifestations and their devastating impact on child development, pregnancy outcome, and productive capacity, the core group of 20 NTDs is responsible for a huge disease burden in the low-income countries of Africa, Asia, and the Americas. The NTDs result in global disabilities almost equivalent to those from malaria, tuberculosis, or HIV/AIDS.[1] Updated estimates from the Global Burden of

Forgotten People, Forgotten Diseases: The Neglected Tropical Diseases and Their Impact on Global Health and Development, Third Edition. Peter J. Hotez.
© 2022 American Society for Microbiology. DOI: 10.1128/9781683673903.ch10

BOX 10.1 NTDs Targeted by Preventive Chemotherapy, or Mass Drug Administration (and the Major Drugs Used for Treatment)

Original list of 7 NTDs (2005–2010)

- Three soil-transmitted helminth infections
  - Ascariasis (albendazole or mebendazole)
  - Trichuriasis (albendazole or mebendazole with ivermectin)
  - Hookworm infection (albendazole or mebendazole)
- Schistosomiasis (praziquantel)
- LF (ivermectin or moxidectin, DEC, albendazole)

- Onchocerciasis (ivermectin or moxidectin)
- Trachoma (azithromycin)

Expanded list of 10 NTDs: Same as above in addition to:

- Scabies (ivermectin or moxidectin)
- Taeniasis (albendazole or praziquantel)
- Yaws (azithromycin)

Disease Study (GBD) 2019 indicate that approximately 1 billion people are infected with at least one NTD, resulting in 25 million DALYs (disability-adjusted life years, i.e., the numbers of healthy life years lost from disability or premature death) and 100,000 to 200,000 deaths.[1]

More than one-half of the NTD global disease burden results from 10 of the most common conditions—ascariasis, trichuriasis, hookworm infection, schistosomiasis, taeniasis, scabies, lymphatic filariasis (LF), trachoma, onchocerciasis, and yaws. These 10 NTDs are the most common infections of poor people in developing countries, occurring in almost all of the "bottom 750 million" who live in extreme poverty. They are also each targets of preventive chemotherapy using a package of medicines (Box 10.1).

Table 10.1 ranks these 10 NTDs by their impact on DALYs, revealing that scabies and hookworm and the other soil-transmitted helminth (STH) infections top the list, followed by schistosomiasis, LF, taeniasis, onchocerciasis, and trachoma. Such DALY estimates may not fully consider all the chronic effects of the NTDs.[2] Examples of these are the mental health and cognitive effects of the NTDs, the long-standing organ damage and growth and developmental delays they cause, and their effects on pregnancy.[3] Some investigators find that as we incorporate these elements our current estimates of disease burden will increase even further.[2]

In addition to their disease burden, the NTDs place an enormous added economic burden on the low-income countries through their impact on child cognition, memory, school performance, and school attendance, as well as on worker productivity and the specific impact on the health of girls and women. Although we are still in the nascent stages of quantifying this economic impact, the information to date suggests that the NTDs represent some of the most important poverty-promoting conditions in developing regions.[3] Along similar lines are the enormous adverse psychological effects stemming from the profound stigma from disfigurement associated with many of these conditions.[3]

An important theme throughout this book has been the concept of preventive chemotherapy, also known as mass drug administration (MDA).

Table 10.1 Ranking of the 10 most prevalent NTDs (targeted by preventive chemotherapy) by DALYs[a]

Rank	Disease	Global disease burden: no. of DALYs	Major drug Rx
1	Scabies	4.84 million	Ivermectin
2	STH infections	1.97 million	Albendazole and ivermectin
3	STH infection: hook-worm disease	0.98 million	
4	STH infection: ascariasis	0.75 million	
5	STH infection: trichuriasis	0.24 million	
6	Schistosomiasis	1.64 million	Praziquantel
7	Lymphatic filariasis	1.63 million	Ivermectin, DEC, and albendazole
8	Taeniasis	1.37 million	Albendazole or praziquantel
9	Onchocerciasis	1.23 million	Ivermectin or moxidectin
10	Trachoma (and yaws)	0.81 million (not determined)	Azithromycin
	Total NTDs targeted by preventive chemotherapy	13.5 million	"Rapid-impact package"
	Total NTDs including venomous animal contact	25.4 million	

[a] Data from GBD 2019 (Global Burden of Disease Collaborative Network, 2020).

We have seen how the simultaneous administration of NTD drugs to large populations over a defined period can produce dramatic health gains, as well as important educational and economic benefits. Proof that this concept would work on large populations was pioneered by the Chinese beginning shortly after the Cultural Revolution, when a part of the nation's salt supply was medicated with diethylcarbamazine (DEC). As a result, by the early 1990s, LF was largely eliminated as a public health problem in the People's Republic of China. Similarly, MDA with DEC or ivermectin was responsible for the elimination of LF from Egypt, many of the islands of the South Pacific, and more than 20 other countries worldwide. MDA has also been effective against other NTDs. MDA with praziquantel has helped to control schistosomiasis in Egypt; MDA with ivermectin has controlled onchocerciasis in West Africa, including elimination of the disease in hyperendemic areas of Mali and Senegal; and MDA, as part of the SAFE strategy (described in chapter 5), has eliminated trachoma as a public health problem in Morocco and now many other countries. Therefore, almost anyone would agree that preventive chemotherapy has significantly cut the global burden of the 10 most prevalent NTDs and has resulted in some of the most extraordinary public health victories in the developing world over the last 3 decades.

In recognition of these public health gains, the World Health Assembly (WHA) has adopted a number of ambitious resolutions over the years to promote the use of MDA for NTDs. Composed of the ministers of health from all 192 member states, the WHA appoints the director-general of the World

Health Organization (WHO) and is the WHO's major decision-making and health policy-setting body. Of relevance to the seven most prevalent NTDs, the WHA has adopted five major resolutions involving MDA.[4] They include the elimination of both LF and trachoma as a public health problem, the establishment of community-based sustainable yearly treatments for onchocerciasis in areas with moderate or high intensity, the regular and periodic treatment of at least 75% of school-age children at risk for STH infections and schistosomiasis, and the establishment of the feasibility of schistosomiasis elimination. Additional WHA targets are in place for multiple-drug treatments for leprosy and for public health intervention efforts against guinea worm, Chagas disease, and human African trypanosomiasis. In 2013, the WHA adopted resolution WHA66.12 as an overarching call for the NTDs, especially through preventive chemotherapy.[4]

To address these WHA targets, several very important international public-private partnerships have been established, each with a commitment to providing technical assistance on NTDs to health ministries in countries in which the diseases are endemic. Table 10.2 gives a summary of the major MDA programs for 8 of the 10 most prevalent NTDs.[5] Working with the WHO, these major partnerships are currently helping to enable access to essential NTD medicines for the poorest people in low-income countries. The major organizations include:

- *Schistosomiasis Control Initiative* (SCI) (https://schistosomiasiscontrolinitiative.org/). London-based SCI also promotes the sustainable control of schistosomiasis and of some STH infections in sub-Saharan Africa by working with health ministries and national control programs to deliver praziquantel and benzimidazole anthelmintics to high-risk groups, especially school-age children.

- *Global Alliance to Eliminate Lymphatic Filariasis* (GAELF) (www.gaelf.org). GAELF coordinates the activities of partners working to deliver DEC, ivermectin, and albendazole, with a focus on political, financial, and technical support. The GAELF Secretariat is currently based at the Liverpool School of Tropical Medicine.

- *International Trachoma Initiative* (ITI) (www.trachoma.org). The ITI is dedicated to the elimination of blinding trachoma through implementation of the SAFE strategy, which includes MDA with azithromycin. The ITI was founded by Pfizer and the Edna McConnell Clark Foundation in 1998. The organization is based in Atlanta at the Task Force for Global Health.

- *Helen Keller International* (HKI) (www.hki.org). HKI was founded in 1915 and is one of the oldest nonprofit organizations dedicated to fighting and treating preventable malnutrition-caused blindness. Headquartered in New York City, HKI maintains large-scale programs to combat trachoma, onchocerciasis, and other causes of blindness. HKI is under the direction of Kathy Spahn.

Table 10.2 Summary of MDA programs for the most prevalent NTDs[a]

Disease	Major initiatives	No. of people treated in 2019	Drug(s) used	Frequency of treatment	Major target populations
STH infections (ascariasis, trichuriasis, and hookworm infection)	Schistosomiasis Control Initiative Children Without Worms Task Force for Global Health Deworm the World Partnership for Child Development GlaxoSmithKline Johnson & Johnson	439 million	Mebendazole or albendazole	Once or twice/yr depending on prevalence or level of transmission	School-age children and some preschool children and women of reproductive age
Schistosomiasis	Schistosomiasis Control Initiative Global Schistosomiasis Alliance Merck KGaA	210 million	Praziquantel	Once/yr or every 2 yr, or twice during primary school depending on prevalence or level of transmission	School-age children
Lymphatic filariasis	Global Alliance to Eliminate Lymphatic Filariasis Task Force for Global Health Mectizan Donation Program Albendazole Donation Program	1.397 billion	DEC or ivermectin (+ albendazole)	Once/yr	Entire at-risk population

Table continues on next page

Table 10.2 (*continued*)

Disease	Major initiatives	No. of people treated in 2019	Drug(s) used	Frequency of treatment	Major target populations
Onchocerciasis	ESPEN Helen Keller International Carter Center Onchocerciasis Elimination Program for the Americas Sightsavers CBM Task Force for Global Health Mectizan Donation Program	302 million	Ivermectin or moxidectin	1–2 times /yr	Entire at-risk population
Trachoma	International Trachoma Initiative of the Task Force for Global Health WHO Alliance for the Elimination of Trachoma Helen Keller International Carter Center Sightsavers CBM Pfizer Inc.	230 million	Azithromycin	Once/yr	Entire at-risk population ≥6 mo of age
Yaws	EMS SA Pharma	0.3 million	Azithromycin	Once/yr	To be determined

a Reported in World Health Organization, 2020.

- *Expanded Special Programme for Elimination of Neglected Diseases* (ESPEN) (www.espen.afro.who.int). Formerly the African Programme for Onchocerciasis Control (APOC), this program represents an expansion of APOC, founded originally during the 1990s to eliminate onchocerciasis. Since 2016 ESPEN has been focused on the major NTDs targeted by preventive chemotherapy in Africa. It is organized under the African Regional Office (AFRO) of the WHO. A cornerstone of the organization is CDTI (community-directed treatment with ivermectin) working through a network of hundreds of thousands of distributors.

- *Task Force for Global Health* (www.taskforce.org). The Task Force houses many of the drug donation programs that work with the partnerships outlined above. They include the Mectizan Donation Program (www.mectizan.org), linked with Merck; Children Without Worms, the donation program for mebendazole (Johnson & Johnson) for STH infections; the Lymphatic Filariasis Support Center; and the ITI. The Task Force is based in Decatur, GA.

- *The UK Coalition against Neglected Tropical Diseases* (www.ntdcoalition.org). The Coalition is a collaborative partnership between organizations in the United Kingdom including the Carter Centre UK, CBM UK, Footwork, LEPRA, Sightsavers, the Centre for Neglected Tropical Diseases of the Liverpool School of Tropical Medicine, London School of Hygiene and Tropical Medicine, Malaria Consortium, ORBIS, Partnership for Child Development, Schistosomiasis Control Initiative, Fred Hollows Foundation, Leprosy Mission England and Wales, and WaterAid.

- *Sightsavers* (www.sightsavers.org). Sightsavers has a mission to eliminate blindness and is a key component of the UK Coalition against Neglected Tropical Diseases. Caroline Harper is the chief executive.

- *Deworm the World-Evidence Action* (www.evidenceaction.org). The mission of Deworm the World is to expand school-based deworming programs; it has so far reached hundreds of millions of children. Deworm the World works closely in partnership with the Abdul Latif Jameel Poverty Action Lab (www.povertyactionlab.org) at Massachusetts Institute of Technology.

- *Partnership for Child Development* (PCD) (https://www.poverty-action.org/organization/partnership-child-development-pcd). The PCD is focused on school health and nutrition and also embraces school-based deworming as a cornerstone of its activities. The PCD is based at Imperial College, London.

- *Carter Center* (www.cartercenter.org). Founded in 1982 by former U.S. President Jimmy Carter and former First Lady Rosalynn Carter, in partnership with Emory University, the Atlanta-based center works to relieve human suffering in more than 70 countries. The Carter Center operates major programs in the control of multiple NTDs, including guinea worm infestation, onchocerciasis, and trachoma. Its success in guinea worm eradication was discussed earlier (in chapter 4).

- *CBM* (www.cbm.org). Originally known as Christian Blind Mission, CBM is an international Christian development organization that leads NTD control efforts in several African countries.

Through these partnerships, MDA has been scaled to an extraordinary level, now reaching more than 2 billion people when adding up the treatments in Table 10.2, including those who are either infected with NTDs or at high risk for acquiring NTDs. However, even this level of activity may not be sufficient to reach the highly ambitious targets for control and elimination set by the WHA. For instance, in 2020 the WHO estimated that we are reaching approximately 60% of the global population who actually require treatment.[5] Therefore, at our current rate of MDA, it could still be many more years before the world's poorest people gain full access to the essential medicines for the NTDs.

The hurdles to increasing the level of MDA by the WHO and the major NTD partnerships are formidable. As pointed out in the beginning of this book, the global health community, especially the Group of Seven (G7) nations, has not supported NTD control activities at nearly the same level as it has HIV/AIDS, malaria, and tuberculosis relief, through the Global Fund, President's Emergency Plan for AIDS Relief (PEPFAR), President's Malaria Initiative (PMI), and other initiatives. To date, the G7 support of NTD control can be measured US$100 million to $200 million range rather than the billions devoted to control of the "big three" diseases, and almost all of these funds are being provided by the United States and United Kingdom.[6]

However, lack of political will has not been the only hurdle to global NTD control. Given that the NTDs occur primarily in the rural (and often remote) areas of the developing world, there is a need to make MDA more accessible to these populations and possibly make it a more efficient enterprise. As we saw earlier (in chapter 5), probably more than any other organization, APOC during the 1990s and the first decade of the 2000s was effective at reaching vulnerable populations. By working through a system of community-based drug distributors, APOC was administering ivermectin to some of the most remote and vulnerable rural populations affected by river blindness. Given that APOC is already doing the "hard part," namely, obtaining access to Africa's hardest-to-reach populations, is there a rationale for APOC and similar organizations to administer not only ivermectin but other NTD drugs as well? Figure 10.1 shows a map of the world illustrating the distribution of the seven most prevalent NTDs (i.e., the three major STH infections plus LF, onchocerciasis, schistosomiasis, and trachoma) currently targeted by MDA. The map illustrates an aggregated distribution of these seven conditions. Thus, whereas none of the most common NTDs are public health problems in North America, Europe, and Central Asia, there are multiple NTDs occurring in the tropical developing world, particularly in sub-Saharan Africa, Southeast Asia, and tropical regions of the Americas. In fact, in many countries of these developing regions, it is common to find five, six, seven, or even more of the most highly prevalent

NTDs occurring in the same geographic region. In other words, the seven most common NTDs are coendemic in these regions. This was the rationale for expanding the remit of APOC to create ESPEN.

To understand better the implications of coendemic NTDs, let us take as an example the nation of Côte d'Ivoire (also known as the Ivory Coast) on the western coast of Africa. Roughly the size of New Mexico, Côte d'Ivoire is a low-income country with a per capita annual income of approximately US$1,600.[7] In addition to high levels of rural poverty and a high rate of infant and under-age-5 child mortality, Côte d'Ivoire has faced recent civil conflict

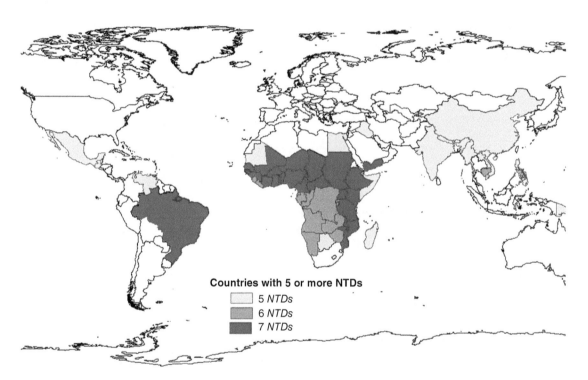

Figure 10.1 Global distribution of the NTDs in the decade of the 2000s. Countries with five, six, and seven NTDs are shown in yellow, orange, and red, respectively. (Map prepared by Sophia Raff based on information summarized in Hotez et al., 2007a.)

between the government holding power in the south and the rebels in the north.[7] In 2002, Giovanna Raso, Juerg Utzinger, and their colleagues at the Swiss Tropical Institute, together with scientists from Côte d'Ivoire's Université de Cocody in Abidjan, conducted a study among 75 randomly selected households in a single rural village where they obtained complete parasitological data on almost all the study participants. They found that three-quarters of the population harbored three or more parasites simultaneously, including hookworms, schistosomes, amebae, and malaria parasites.[8] Practically speaking, this finding means that in rural Côte d'Ivoire, as in many of the areas illustrated in red and orange in Fig. 10.1, people living in rural poverty are *polyparasitized*; i.e., they suffer from multiple NTDs at any given time. In a follow-up study among schoolchildren in Côte d'Ivoire, coinfections with hookworm and schistosomes were found to be particularly common, a finding subsequently confirmed in Brazil.[8] Indeed, the Brazilian studies suggest that infection with hookworm may predispose to schistosomiasis, so that there are synergistic effects as well as additive ones. Thus, investigations in many developing countries have shown that polyparasitism is common, not only because schoolchildren frequently have all three major STH infections (the unholy trinity of ascariasis, trichuriasis, and hookworm infection) but also because they have additional coinfections with schistosomiasis, LF, onchocerciasis, and even trachoma.

It is important to understand the full extent of NTD coinfections across sub-Saharan Africa and elsewhere. Accordingly, efforts are under way to employ satellite mapping and geographic information systems in order to overlay the prevalence of specific NTDs with maps of climate and vegetation as a means of predicting the global extent of coinfection and geographic overlap.[9] By folding in behavioral, demographic, and socioeconomic factors, it should become possible to develop complete risk maps for identifying large-scale patterns of potential overlap.[9] However, the evidence emerging to date already indicates that in the poorest rural areas of the developing world, individuals are, more frequently than not, simultaneously infected with multiple NTDs, typically 1 or more of the 10 most common NTDs.

The facts that the 10 NTDs being targeted for preventive chemotherapy cluster in people and that polyparasitism is common mean that it may be possible to bundle some of the NTD drugs together in a highly cost-effective MDA package. Beginning in 2003, those of us concerned about the plight of polyparasitized populations began to reassess the disease burden associated with the NTDs and opportunities for control. Through a series of meetings held in Europe and the United States, a group of NTD experts who headed partnerships involved in the control of specific NTDs, initially including David Molyneux (GAELF), Alan Fenwick (SCI), and myself, as well as Vinand Nantulya (Global Fund to Fight AIDS, Tuberculosis, and Malaria), Jacob Kumaresan (then at the ITI), Eric Ottesen (Task Force for Global Health), Frank Richards (Carter Center), Jeffrey and Sonia Ehrlich Sachs (Earth Institute at Columbia University), and Lorenzo Savioli (then-Director, Department of NTDs, WHO), discussed the situation. We identified the following important elements of polyparasitism and opportunities for integrated control.

1. The aggregate NTD health and economic burden is enormous, and the need for a global effort on these diseases is as urgent as the need for major HIV/AIDS and malaria initiatives such as the Global Fund and PEPFAR, to name a few.[10]
2. Because of the high degree of geographic overlap and coendemicity of the NTDs and the high prevalence of polyparasitism, the most prevalent NTDs—the three STH infections (ascariasis, hookworm

infection, and trichuriasis), scabies, schistosomiasis, LF, trachoma, and onchocerciasis—could be simultaneously targeted with a package of drugs comprising albendazole or mebendazole (targeting primarily the STHs), praziquantel (for schistosomiasis), ivermectin or DEC (for LF and onchocerciasis), and azithromycin (for trachoma).[11] The package of drugs for integrated NTD control has been named by some the "rapid-impact package" because the drugs can be quickly deployed by community-based drug distributors.[11] Safety data are in hand to support the coadministration of these drugs.[11]

3. Integrated NTD control through the rapid-impact package would be expected to have a number of health impacts, especially for children and women. Among them would be worm burden reductions and reductions in worm-associated anemia from hookworm infection and schistosomiasis; improvements in child growth and development, pregnancy outcome (both in terms of neonatal birth weight and maternal morbidity and mortality), and worker productivity; and prevention of blindness, chronic disability, and disfigurement.[12] In addition, new findings indicate that preventive chemotherapy produces important collateral benefits for additional helminth infections, including strongyloidiasis, taeniasis, and oesophagostomiasis. There are also effects on ectoparasitic skin infestations such as scabies and pediculosis (lice). The reduction of skin disease from both onchocerciasis and ectoparasites could also prevent secondary complications from cutaneous bacterial infections.[12] Still another benefit is the finding that yearly azithromycin treatments for trachoma control also had an impact on child mortality reduction, possibly through effects on respiratory or diarrheal pathogens.[12] Thus, the original list of 7 NTDs targeted by preventive chemotherapy was extended to 10 conditions.

4. The ability of integrated NTD control to reduce anemia is of particular importance. Hookworm infection, trichuriasis, and schistosomiasis all cause anemia, and an important clinical consequence of polyparasitism is enhancement of anemia, particularly in school-age children and pregnant women.[13] Anemia in school-age children results in decreased motor activity, social inattention, and decreased school performance and increases susceptibility to infection.[14] Anemia is also associated with increased maternal morbidity and low neonatal birth weight and accounts for 3.7 and 12.8% of maternal deaths during pregnancy and childbirth in Africa and Asia, respectively.[14] To date, studies have shown that hookworm burden reduction through deworming with albendazole improves childhood iron status and reduces anemia in pregnant women, resulting in reductions in the frequency of low birth weight and infant and maternal mortality.[15]

5. In addition to improving health, the integration of NTD control would improve child cognition and educational performance and attendance, as well as promote economic development in the poorest regions on Earth.[16]

6. Through integrated control with the rapid-impact package, health, educational, and economic improvements could be achieved at remarkably low cost. Ivermectin and azithromycin are donated free of charge and for as long as needed by Merck and Pfizer, respectively, while GlaxoSmithKline donates albendazole for LF and STH control, Johnson & Johnson donates the mebendazole required for the global control of STH infection, Eisai donates DEC, and Merck KGaA donates much of the praziquantel required for schistosomiasis control. In 2012, the major pharmaceutical companies reaffirmed or increased their commitment through the London Declaration on NTDs.[17] Therefore, the costs of purchase and delivery of the drugs comprising the rapid-impact package are extremely modest. Initial estimates suggested that the package could be administered for US$0.40 per person, with other estimates ranging up to US$0.79 per person.[18] A subsequent study by the Carter Center has shown that integrating NTD control with the rapid-impact package (albendazole, praziquantel, and ivermectin) can reduce the costs by 41% compared to the costs of MDA with each drug separately and that the package can be administered for far less than US$0.50 per person.[18] As shown in Fig. 10.2, the costs of administering the rapid-impact package are extremely small compared to the annual per-person costs of antiretroviral therapy for HIV/AIDS, direct observed therapy for tuberculosis, or even antimalarial drugs and bed nets.

On a macroeconomic scale, the low cost of integrated NTD control means that the health and education of entire populations can be improved for extremely modest sums of money. For instance, the entire at-risk population of rural sub-Saharan Africa (500 million people) could be treated annually

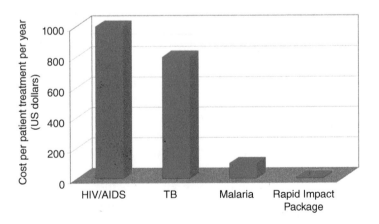

Figure 10.2 Range of treatment costs per person per year for HIV/AIDS, tuberculosis (TB), malaria, and the rapid-impact package for integrated NTD control. The rapid-impact package costs as little as US$0.50 per person for packaged intervention (Hotez et al., 2007a). (Modified from Molyneux DH, et al. 2005. "Rapid-impact interventions": how a policy of integrated control for Africa's neglected tropical diseases could benefit the poor. *PLoS Med* 2(11):e336. © 2005 Molyneux et al., CC BY 4.0.)

for between US$200 million and $400 million.[18] Therefore, for US$1 billion to $2 billion over a period of 5 years, up to 500 million at-risk people in sub-Saharan Africa could be blanketed with NTD drugs. Compared to an estimated US$3 billion annually for malaria control (which is still very reasonable),[19] the cost of NTD control is modest. Therefore, according to these estimates, integrated NTD control represents a "best buy" in public health. Harvard's David Canning has further pointed out that integrated NTD control also represents a priority investment in human capital.[20]

Administering the rapid-impact package requires recognition that not all groups should receive the same drugs and at the same time.[21] For instance, the drugs for soil-transmitted infections and schistosomiasis—albendazole, mebendazole, and praziquantel—are intended primarily for school-age (and some preschool) children, whereas ivermectin and DEC are administered to adult populations and to some children who have reached a specified age or height. However, through the Deworm3 Project (highlighted in chapter 2) there are efforts to extend albendazole or mebendazole treatment to adults. Usually, DEC is not administered at all for LF control in sub-Saharan Africa, because it is toxic for patients who are also infected with onchocerciasis, whereas ivermectin is safe for both LF and onchocerciasis in Africa. Therefore, DEC is primarily reserved for Asian LF and for LF in the Americas. In order to cover many of these contingencies and to tailor the rapid-impact package drugs to situations where some but not all the NTDs are endemic, the WHO has developed and issued a set of detailed guidelines and algorithms for administering the major NTD drugs to polyparasitized populations.[21]

Initially, in order to raise awareness about the NTDs and the promise of integrated NTD control, the Global Network for Neglected Tropical Diseases was established in 2006 to work with the health ministries in countries where the diseases are endemic, the major public-private partnerships committed to NTD control listed above, the WHO and its regional offices, and the World Bank and other international agencies. Its founding mission was to help those who suffer from and are at risk for the NTDs through effective collaboration, scaling up of existing disease interventions, and development of a new generation of improved control tools.[22] The vision is a world free of NTDs in which healthy people can develop fully, learn effectively, raise families, and be productive members of their communities.

Launched at the Clinton Global Initiative in 2006, the Global Network focuses its activities primarily on advocacy, resource mobilization, bilateral engagement (parliaments and ministries of health), and international development banks in order to achieve the overall goal of increasing access to essential medicines for the estimated 1.4 billion people affected by the major NTDs (estimate at that time) targeted through preventive chemotherapy. Doing so would improve childhood development, school attendance and performance, pregnancy outcomes, and worker productivity among the world's poorest populations. Moreover, reduction of the health, educational, and economic burden of the NTDs will directly address the Millennium Development Goals for sustainable poverty reduction in developing countries. To promote this advocacy agenda, the Global Network inaugurated a website devoted to NTDs, and articles about its activities have appeared in the print and electronic media, including a lead editorial by *New York Times* columnist Nicholas Kristof.[23] Actress Alyssa Milano also committed her time and energy (and has provided a financial commitment) to the Global Network by agreeing to serve as its inaugural global goodwill ambassador. In April 2012, His Excellency John A. Kufuor (president of the Republic of Ghana, 2001 to 2009) joined the Global Network as its NTD Special Envoy. Through this role, Mr. Kufuor served to elevate the fight against NTDs among development partners to encourage them to incorporate NTD control into existing global health or cross-sectoral development programs. The Global Network launched END7, an international campaign dedicated to raising public awareness and mobilizing the resources necessary to control and eliminate the most common NTDs. Since its launch, the campaign has engaged with global health partners including the WHO, corporate partners including prominent entertainment companies and artists, and leading actors, musicians, and activists to help raise awareness about NTDs and earn support from policymakers, governments, and the general public. The Global Network, the WHO, and the World Bank form a full and comprehensive partnership for the control and elimination of the major NTDs.

In association with the 2012 London Declaration for NTDs, national programs of integrated NTD control have begun in more than 20 African and Asian countries, as well as in Haiti (www.neglecteddiseases.gov). Funding for these early NTD integration efforts has come primarily from an annual congressional appropriation of approximately US$100 million administered through the NTD Program of the U.S. Agency for International Development (USAID), which operates through the contracting organizations RTI International and FHI 360. Additional large-scale support for NTD control is provided by the Department for International Development (DFID) of the United Kingdom. The END Fund is a philanthropic fund, initially administered by Geneva Global and the British investment concern Legatum, to support the effective delivery of preventive chemotherapy interventions that reduce the prevalence of the most common NTDs in sub-Saharan Africa. Currently the END Fund (www.end.org) is headquartered in New York City under the direction of Ellen Agler.[23] Since its start in 2012, the END Fund has treated more than 140 million people and mobilized more than US$100 million in funding, while training almost 1 million health workers. It also provides urgently required simple surgeries to those suffering from LF and blinding trachoma.[23] In parallel, the Carter Center is assisting in NTD control in Nigeria and elsewhere. These nascent efforts at integrated control are expected to produce dramatic gains for the poorest people in these countries. Monitoring and evaluation of these programs will include both measuring process indicators, including the number of people actually treated, and assessing reductions in NTD prevalence and worm intensity, improvements in anemia and other nutritional parameters, and catch-up growth in children. A high-level forum sponsored by the Global Network in 2006 revealed a number of exciting benefits from integrated NTD control, including the strengthening of community-based health systems and the African health workforce by promoting the activities of community drug distributors from ESPEN and other control programs. Because schoolteachers can be trained to administer deworming medicines, Africa's schools can also be developed as potent allies for building health systems there.[24] As NTD integration progresses, an important component of monitoring and evaluation will be a careful appraisal of its cost-effectiveness and efforts to determine how NTD control strengthens health systems.[24] Efforts to answer many of these operational questions are being funded by the Bill & Melinda Gates Foundation, leading to annual meetings for the Coalition for Operational Research for NTDs (COR-NTDs), headed initially by Julie Jacobson and more recently by Katey Owen.

Following the sunset of the Global Network in 2012, a new organization, Uniting to Combat Neglected Tropical Diseases, was formed specifically around London Declaration elimination targets and goals (https://unitingto combatntds.org/). Based in London, Uniting is a partnership that works to support the WHO in order to meet both the Sustainable Development Goals (SDGs) for health as they relate to the NTDs and the actual London Declaration targets. Central to this effort is a new 2030 roadmap released by the WHO in 2020 and led by Dr. Mwelecele Ntuli Malecela, Director of the WHO Department of Control of NTDs.[24] It emphasizes horizontal integration of programs directed at controlling NTDs, similar to the broadening of the remit of APOC to form ESPEN. The approach also works toward shifting ownership away from the G7 nations and nongovernmental development organizations in favor of local and national ownership in the countries in which the diseases are endemic. Endorsed by the 73rd WHA, the roadmap aspires to achieve specific 10-year milestones (2021 to 2030), highlighted in Box 10.2.

The early days of integrated NTD control revealed important challenges and hurdles.[24] They included (i) a requirement for improved or simplified strategies of surveillance and rapid disease mapping in order to identify the best algorithm for administering an optimal combination of NTD drugs; (ii) inadequate access to certain NTD drugs; (iii) inadequate data on patient compliance and on safe and efficacious administration of certain combinations of NTD drugs; and (iv) lack of consistent policies on data collection, management, and reporting, including the reporting of adverse events. We also have not looked carefully at the specter of possible emerging drug resistance, especially given the likelihood that use of the NTD drugs will become widespread in Africa and elsewhere. This possibility is discussed in more detail in chapter 11. Finally, we are facing the enormous hurdle of effecting integrated NTD control

---

BOX 10.2 Overarching Goals of the 2021–2030 WHO Roadmap for NTDs[24]

1. Reduce by 90% the number of people requiring treatment for NTDs.

2. Eliminate at least one NTD in 100 countries.

3. Eradicate two diseases (dracunculiasis and yaws).

4. Reduce by 75% the DALYs related to NTDs.

in Africa's fragile conflict-ridden and post-conflict nation-states. Because of long interruptions in public health control measures caused by the collapse of their health care infrastructures, countries such as Central African Republic, Chad, Côte d'Ivoire, the Democratic Republic of the Congo, northern Nigeria, and South Sudan have some of the world's highest NTD prevalence and intensity rates. This issue will be addressed in chapter 12.

Ultimately, as lessons are learned and if the major operational research questions are addressed, integrated control should, in time, become more efficient and streamlined. Assisting in this effort is the exciting possibility of expanding geographic information and remote sensing efforts in order to map large areas of sub-Saharan Africa for NTDs.[9] Still another exciting opportunity is the prospect of linking NTD control to the control of malaria and HIV/AIDS. For example, while working with community ivermectin drug distributors on LF and onchocerciasis control projects in central Nigeria, Frank Richards of the Carter Center and his colleagues from the Centers for Disease Control and Prevention observed that in areas where MDA was practiced, the ownership and use of antimalarial bed nets dramatically increased, as much as 9-fold.[25] Since insecticide-treated nets are considered a major tool in the fight against malaria, it is worth exploring whether there might be other entry points for integrated NTD control to embrace malaria control as a part of its mission or vice versa. Indeed, there is a high degree of geographic overlap and coendemicity of malaria with the major NTDs. For example, Fig. 10.3 shows a map of Africa demonstrating the geographic overlap between malaria and hookworm infection.[26] Studies by Simon Brooker and his colleagues at the London School of Tropical Medicine and Hygiene indicate that of the 179 million school-age children living Africa, approximately 50 million are infected with hookworm, and 90% of these children are at risk for coincident infection with malaria.[26]

The observation of NTD-malaria coendemicity has important implications. Malaria is a leading cause of morbidity and mortality in sub-Saharan Africa, and a significant component of malaria's disease burden results from the severe anemia that it causes in both children and pregnant women. If we now superimpose on these two vulnerable malaria-infected populations the blood loss and iron deficiency anemia of hookworm infection, and possibly as well the anemia-causing elements of schistosomiasis, the result is a "perfect storm" of anemia in areas of rural poverty in sub-Saharan Africa. In Kenya, Brooker and his colleagues were able to show that in both preschool children and school-age children, coinfection with malaria and heavy hookworm was associated with significantly worse malaria than were single infections with either malaria or hookworm.[27] Moreover, it is believed that most of the estimated 7.5 million pregnant women in Africa with hookworm are exposed to malaria sometime during their pregnancy, so that coinfection would also exacerbate anemia in this vulnerable population.[28] The term "agriculture-related anemias" has been used to describe the high prevalence of anemia in rural areas of developing countries resulting from the cumulative morbidities of NTDs and malaria and nutritional deficiencies, as well as some common genetic causes of anemia such as sickle-cell diseases and thalassemias.[29]

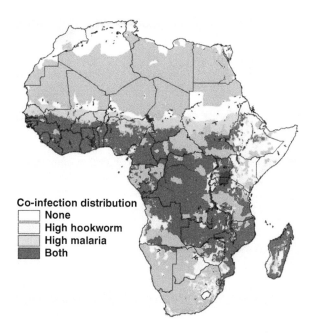

Figure 10.3 Geographic overlap of moderate to high hookworm infection prevalence (>20% prevalence of infection among school-age children) and transmission of falciparum malaria (based on a map of climatic suitability for malaria transmission adjusted for urbanization). (Modified from Brooker S, et al. 2006. The co-distribution of *Plasmodium falciparum* and hookworm among African school-children. *Malar J* 5:99. CC BY 2.0.)

The public health impact of the additive anemia resulting from overlapping NTDs both with and without malaria was described earlier. Briefly, in young children severe anemia is a leading cause of death, while in older children anemia results in impaired physical growth and cognition and is an important reason why hookworm is such an important global pathogen. Similarly, in pregnant women, anemia is a major contributory factor in low birth weight and even in the death of the mother. Moreover, the morbidities resulting from malaria and helminth coinfections may be synergistic as well as additive. Previously, Pierre Druilhe and his colleagues at the Institut Pasteur in Paris accumulated evidence indicating that selected NTDs, especially schistosomiasis and hookworm infection, may, by modulating the immune system, increase the likelihood of infection with malaria or worsen the number or severity of clinical malaria attacks,[30] although some investigators have not been able to replicate such studies.

Therefore, an anticipated benefit of linking malaria control and NTD control in Africa would be the major health gains that could result from further reductions in anemia, greater than what would result from NTD control alone, as well as the possible provision of a back-door mechanism to help achieve malaria control. Investigators associated with the Global Network and others have identified a number of possible entry points for linking malaria control with NTD control.[31] They include both mosquito vector control measures and simultaneous administration of antimalarial and NTD drugs. With regard to the former in rural sub-Saharan Africa, LF is transmitted by the same *Anopheles* species that transmits malaria parasites, so insecticide-treated bed nets used to control malaria can also be used to prevent LF.[24,31] Moreover, anthelmintic drugs used during pregnancy to reduce anemia from hookworm could be used alongside a recommended regimen of antimalarial drugs known as intermittent preventive therapy (IPT) of pregnancy, in which administration of a curative regimen of sulfadoxine-pyrimethamine after 20 weeks of pregnancy is followed by a second dose at least 1 month later.[27,31] IPT is now a WHO-recommended strategy for the prevention of malaria-related maternal anemia and low birth weight; it is also being investigated as an approach for malaria control in children.[31]

Linking malaria control with NTD control is expected to provide for more efficient use of the limited funds available for vulnerable populations, while at the same time maximizing involvement of the community. Therefore, I feel that the major malaria experts and policymakers need to recognize the potential for enhanced access through NTD control measures, particularly because it could be achieved at minimal cost. Jeffrey Sachs and I previously estimated that the cost of NTD control would add only 10% to estimated malaria control costs.[32] This is a very modest investment, given the return in terms of anemia reduction and enhanced bed net use and other operational synergies.

In addition to its contribution to malaria prevention, there are reasons to believe that NTD control could have an impact on the global HIV/AIDS epidemic.[33] For instance, Zvi Bentwich and his colleagues in Israel have shown that parasitic worms affect the host immune system in such a manner that they cause people already infected with HIV/AIDS to have larger amounts of virus in their system.[33] These studies have been confirmed in baboons coinfected with schistosomes and HIV.[33] Further, as described in chapter 3, there is abundant evidence that in women, urinary schistosomiasis also affects the genital tract. Studies conducted by a group of Norwegian scientists led by Eyrun Kjetland and a group of U.S. scientists led by Jennifer Downs have shown that such schistosome-infected women have a 3- to 4-fold increase in risk of contracting the AIDS virus. Beyond its obvious benefit for the health of girls and women, schistosomiasis control has been proposed as a back-door approach to HIV/AIDS control if we could only link programs such as PEPFAR and the Global Fund with the USAID NTD Program and the DFID's programs.[33] Additional investigations have shown that worms also increase the likelihood of virus passing from mother to baby.[33] Therefore, it may turn out that NTD control may be a key not only for malaria control but for HIV/AIDS control as well.

The important coendemic and synergistic effects of the NTDs on malaria and HIV/AIDS in Africa, and possibly elsewhere, suggest that an important opportunity is being overlooked in the failure to embrace NTD control as part of the global agenda to fight the big three diseases. As I indicated in the beginning of this book, the G7 nations have pumped billions of dollars into large-scale treatment programs for the big three. The largest funding instruments are the Global Fund to Fight AIDS, Tuberculosis, and Malaria and PEPFAR. The coendemicity and the operational synergies between AIDS, malaria, and the NTDs argue strongly for including low-cost and cost-effective NTD control measures in the next round of the Global Fund, as well as large-scale bilateral initiatives from the U.S. government such as PEPFAR and PMI. As an alternative scenario, we could establish a stand-alone NTD drug fund in order to ensure that the world's poorest people receive access to these essential medicines. I believe strongly that these topics deserve discussion at the next G7 or G20 summit. For too long the NTDs have been the forgotten diseases among forgotten people living in the remote and rural areas of the developing world. I am hopeful, however, that through these advocacy organizations, the coming years could be remembered for catapulting the NTDs to the attention of global health and poverty advocates everywhere.

SUMMARY POINTS   Uniting to Combat NTDs

- The 10 most prevalent NTDs targeted by preventive chemotherapy—ascariasis, trichuriasis, hookworm infection, schistosomiasis, scabies, LF, taeniasis, trachoma, onchocerciasis, and yaws—exhibit a high degree of geographic overlap and coendemicity, especially in sub-Saharan Africa.

- In areas of coendemicity, populations are frequently polyparasitized, creating the opportunity to simultaneously target multiple NTDs with a package of drugs.

- A rapid-impact package of drugs—comprising either albendazole or mebendazole; DEC, ivermectin, or moxidectin; praziquantel; and azithromycin—has been designed to target the 10 most common NTDs. The drugs are given in different combinations according to WHO preventive chemotherapy guidelines.

- Because many of the drugs are donated, the rapid-impact package is both extremely inexpensive and highly cost-effective and can be administered for approximately US$0.40 per person per year. It represents one of the best buys in public health because of its health, educational, and economic impact.

- The Global Network for Neglected Tropical Diseases initiated efforts to work with a variety of partners to raise the awareness, political will, and funding necessary to control and eliminate the 10 NTDs. Since 2012, under the auspices of the London Declaration, the privately run END Fund and a partnership known as Uniting to Combat Neglected Tropical Diseases have worked closely with the WHO Department of Control of NTDs to accelerate preventive chemotherapy. These organizations operate under WHA resolutions and the new WHO 2021–2030 Roadmap.

- Integrated control of the NTDs is under way across sub-Saharan Africa.

- The NTDs also exhibit coendemicity with malaria and HIV/AIDS. There are both theoretical and practical reasons to link NTD control programs with malaria and HIV/AIDS control programs, including the Global Fund.

Notes

1. Based on information from Hotez et al., 2006; and Hotez et al., 2007a. Updated numbers are from http://www.healthdata.org/results/gbd_summaries/2019/neglected-tropical-diseases-and-malaria-level-2-cause (Global Burden of Disease Collaborative Network, 2020). To obtain the specific numbers for the NTDs (16 million to 25 million DALYs), go to the detailed notes in Table 1.5 in chapter 1.

2. For the importance of incorporating the chronic effects of schistosomiasis and other NTDs into DALY estimates, see King et al., 2005; and King, 2007.

3. Information on the economic impact is summarized in Hotez and Ferris, 2006; Hotez et al., 2007a; Adelman et al., 2012: Hotez et al., 2009; Lenk et al., 2016; and Kirigia and Mburugu, 2017. Information on the mental health aspects of NTDs is in Hotez, 2014; and Bailey et al., 2019.

4. A summary of the WHA resolutions on NTDs is found in https://www.who.int/neglected_diseases/mediacentre/resolutions/en/. WHA66.12 is described in https://www.who.int/neglected_diseases/WHA_66_seventh_day_resolution_adopted/en/.

5. Information on the number of people reached through MDA/preventive chemotherapy can be found in World Health Organization, 2020.

6. Hotez, 2010.

7. Statistical information on Côte d'Ivoire is found in Wright, 2006, p. 605, and at https://www.cia.gov/the-world-factbook/countries/cote-divoire/.

8. Information on polyparasitism in Côte d'Ivoire, Brazil, and elsewhere is found in Raso et al., 2004; Raso et al., 2006; Fleming et al., 2006; and McKenzie, 2005.

9. Information is found in Brooker and Utzinger, 2007; and Magalhaes et al., 2011.

10. The combined disease burden of the NTDs is found in Hotez et al., 2006; and Hotez et al., 2007a.

11. The rationale for the rapid-impact package is in Molyneux et al., 2005. The definition of the rapid-impact package is in Hotez et al., 2007a; and Hotez, 2009. The original targets were seven conditions: ascariasis, trichuriasis, hookworm infection, schistosomiasis, LF, onchocerciasis, and trachoma. Only later were scabies, taeniasis, and yaws identified as targets for the same package (described in Hotez et al., 2019c) to bring the total to 10.

12. Information on the potential health and economic impacts of the rapid-impact package is found in Hotez et al., 2006; Hotez et al., 2009b; Hotez, 2009; and Porco et al., 2009. Information on the extended or collateral benefits of preventive chemotherapy is in Hotez et al., 2019c.

13. Information is in Hotez et al., 2006.

14. Information on the global impact of anemia is in Zimmermann and Hurrell, 2007.

15. Information on the impact of deworming on anemia rates in hookworm-infected children and in pregnancy is in Torlesse and Hodges, 2000; Christian et al., 2004; Brooker et al., 2008; and Smith and Brooker, 2010.

16. Information on the impact of deworming and anemia reduction on education and economic development is in Miguel and Kremer, 2004; World Bank, 2003; Bleakley, 2007; and Baird et al., 2012.

17. Information about the London Declaration can be found at https://unitingtocombatntds.org/resource-hub/who-resources/london-declaration-neglected-tropical-diseases/.

18. The costs of integrated NTD control are provided in Fenwick et al., 2005; Brady et al., 2006; Hotez et al., 2007a; and Evans et al., 2011.

19. The US$3 billion annual estimate for malaria control in Africa is from Teklehaimanot et al., 2007; and Sachs and Hotez, 2006.

20. Canning, 2006.

21. World Health Organization, 2006.

22. Information about the Global Network and its partners is in Hotez et al., 2007a.

23. Comments from Nick Kristof are in Kristof, 2007. Information about the END Fund is at www.end.org, in addition to a recent book, Agler, 2019. Information about the simple surgeries for NTDs is in Karun et al., 2017.

24. An overview of the WHO 2021–2030 roadmap can be found at https://www.who.int/news/item/12-11-2020-neglected-tropical-diseases-world-health-assembly-endorses-bold-new-road-map-targets-for-2030. The challenges of integration are detailed in Hotez et al., 2007a; and Hotez et al., 2007b.

25. Studies on the integration of insecticide-treated net distribution with MDA are described in Blackburn et al., 2006.

26. The geographic overlap between malaria and hookworm infection is in Brooker et al., 2006.

27. Information is from Brooker et al., 2007.

28. The impact of hookworm infection and malaria on pregnancy in Africa is described in Crompton, 2000; Guyatt and Snow, 2004; and Torlesse and Hodges, 2000.

29. The "perfect storm" of anemia and the agricultural anemias is described in Hotez et al., 2006; and Fleming, 1994.

30. The synergistic effects between NTDs and malaria are summarized in Druilhe et al., 2005.

31. The entry points between NTD control and malaria control are summarized in Hotez et al., 2006; and Brooker et al., 2007. Relevant specific papers on IPT for malaria include Greenwood, 2006; and Ntab et al., 2007.

32. Sachs and Hotez, 2006.

33. The impact of NTDs on the HIV/AIDS epidemic is described in the following papers: Borkow and Bentwich, 2006; Secor, 2006; Walson et al., 2009; Hotez et al., 2009a; Hotez et al., 2011; Gerns et al., 2012; Kjetland et al., 2006; Hotez et al., 2019b; and Engels et al., 2020.

References

Adelman C, Norris J, Spantchak Y, Marano K. 2012. *Social and Economic Impact Review on Neglected Tropical Diseases.* Hudson Institute, Washington, DC.

Agler E, Crigler M. 2019. *Under the Big Tree: Extraordinary Stories from the Movement to End Neglected Tropical Diseases.* Johns Hopkins University Press, Baltimore, MD.

Bailey F, Eaton J, Jidda M, van Brakel WH, Addiss DG, Molyneux DH. 2019. Neglected tropical diseases and mental health: progress, partnerships, and integration. *Trends Parasitol* 35:23–31.

Baird S, Hicks JH, Kremer M, Miguel E. 2016. Worms at work: long-run impacts of a child health investment. *Q J Econ* 131:1637–1680.

Blackburn BG, Eigege A, Gotau H, Gerlong G, Miri E, Hawley WA, Mathieu E, Richards F. 2006. Successful integration of insecticide-treated bed net distribution with mass drug administration in Central Nigeria. *Am J Trop Med Hyg* 75:650–655.

Bleakley H. 2007. Disease and development: evidence from hookworm eradication in the American South. *Q J Econ* 122:73–117.

Borkow G, Bentwich Z. 2006. HIV and helminth co-infection: is deworming necessary? *Parasite Immunol* 28:605–612.

Brady MA, Hooper PJ, Ottesen EA. 2006. Projected benefits from integrating NTD programs in sub-Saharan Africa. *Trends Parasitol* 22:285–291.

Brooker S, Akhwale W, Pullan R, Estambale B, Clarke SE, Snow RW, Hotez PJ. 2007. Epidemiology of plasmodium-helminth co-infection in Africa: populations at risk, potential impact on anemia, and prospects for combining control. *Am J Trop Med Hyg* 77(Suppl):88–98.

Brooker S, Clements AC, Hotez PJ, Hay SI, Tatem AJ, Bundy DA, Snow RW. 2006. The co-distribution of *Plasmodium falciparum* and hookworm among African schoolchildren. *Malar J* 5:99.

Brooker S, Hotez PJ, Bundy DA. 2008. Hookworm-related anaemia among pregnant women: a systematic review. *PLoS Negl Trop Dis* 2:e291.

Brooker S, Utzinger J. 2007. Integrated disease mapping in a polyparasitic world. *Geospat Health* 1:141–146.

Canning D. 2006. Priority setting and the 'neglected' tropical diseases. *Trans R Soc Trop Med Hyg* 100:499–504.

Christian P, Khatry SK, West KP Jr. 2004. Antenatal anthelmintic treatment, birthweight, and infant survival in rural Nepal. *Lancet* 364:981–983.

Crompton DW. 2000. The public health importance of hookworm disease. *Parasitology* 121(Suppl):S39–S50.

Druilhe P, Tall A, Sokhna C. 2005. Worms can worsen malaria: towards a new means to roll back malaria? *Trends Parasitol* 21:359–362.

Engels D, Hotez PJ, Ducker C, Gyapong M, Bustinduy AL, Secor WE, Harrison W, Theobald S, Thomson R, Gamba V, Masong MC, Lammie P, Govender K, Mbabazi PS, Malecela MN. 2020. Integration of prevention and control measures for female genital schistosomiasis, HIV and cervical cancer. *Bull World Health Organ* 98:615–624.

Evans D, McFarland D, Adamani W, Eigege A, Miri E, Schulz J, Pede E, Umbugadu C, Ogbu-Pearse P, Richards FO. 2011. Cost-effectiveness of triple drug administration (TDA) with praziquantel, ivermectin and albendazole for the prevention of neglected tropical diseases in Nigeria. *Ann Trop Med Parasitol* 105:537–547.

Fenwick A, Molyneux D, Nantulya V. 2005. Achieving the Millennium Development Goals. *Lancet* 365:1029–1030.

Fleming AF. 1994. Agriculture-related anaemias. *Br J Biomed Sci* 51:345–357.

Fleming FM, Brooker S, Geiger SM, Caldas IR, Correa-Oliveira R, Hotez PJ, Bethony JM. 2006. Synergistic associations between hookworm and other helminth species in a rural community in Brazil. *Trop Med Int Health* 11:56–64.

Gerns HL, Sangaré LR, Walson JL. 2012. Integration of deworming into HIV care and treatment: a neglected opportunity. *PLoS Negl Trop Dis* 6:e1738.

Global Burden of Disease Collaborative Network. 2020. Global Burden of Disease Study 2019 (GBD 2019). GBD Results Tool. Institute for Health Metrics and Evaluation, Seattle, WA.

Greenwood B. 2006. Review: intermittent preventive treatment—a new approach to the prevention of malaria in children in areas with seasonal malaria transmission. *Trop Med Int Health* 11:983–991.

Guyatt HL, Snow RW. 2004. Impact of malaria during pregnancy on low birth weight in sub-Saharan Africa. *Clin Microbiol Rev* 17:760–769.

Hotez PJ. 2009. Mass drug administration and integrated control for the world's high-prevalence neglected tropical diseases. *Clin Pharmacol Ther* 85:659–664.

Hotez PJ. 2010. Neglected tropical disease control in the "post-American world". *PLoS Negl Trop Dis* 4:e812.

Hotez PJ. 2014. Neglected infections of poverty in the United States and their effects on the brain. *JAMA Psychiatry* 71:1099–1100.

Hotez PJ, Biritwum NK, Fenwick A, Molyneux DH, Sachs JD. 2019a. Ghana: accelerating neglected tropical disease control in a setting of economic development. *PLoS Negl Trop Dis* 13:e0007005.

Hotez PJ, Engels D, Gyapong M, Ducker C, Malecela MN. 2019b. Female genital schistosomiasis. *N Engl J Med* 381:2493–2495.

Hotez PJ, Fenwick A, Kjetland EF. 2009a. Africa's 32 cents solution for HIV/AIDS. *PLoS Negl Trop Dis* 3:e430.

Hotez PJ, Fenwick A, Molyneux DH. 2019c. Collateral benefits of preventive chemotherapy—expanding the war on neglected tropical diseases. *N Engl J Med* 380:2389–2391.

Hotez PJ, Fenwick A, Ray SE, Hay SI, Molyneux DH. 2018. "Rapid impact" 10 years after: the first "decade" (2006-2016) of integrated neglected tropical disease control. *PLoS Negl Trop Dis* 12:e0006137.

Hotez PJ, Fenwick A, Savioli L, Molyneux DH. 2009b. Rescuing the bottom billion through control of neglected tropical diseases. *Lancet* 373:1570–1575.

Hotez PJ, Ferris MT. 2006. The antipoverty vaccines. *Vaccine* 24:5787–5799.

Hotez PJ, Mistry N, Rubinstein J, Sachs JD. 2011. Integrating neglected tropical diseases into AIDS, tuberculosis, and malaria control. *N Engl J Med* 364:2086–2089.

Hotez PJ, Molyneux DH, Fenwick A, Kumaresan J, Sachs SE, Sachs JD, Savioli L. 2007a. Control of neglected tropical diseases. *N Engl J Med* 357:1018–1027.

Hotez PJ, Molyneux DH, Fenwick A, Ottesen E, Ehrlich Sachs S, Sachs JD. 2006. Incorporating a rapid-impact package for neglected tropical diseases with programs for HIV/AIDS, tuberculosis, and malaria. *PLoS Med* 3:e102.

Hotez P, Raff S, Fenwick A, Richards F Jr, Molyneux DH. 2007b. Recent progress in integrated neglected tropical disease control. *Trends Parasitol* 23:511–514.

Karun V, Hotez PJ, Rosengart TK. 2017. Global surgery and the neglected tropical diseases. *PLoS Negl Trop Dis* 11:e0005563.

King CH. 2007. Lifting the burden of schistosomiasis—defining elements of infection-associated disease and the benefits of anti parasite treatment. *J Infect Dis* 196:653–655.

King CH, Dickman K, Tisch DJ. 2005. Reassessment of the cost of chronic helmintic infection: a meta-analysis of disability-related outcomes in endemic schistosomiasis. *Lancet* **365**:1561–1569.

Kirigia JM, Mburugu GN. 2017. The monetary value of human lives lost due to neglected tropical diseases in Africa. *Infect Dis Poverty* **6**:165.

Kjetland EF, Ndhlovu PD, Gomo E, Mduluza T, Midzi N, Gwanzura L, Mason PR, Sandvik L, Friis H, Gundersen SG. 2006. Association between genital schistosomiasis and HIV in rural Zimbabwean women. *AIDS* **20**:593–600.

Kristof N. 2007. Attack of the worms. *The New York Times* 2007(July 2).

Lammie PJ, Fenwick A, Utzinger J. 2006. A blueprint for success: integration of neglected tropical disease control programmes. *Trends Parasitol* **22**:313–321.

Lenk EJ, Redekop WK, Luyendijk M, Rijnsburger AJ, Severens JL. 2016. Productivity loss related to neglected tropical diseases eligible for preventive chemotherapy: a systematic literature review. *PLoS Negl Trop Dis* **10**:e0004397.

Lo NC, Addiss DG, Hotez PJ, King CH, Stothard JR, Evans DS, Colley DG, Lin W, Coulibaly JT, Bustinduy AL, Raso G, Bendavid E, Bogoch II, Fenwick A, Savioli L, Molyneux D, Utzinger J, Andrews JR. 2017. A call to strengthen the global strategy against schistosomiasis and soil-transmitted helminthiasis: the time is now. *Lancet Infect Dis* **17**:e64–e69.

Magalhães RJ, Clements AC, Patil AP, Gething PW, Brooker S. 2011. The applications of model-based geostatistics in helminth epidemiology and control. *Adv Parasitol* **74**:267–296.

McKenzie FE. 2005. Polyparasitism. *Int J Epidemiol* **34**:221–222, author reply 222–223.

Miguel E, Kremer M. 2004. Worms: identifying impacts on education and health in the presence of treatment externalities. *Econometrica* **72**:159–217.

Molyneux DH. 2014. Neglected tropical diseases: now more than just 'other diseases'—the post-2015 agenda. *Int Health* **6**:172–180.

Molyneux DH, Hotez PJ, Fenwick A. 2005. "Rapid-impact interventions": how a policy of integrated control for Africa's neglected tropical diseases could benefit the poor. *PLoS Med* **2**:e336.

Murray CJ, et al. 2012. Disability-adjusted life years (DALYs) for 291 diseases and injuries in 21 regions, 1990-2010: a systematic analysis for the Global Burden of Disease Study 2010. *Lancet* **380**:2197–2223.

Ntab B, Cissé B, Boulanger D, Sokhna C, Targett G, Lines J, Alexander N, Trape JF, Simondon F, Greenwood BM, Simondon KB. 2007. Impact of intermittent preventive anti-malarial treatment on the growth and nutritional status of preschool children in rural Senegal (west Africa). *Am J Trop Med Hyg* **77**:411–417.

Porco TC, Gebre T, Ayele B, House J, Keenan J, Zhou Z, Hong KC, Stoller N, Ray KJ, Emerson P, Gaynor BD, Lietman TM. 2009. Effect of mass distribution of azithromycin for trachoma control on overall mortality in Ethiopian children: a randomized trial. *JAMA* **302**:962–968.

Raso G, Luginbühl A, Adjoua CA, Tian-Bi NT, Silué KD, Matthys B, Vounatsou P, Wang Y, Dumas ME, Holmes E, Singer BH, Tanner M, N'goran EK, Utzinger J. 2004. Multiple parasite infections and their relationship to self-reported morbidity in a community of rural Côte d'Ivoire. *Int J Epidemiol* **33**:1092–1102.

Raso G, Vounatsou P, Singer BH, N'Goran EK, Tanner M, Utzinger J. 2006. An integrated approach for risk profiling and spatial prediction of *Schistosoma mansoni*-hookworm coinfection. *Proc Natl Acad Sci USA* **103**:6934–6939.

Sachs JD, Hotez PJ. 2006. Fighting tropical diseases. *Science* **311**:1521.

Secor WE. 2006. Interactions between schistosomiasis and infection with HIV-1. *Parasite Immunol* **28**:597–603.

Smith JL, Brooker S. 2010. Impact of hookworm infection and deworming on anaemia in non-pregnant populations: a systematic review. *Trop Med Int Health* **15**:776–795.

Teklehaimanot A, Sachs JD, Curtis C. 2007. Malaria control needs mass distribution of insecticidal bednets. *Lancet* **369**:2143–2146.

Torlesse H, Hodges M. 2000. Anthelminthic treatment and haemoglobin concentrations during pregnancy. *Lancet* **356**:1083.

Walson JL, Herrin BR, John-Stewart G. 2009. Deworming helminth co-infected individuals for delaying HIV disease progression. *Cochrane Database Syst Rev* **8**:CD006419.

Webster JP, Molyneux DH, Hotez PJ, Fenwick A. 2014. The contribution of mass drug administration to global health: past, present and future. *Philos Trans R Soc Lond B Biol Sci* **369**:20130434.

World Bank. 2003. School deworming at a glance. World Bank, Washington, DC.

World Health Organization. 2006. *Preventive Chemotherapy in Human Helminthiasis*. World Health Organization, Geneva, Switzerland.

World Health Organization. 2020. Neglected tropical diseases: impact of COVID-19 and WHO's response. *Wkly Epidemiol Rec* **95**:461–476.

Wright JW (ed). 2006. *The 2006 New York Times Almanac*. Penguin, New York, NY.

Zimmermann MB, Hurrell RF. 2007. Nutritional iron deficiency. *Lancet* **370**:511–520.

11 Future Trends in Control of Neglected Tropical Diseases and the Antipoverty Vaccines

We've begun to share the promise of modern medicine with those suffering from river blindness, sleeping sickness, and other age-old diseases. A handful of innovative, compassionate people around the world have joined forces to develop new drugs for neglected diseases.

SENATOR EDWARD KENNEDY, 2006

The main obstacle to responding to the needs of those suffering is insufficient incentive for companies to produce drugs that treat and prevent neglected tropical diseases.

SENATOR SAM BROWNBACK, 2007

A scientist who is also a human being cannot rest while knowledge which might reduce suffering rests on the shelf.

ALBERT B. SABIN

The most prevalent neglected tropical diseases (NTDs) (in order of prevalence, they are ascariasis, trichuriasis, hookworm infection, scabies, schistosomiasis, lymphatic filariasis [LF], onchocerciasis, taeniasis, trachoma, and yaws) targeted for preventive chemotherapy, together with dengue and the food-borne trematode infections, are the most common infections of the world's poorest people. As a group, they cause disease and disability on a scale that almost rivals HIV/AIDS or malaria. These conditions also represent a major cause of economic underdevelopment, as they currently block the escape from poverty for the poorest people on our planet. Because integrated NTD control through widespread use of the rapid-impact package of drugs is expected either to reduce the morbidity and disease burdens of these 10 highly prevalent NTDs or in some cases to result in their elimination, it is likely that this approach will become an important ally in the fight for sustainable poverty reduction.

Forgotten People, Forgotten Diseases: The Neglected Tropical Diseases and Their Impact on Global Health and Development, Third Edition. Peter J. Hotez.
© 2022 American Society for Microbiology. DOI: 10.1128/9781683673903.ch11

In the previous chapter, I outlined some of the challenges facing large-scale deployment of the rapid-impact package, including the possibility of emerging drug resistance. My concerns about resistance to drugs and other chemical agents stem partly from what we know about the mixed legacy of mass drug administration in the fight against malaria that began shortly after World War II. Widespread use of the insecticide DDT (dichlorodiphenyltrichloroethane) was instrumental in eliminating malaria from much of southern Europe and helped to stimulate economic growth in Greece, Italy, and Spain during the late 1940s and into the 1950s.[1] On the other hand, in some regions of intense malaria endemicity, such as India and sub-Saharan Africa, this same approach has failed.[2] In India throughout the 1950s, mass administration of the antimalarial drug chloroquine, together with intensive spraying of DDT, resulted in the reduction of malaria incidence from an estimated 75 million cases to only 50,000 cases by 1961. As a result of these successes, by the early 1960s there was considerable talk within the global health community and among policymakers about the eventual eradication of malaria in India and, possibly, worldwide. However, the emergence of chloroquine and DDT resistance almost completely reversed these early successes, so that by 1977 there were again tens of millions of malaria cases in India.[2]

This sobering lesson requires us to give serious consideration to the possibility that in areas of high parasite transmission, resistance to some of the drugs being used in the rapid-impact package could also emerge. With an ambitious global agenda to scale up integrated interventions for more than 1 billion people where multiple NTDs are coendemic, we need to carefully consider the specter of resistance and other forms of drug failure. Because helminth parasites have much longer replication times than do viruses, bacteria, or even a protozoan organism such as the malaria parasite, we can expect that resistance will probably not emerge as rapidly.[3] However, this observation alone is not an excuse for complacency.

Of the six possible different drugs contained in the rapid-impact package—azithromycin; praziquantel; albendazole or mebendazole; and diethylcarbamazine (DEC), ivermectin, or moxidectin—the agents albendazole and mebendazole particularly stand out as drugs that could induce resistance. Both albendazole and mebendazole belong to the benzimidazole class of compounds. Benzimidazoles similar to albendazole and mebendazole are used widely to deworm ruminant livestock of their soil-transmitted helminths. The gastrointestinal helminth infections of sheep and cattle are both important veterinary public health and economic problems and are a major reason why it is difficult and sometimes impossible to maintain these animals in an acceptable state of health in many subtropical and tropical countries. For decades, livestock producers in the Southern Hemisphere, especially in South America, South Africa, Australia, and New Zealand, have relied heavily on frequent and periodic treatments with benzimidazole anthelmintics to ensure that their animals harbor low worm burdens throughout the year. Today, however, benzimidazole resistance is widespread in many parts of the Southern Hemisphere and elsewhere, and in such regions these agents are no longer effective.

Resistance to benzimidazoles now actually thwarts livestock production in many areas of the world.[4]

Could similar problems emerge among human helminth parasites? For most, if not all, parasitic nematode species, resistance to benzimidazoles results from a point mutation in the parasite genes encoding a protein known as beta-tubulin. Presumably, the benzimidazoles exert their drug effect by binding to beta-tubulin contained in the worm, so that a mutation reduces binding and prevents the drug from acting on the parasite. For the veterinary nematodes, it has been determined that a specific point mutation in the beta-tubulin gene can produce an amino acid change of a phenylalanine to a tyrosine residue in either of two positions in the beta-tubulin gene product, which appears to be sufficient to reduce benzimidazole binding and cause resistance.[5] Recently, this benzimidazole resistance-associated mutation was also found in *Wuchereria bancrofti*, the major cause of LF,[5] possibly in association with scaled-up use of albendazole and ivermectin combinations. At the same time, although the resistant phenotype has not yet been demonstrated for hookworm beta-tubulins, there is evidence demonstrating that mebendazole often is ineffective against hookworms in some parts of the world, especially in areas where the drug has been used frequently.[5] In a systematic review and meta-analysis published in *JAMA* in 2008, Jennifer Keiser and Juerg Utzinger of the Swiss Tropical and Public Health Institute showed that the overall cure rate for mebendazole against hookworm infection is only 15%, with quantitative reductions of the eggs of the major human hookworm, *Necator americanus*, ranging between 0 and 68%.[5] Moreover, in one study it was shown that the efficacy of mebendazole actually worsens with frequent and increasing use.[5] These observations indicate that drug failure is common when mebendazole is used for the treatment of hookworm infection and suggest the need to investigate the possibility of emerging drug resistance. It has been further noted that the effectiveness of albendazole for trichuriasis has also declined in recent years, while ivermectin efficacy against onchocerciasis may have also diminished in Ghana, leading some groups to propose combining drugs—such as albendazole together with ivermectin—to promote antiparasitic efficacy or forestall resistance.[5,6]

The occurrence of sporadic instances of drug failure or resistance in treatment of helminth infections in Africa (and possibly elsewhere) should not deter us from aggressively working to widely deploy rapid-impact packages in developing countries. Drug resistance to helminths does not evolve as rapidly as it does for bacterial and viral pathogens, and in reality, what choice is there? In my opinion, it would be morally unacceptable to withhold essential medicines from the world's poorest people because of theoretical concerns about emerging drug resistance. Instead, the best possible approach is to get the existing rapid-impact drugs out to the people who urgently need them, but at the same time recognize the urgency of monitoring for the possible emergence of new pockets of drug resistance. In this regard, a new single-nucleotide polymorphism technique has been developed that could be used to detect benzimidazole resistance among human-parasitic nematodes.[7] In response, the

Gates Foundation has launched the "Starworms" (Stop Anthelmintic Resistant Worms; www.starworms.org) project to monitor potential drug resistance based on single-nucleotide polymorphisms.[7]

Concurrently, we need to continue research and development efforts for a new pipeline of anthelmintic drugs. As new drugs are developed, they could be folded into the rapid-impact package when necessary. To date, there are several drugs that should be undergoing additional development and testing in the eventual likelihood that they could be folded into the rapid-impact package. Two such are flubendazole and moxidectin, promising new drugs that can be used to treat filarial worm infections, i.e., onchocerciasis and LF. Both were originally developed as veterinary anthelmintic drugs.[8] Flubendazole is now licensed in Europe as a treatment for human intestinal nematodes but requires extended studies to determine if it is effective as a so-called macrofilaricide that can target the adult worms so that fewer rounds of ivermectin are required. Similarly, moxidectin was recently licensed by the Food and Drug Administration (FDA) in the United States, as was an anti-fluke medication known as triclabendazole.[8] In parallel, Raffi Aroian at the University of Massachusetts is exploring a novel recombinant *Bacillus thuringiensis* Cry5B protein as an anthelmintic, while a new generation of anti-*Wolbachia* therapies, including antibiotics being developed by Mark Taylor and his colleagues at Liverpool School of Tropical Medicine, that target filarial endosymbionts required for effective parasite replication and survival could eventually offer an entirely new approach to the treatment of both onchocerciasis and LF.[9]

A major obstacle to the development and clinical testing of flubendazole, moxidectin, new anti-*Wolbachia* compounds, or really any NTD drug is the total absence of a viable commercial market and therefore the absence of incentive for pharmaceutical companies to embark on anthelmintic drug development projects. Although the major pharmaceutical companies have been willing to donate selected products free of charge in order to combat NTDs in developing countries, it is difficult for a publicly held company responsible to shareholders to take the next step and commit precious resources toward extensive research into and development of new NTD products. As a result, just a few years ago we were faced with a pathetic situation in which out of a total of 1,556 new chemical entities marketed between 1975 and 2004, only 21 drugs were for tropical diseases.[10] Of these, 11 drugs were for malaria or tuberculosis (TB), which have some North American and European markets. This left a total of 10 products developed specifically for NTDs over that 30-year period, or roughly 0.6% of the total products developed. The Global Forum for Health Research has coined the term "10/90 gap" to describe the observation that only 10% of worldwide expenditure on medical research and development is devoted to the problems that primarily affect the poorest 90% of the world's population. In the case of the NTDs, it is more like a 1/99 or a 1/199 gap!

Up until recently, most of the new drugs for NTDs were created by the large, multinational pharmaceutical companies. In at least two cases, the drugs were discovered and developed for a lucrative market for livestock antiparasitic drugs and then later the company supported the development of these drugs as human anthelmintics. This includes the original investment by Merck in ivermectin as a veterinary product and its subsequent, welcomed support of important clinical testing in Africa in order to demonstrate ivermectin's efficacy against onchocerciasis, as well as the discovery and development of albendazole as a veterinary product by the Animal Health division of Smith Kline & French, before it was subsequently developed by the company now known as GlaxoSmithKline (GSK) as a human anthelmintic in 1987. In addition, Bayer AG in Germany developed praziquantel and nifurtimox, Pfizer developed oxamniquine (although this drug has been largely replaced by praziquantel), and Sanofi-Aventis took on the industrial development and distribution of all three major drugs for human African trypanosomiasis (HAT)—eflornithine, melarsoprol, and pentamidine.

I speak about this problem in the past tense since I think we are at the beginning of a paradigm shift and a new vision for developing new NTD drugs. Given the complete lack of commercial incentive for developing NTD drugs, it is amazing to me that the pharmaceutical world is willing to take on product development for these conditions. But there are other incentives based on longer-term business considerations, including corporate social responsibilities and ethical concerns, the strategic consideration of securing better access to developing-country markets and researchers from developing countries, and company reputation.[11]

For instance, while attending the launch in 2006 of Scientists Without Borders, a nonprofit venture previously launched by the New York Academy of Sciences, I was impressed by an evening speech delivered by Jean-Pierre Garnier, the chief executive officer of GSK. He indicated that profit alone was not sufficient motivation for his tens of thousands of employees to come to work every day. Instead, the employees are partly driven by the knowledge that their company is a leader in helping people in need everywhere. Later, his successor Andrew Witty has expanded the company's commitment to neglected diseases by increasing its donations of albendazole to target both soil-transmitted helminth infections and LF; advancing a new malaria vaccine toward licensure; and establishing a new facility in Tres Cantos, Spain, to allow company scientists to work in product development partnerships on drugs for neglected diseases (https://www.openlabfoundation.org/). GSK also acquired a global health vaccine institute from Novartis (as part of a larger corporate merger) and has continued this activity. Known as the GSK Vaccines Institute for Global Health, it was launched under the visionary leadership of Rino Rappuoli, with Allen Saul as its inaugural director, emphasizing bacterial enteric pathogen vaccines. In parallel, Merck & Co., in collaboration with the Wellcome Trust, created the Hilleman Laboratories in India for new vaccines (https://www.hillemanlabs.org/), named after Dr. Maurice Hilleman, a pioneer in vaccinology and an important mentor for me. But it does not stop there—Sanofi Pasteur is taking a leading role in developing and commercializing the first human dengue vaccine, and now Takeda and Merck & Co. are also developing dengue vaccines. Merck & Co. also led the development of an innovative Ebola vaccine that helped to halt the spread of this disease beyond the Democratic Republic of the Congo.[11] Janssen Global Public Health formed in 2014 as part of Johnson & Johnson, and Pfizer Inc. has maintained its commitment to global health research and development. I'm sure there are more examples, but these are the ones I am most familiar with.

In parallel with these new activities by major pharmaceutical companies, several advocacy organizations have recently been established or expanded in order to track research and development activities for NTDs and big three diseases. They include BIO Ventures for Global Health (www.bvgh.org) and Policy Cures, an Australian policy think tank founded by Mary Moran, formerly of the London School of Economics and Political Science. Policy Cures produces an important G-FINDER (Global Funding of Innovation for Neglected Diseases) report and public search tool that provides up-to-date information on public and private research and development funding for multiple NTDs and big three diseases. Their analysis of the NTD drug pipeline has identified some interesting factors that could stimulate a multinational company to consider a program of drug and vaccine development for NTDs.[12]

Despite an emerging track record of developing new drugs for NTDs, it is unreasonable to think that there will be a significant improvement in research and development capacity for new agents if we rely on multinational companies alone. Even with enticements and entitlements for developing NTD products, such as the innovative prospect of advance market commitments proposed by Harvard's Michael Kremer and Massachusetts Institute of Technology's Rachel Glennerster or a very timely amendment for priority review vouchers (https://sites.fuqua.duke.edu/priorityreviewvoucher/) passed by the Senate, the Brownback-Brown Elimination of Neglected Diseases Amendment of the Food and Drug Administration Revitalization Act,[13] it is unlikely that these so-called pull mechanisms will by themselves ensure a robust pipeline of new NTD drugs or other health products such as vaccines and diagnostics. Duke University's Gavin Yamey has identified more than 600 potential neglected disease product candidates (including those for HIV/AIDS, malaria, and TB) that were in development as of 2017, and estimated that there is a US$1.5 billion to US$2.8 billion funding gap to advance just a small fraction of these.[11] I believe that to create a truly robust pipeline of NTD drugs, we also need to look to either smaller biopharmaceutical companies such as AEterna Zentaris, based in Frankfurt, Germany, which is producing miltefosine (Impavido), the first oral agent for visceral leishmaniasis; or a new generation of NTD product development partnerships (PDPs) (also known as public-private partnerships). PDPs are defined by Mary Moran as "public-health-driven not-for-profit organizations that drive neglected-disease drug development in conjunction with industry groups."[11] These PDPs are either exploiting just-finished genome sequencing projects for protozoan parasites,[14] in order to identify potential drug targets, or embarking on high-throughput screening and other traditional approaches to drug development and clinical testing.[15] Many of these PDPs are supported partly by the Bill & Melinda Gates Foundation. PDPs are currently driving the development and clinical testing of several new antiprotozoan drugs for leishmaniasis and HAT, including fexi-nidazole and nifurtimox-eflornithine.[15] A partial listing of at least one dozen PDPs is shown in Table 11.1. Activities are also in place to develop vaccines to combat the NTDs. I coined the term "antipoverty vaccines" to emphasize the poverty-promoting features of the NTDs and the impact that such vaccines would have not only on improving global public health but also on preventing the socioeconomic consequences of these conditions.[16] The vaccine PDPs are developing new antipoverty vaccines for enteric infections, malaria, TB, and several key NTDs, including helminth infections—schistosomiasis, hookworm infection, and onchocerciasis—protozoan infections, and other neglected conditions.[17] Many of these PDPs have manufacturing partners in India, Brazil, and other low- and middle-income countries, and in modeling studies were shown to be highly cost-effective or even cost-saving.[17]

Most of the NTD vaccines use recombinant protein technology building on genomic and "reverse vaccinology" approaches based on completed genome projects. For example, schistosomiasis, onchocerciasis, and Chagas disease vaccines focus on surface or subsurface expressed proteins from the parasite that interact with human blood and tissues and might be targeted by the immune system, whereas the human hookworm vaccine targets enzymes required by the parasite for blood feeding and nutrition. A leishmaniasis vaccine is based on parasite epitopes that might stimulate host T-cell responses.

One of the great challenges that faces all PDPs is the requirement of following industry practices and engaging federal regulatory bodies such as the FDA while conducting these activities in the nonprofit sector.[23] For instance, to develop the human hookworm vaccine, the Texas Children's Center for Vaccine Development (CVD), co-directed by me and my science partner for the last 20 years, Dr. Maria Elena Bottazzi, conducts process development consisting of scale-up fermentation and protein purification; formulation and potency testing; an extensive documentation system that includes standard operating procedures, protocols, and batch production records; and rigorous quality control and quality assurance standards. The Texas Children's CVD also uses these elements to detail steps for the transfer of these technologies to a manufacturer that can reproduce the process under so-called current good manufacturing practices.[18] Following a careful analysis of the manufactured product through lot release and stability testing, it is necessary to submit specific information about the product to an appropriate and internationally recognized regulatory agency (such as the FDA) in order to obtain permission for beginning "first in humans" testing of a new product. Such phase 1 and 2 studies are conducted under current good clinical practices.

In the preface to this book, I mentioned that after some 40 years of conducting hookworm research and heading a hookworm research team that in the last few years has successfully developed hookworm antigens as recombinant products for clinical trials, I have begun to realize that this activity is, in some respects, just a beginning exercise. I make this statement because of a terrible record of getting high-technology products such as recombinant vaccines into the hands of vaccinators in developing countries. For instance, it required some 30 years after its discovery for the hepatitis B

Table 11.1 Major PDPs for NTDs and related neglected diseases

PDP	Class(es) of product	Major disease(s)	Location of headquarters	Website
Drugs for Neglected Diseases initiative (DNDi)	New drugs and new chemical entities (NCEs)	HAT, Chagas disease, leishmaniasis, malaria, pediatric HIV, hepatitis C, and COVID-19	Geneva, Switzerland	https://dndi.org/
Medicines for Malaria Ventures (MMV)	New drugs and NCEs	Malaria	Geneva, Switzerland	https://www.mmv.org/
TB Alliance	New drugs and NCEs	TB	New York, NY	https://www.tballiance.org/
Innovative Vector Control Consortium (IVCC)	New insecticides and anti-vector technologies	Vector-borne diseases	Liverpool, UK	https://www.ivcc.com/
Foundation for Innovative New Diagnostics (FIND)	Diagnostics	NTDs, malaria, hepatitis C, and COVID-19	Geneva, Switzerland	https://www.finddx.org/
Program for Appropriate Technology in Health (PATH)	Diagnostics and vaccines	Malaria, HIV, TB, NTDs, and COVID-19	Seattle, WA	https://www.path.org/
International Vaccine Institute (IVI)	Vaccines	NTDs, infectious diseases, and COVID-19	Seoul, South Korea	https://www.ivi.int/
Infectious Diseases Research Institute (IDRI)	Vaccines and adjuvant technologies	NTDs and infectious diseases	Seattle, WA	http://www.idri.org/
HDT Bio	Vaccines and adjuvant technologies	NTDs and infectious diseases	Seattle, WA	https://www.hdt.bio/
Texas Children's Center for Vaccine Development	Vaccines	NTDs and COVID-19	Houston, TX	https://www.texaschildrens.org/departments/vaccine-development
Hilleman Laboratories	Vaccines	Enteric infectious diseases	New Delhi, India	https://www.hillemanlabs.org/
GSK Vaccine Institute for Global Health	Vaccines	Enteric infectious diseases	Siena, Italy	https://www.gsk.com/en-gb/responsibility/inside-the-gvgh/
International AIDS Vaccine Initiative (IAVI)	Vaccines	HIV and TB	New York, NY	https://www.iavi.org/
Centro de Desenvolvimento Tecnológico em Saúde (CDTS-FIOCRUZ)	Drugs, diagnostics, and vaccines	Neglected and infectious diseases	Rio de Janeiro, Brazil	https://www.cdts.fiocruz.br/

vaccine to be used widely in some developing countries. Today, it is still not used in many low-income countries in Africa and elsewhere.[19] During the past 30 years, countless thousands of people, particularly in East Asia, have unnecessarily suffered the consequences of chronic hepatitis B infection, including liver failure and liver cancer. One of the major reasons for this lag was cost. When it was first produced, hepatitis B vaccine, a recombinant vaccine produced in genetically engineered yeast, was expensive (over US$100). The high price of the original recombinant hepatitis B vaccine allowed the manufacturer to recover its research, development, and manufacturing costs.[19] Only now are public sector manufacturers in developing countries, such as the Instituto Butantan (São Paulo, Brazil), producing hepatitis B vaccine for less than US$1. I worry that decades also will pass before the new human papillomavirus vaccine to prevent cervical cancer comes down from approximately US$360 for the full series of shots to a price suitable for poor women in the developing world. If 30 years is again required, we are facing the prospect of hundreds of thousands of poor women (who do not have access to routine Pap smear screening) needlessly dying of cervical cancer.

Therefore, an important rationale for my starting the Human Hookworm Vaccine Initiative, now a program of our Texas Children's CVD, was to produce a vaccine that would be low-cost from the very beginning. For it to be cost-effective, we estimate that the human hookworm vaccine (likely a "bivalent product," meaning that it will contain at least two recombinant antigens, each encoding a larval- and adult-stage antigen) will need to be produced for less than US$1 to $2. This requirement compels us to build in low-cost manufacturing processes from the very beginning, including the use of inexpensive expression hosts, such as bacteria or yeast, rather than the more expensive mammalian or insect cell lines. It also requires us to produce these products in high volume and to use low-cost column resins to purify the recombinant antigens.

Each of our vaccine programs, including hookworm, schistosomiasis, Chagas disease, and most recently COVID-19, represents a key component of what is referred to now as global access. The NTD PDPs, whether they focus on drug, vaccine, or diagnostic test development, must consider from day one how their products might be used for large numbers of impoverished people. The Texas Children's CVD develops detailed global access roadmaps in order to plan how its vaccines will be developed and tested in resource-poor settings. For us, an important component of global access is partnering with vaccine manufacturers in middle-income countries where hookworm is endemic. Several of these countries, such as Brazil, China, Cuba, India, Indonesia, Iran, Senegal, and others, have some degree of sophistication in manufacturing health products, including vaccines. Brazil's Carlos Morel and his colleagues refer to such countries as IDCs (innovative developing countries), with innovation defined by quantitatively measuring peer-reviewed papers, international patents, and biotechnology capacity.[20] To ensure global access, the Texas Children's CVD works to transfer the technology for its human hookworm vaccine to IDC vaccine producers. I believe that partnerships with IDCs and low-income countries represent one of the very best opportunities to promote

Southern Hemispheric ownership and to ensure global access to new NTD drugs, vaccines, and diagnostics.

With genome sequencing completed for a large number of NTD pathogens, including the agents that cause Chagas disease, HAT, leishmaniasis, leprosy, leptospirosis, LF, schistosomiasis, and trachoma, it should be theoretically possible to mine bioinformatics databases in order to develop a large number of antipoverty vaccines in the coming decade. The financier Mike Milken has helped to popularize the phrase "financial innovation" (www.milkeninstitute. org). We have the technology in hand to develop new antipoverty drugs, vaccines, and diagnostics, but it is financial innovation that is most needed in order to promote institutions for conducting scale-up process development, manufacturing, and clinical testing and for securing global access to these new products. In the next chapter, I make a case for the United States apportioning a modest percentage of its Global Health Initiative to support research and development.[21] At the 59th World Health Assembly, the world's health ministers called for increased access to innovation and intellectual property as an important component of the global health agenda.[22] The G7 and G20 nations need to do more to finance research, development, and clinical testing of new health products for NTDs and to support resolutions for addressing the 1/199 gap.

In parallel, we need to do a better job of developing health system infrastructures so that these new drugs, vaccines, and diagnostics can be folded into NTD rapid-impact packages once they are developed. The control and elimination of the NTDs as a means for sustainably reducing poverty and meeting Millennium Development Goal targets will depend on how successful we are at scaling up integrated NTD control through providing access to existing essential medicines together with access to innovation, so that new products can be incorporated into a new-generation rapid-impact package combining multidimensional health products including drugs, vaccines, and diagnostics. Ultimately, this comprehensive health package will need to be linked with malaria and HIV/AIDS control measures in a holistic and integrated infectious disease control initiative. A roadmap for advancing this agenda at the scientific and technical level would not be simple, but it is almost definitely feasible. Our far greater challenges are financial innovation and global political will.

SUMMARY POINTS   **Future Trends and the Antipoverty Vaccines**

- Because widespread use of the rapid-impact package of drugs is expected to reduce the morbidity and disease burdens of the 10 most prevalent NTDs, it is likely that this approach will become an important ally in the fight for sustainable poverty reduction.
- Of the six different drugs contained in the rapid-impact package, albendazole, mebendazole, and ivermectin are potentially vulnerable to drug resistance and other mechanisms of drug failure. The Starworms initiative, supported by the Bill & Melinda Gates Foundation, is monitoring for emerging anthelmintic drug resistance.
- As alternative agents for the rapid-impact package, flubendazole, moxidectin, and

the anti-*Wolbachia* therapies are under development.

- A major obstacle to the development and clinical testing of NTD drugs is the total absence of commercial incentive.
- In the last few years, some of the multinational pharmaceutical companies have stepped up to establish facilities for producing new NTD drugs and vaccines, although not much of this effort is directed toward the 20 major NTDs.
- To create a truly robust pipeline of NTD drugs, we need to look to either small biopharmaceuticals or a new generation of NTD PDPs. Many of the NTD PDPs are funded in part by the Gates Foundation.
- Within the last decade, several important NTD PDPs have been established for research, development, and clinical testing of new drugs for Chagas disease, HAT, and leishmaniasis.
- A few PDPs have also successfully developed NTD vaccines, also known as antipoverty vaccines, for clinical testing. These include new vaccines for schistosomiasis, leishmaniasis, hookworm infection, and Chagas disease, now in clinical trials in developing countries. These PDPs are producing low-cost vaccines from the outset. The major multinational pharmaceutical companies are developing and testing vaccines for dengue.

- An important component of global access for health products in developing countries is partnering with manufacturers in middle-income countries where hookworm is endemic. Several of these countries, such as Brazil, China, Cuba, India, Indonesia, and others, have some degree of sophistication in manufacturing health products, including vaccines.
- With genome sequencing completed for a large number of NTD pathogens, including the agents that cause Chagas disease, HAT, leishmaniasis, leprosy, leptospirosis, LF, schistosomiasis, and trachoma, it should be theoretically possible to mine such databases in order to develop a large number of antipoverty vaccines in the coming decade.
- We have the technology in hand to develop new antipoverty vaccines, but it is financial innovation that is most needed in order to promote institutions for conducting scale-up process development, manufacturing, and clinical testing and for securing global access to these new products.
- The global control and elimination of the NTDs will require mechanisms to introduce new health products, e.g., vaccines, drugs, and diagnostics, together with existing drugs, in order to create new-generation rapid-impact packages. This approach will be necessary for sustainably reducing poverty and for meeting Sustainable Development Goal targets.

Notes

1. The impact of malaria control on economic development in southern Europe is described in Gallup and Sachs, 2001.

2. The history of drug resistance in global malaria eradication efforts is described in Harrison, 1978, p. 242–254; and Hotez, 2004.

3. Some of the mathematical concepts regarding drug resistance, pathogen transmission, and population genetics are detailed in Anderson, 1999; and Austin et al., 1997.

4. Drug resistance and its impact on livestock production are described in Conder and Campbell, 1995.

5. Detection of the benzimidazole-resistant phenotype in *W. bancrofti* and the evidence for mebendazole drug failure or resistance to benzimidazoles by hookworms are described or summarized in Keiser and Utzinger, 2008; Prichard, 2007; Schwab et al., 2007; Schwab et al., 2005; Bennet and Guyatt, 2000; Albonico et al., 2004a; Albonico et al., 2004b; Albonico et al., 2003; and Albonico et al., 2006.

6. Declining albendazole efficacy and use of combination therapies are discussed in Moser et al., 2017; and Moser et al., 2019. Evidence for emerging ivermectin resistance against *Onchocerca volvulus* is found in Osei-Atweneboana et al., 2007; and Hotez, 2007.

7. Prichard, 2007. Information about Starworms is in Vlaminck et al., 2020.

8. Information is found in Mackenzie and Geary, 2011; Bockarie and Deb, 2010; and de Moraes et al., 2020.

9. Information about the *B. thuringiensis* Cry5B protein is in Hu et al., 2018. Information about the anti-*Wolbachia* approach is in Taylor et al., 2019.

10. Information is from Chirac and Torreele, 2006.

11. Moran et al., 2011. The story about how the Ebola vaccine was developed and tested in the Democratic Republic of the Congo has not yet been fully explained, but some of it can be found in Wolf et al., 2020.

12. Information on G-FINDER can be found at https://gfinder.policycuresresearch.org/.

13. The advanced market commitment concept is explained in Kremer and Glennerster, 2004. The estimates from Duke University are in Young et al., 2018.

14. The comparison of the genomes for the two human trypanosomes and leishmania is in El Sayed et al., 2005.

15. The PDP/public-private partnership activities for developing new drugs to combat kinetoplastid infections are described in Caridha et al., 2019; and Neau et al., 2020.

16. Hotez and Ferris, 2006; Hotez, 2011a; and Hotez, 2011c.

17. Information about the specific NTD vaccines can be found in Hotez et al., 2016; Diemert et al., 2017; Jones et al., 2018; Lustigman et al., 2018; Bartsch et al., 2019; Hotez et al., 2019; and Keitel et al., 2019.

18. Information on product development challenges can be found in Bottazzi and Hotez, 2019.

19. The problems of vaccine global access are described in Mahoney and Maynard, 1999; and Mahoney et al., 2007.

20. Morel et al., 2005.

21. Hotez, 2011b.

22. Information is found at https://apps.who.int/gb/ebwha/pdf_files/WHA59/A59_17-en.pdf.

References

Albonico M, Bickle Q, Ramsan M, Montresor A, Savioli L, Taylor M. 2003. Efficacy of mebendazole and levamisole alone or in combination against intestinal nematode infections after repeated targeted mebendazole treatment in Zanzibar. *Bull World Health Organ* **81**:343–352.

Albonico M, Engels D, Savioli L. 2004a. Monitoring drug efficacy and early detection of drug resistance in human soil-transmitted nematodes: a pressing public health agenda for helminth control. *Int J Parasitol* **34**:1205–1210.

Albonico M, Montresor A, Crompton DW, Savioli L. 2006. Intervention for the control of soil-transmitted helminthiasis in the community. *Adv Parasitol* **61**:311–348.

Albonico M, Wright V, Bickle Q. 2004b. Molecular analysis of the β-tubulin gene of human hookworms as a basis for possible benzimidazole resistance on Pemba Island. *Mol Biochem Parasitol* **134**:281–284.

Anderson RM. 1999. The pandemic of antibiotic resistance. *Nat Med* **5**:147–149.

Austin DJ, Kakehashi M, Anderson RM. 1997. The transmission dynamics of antibiotic-resistant bacteria: the relationship between resistance in commensal organisms and antibiotic consumption. *Proc Biol Sci* **264**:1629–1638.

Bartsch SM, Bottazzi ME, Asti L, Strych U, Meymandi S, Falcón-Lezama JA, Randall S, Hotez PJ, Lee BY. 2019. Economic value of a therapeutic Chagas vaccine for indeterminate and Chagasic cardiomyopathy patients. *Vaccine* **37**:3704–3714.

Bartsch SM, Hotez PJ, Hertenstein DL, Diemert DJ, Zapf KM, Bottazzi ME, Bethony JM, Brown ST, Lee BY. 2016. Modeling the economic and epidemiologic impact of hookworm vaccine and mass drug administration (MDA) in Brazil, a high transmission setting. *Vaccine* **34**:2197–2206.

Bennett A, Guyatt H. 2000. Reducing intestinal nematode infection: efficacy of albendazole and mebendazole. *Parasitol Today* **16**:71–74.

Bockarie MJ, Deb RM. 2010. Elimination of lymphatic filariasis: do we have the drugs to complete the job? *Curr Opin Infect Dis* **23**:617–620.

Bottazzi ME, Hotez PJ. 2019. "Running the Gauntlet": formidable challenges in advancing neglected tropical diseases vaccines from development through licensure, and a "Call to Action". *Hum Vaccin Immunother* **15**:2235–2242.

Caridha D, Vesely B, van Bocxlaer K, Arana B, Mowbray CE, Rafati S, Uliana S, Reguera R, Kreishman-Deitrick M, Sciotti R, Buffet P, Croft SL. 2019. Route map for the discovery and preclinical development of new drugs and treatments for cutaneous leishmaniasis. *Int J Parasitol Drugs Drug Resist* **11**:106–117.

Chirac P, Torreele E. 2006. Global framework on essential health R&D. *Lancet* **367**:1560–1561.

Conder GA, Campbell WC. 1995. Chemotherapy of nematode infections of veterinary importance, with special reference to drug resistance. *Adv Parasitol* **35**:1–84.

de Moraes J, Geary TG. 2020. FDA-approved antiparasitic drugs in the 21st century: a success for helminthiasis? *Trends Parasitol* **36**:573–575.

Diemert DJ, Freire J, Valente V, Fraga CG, Talles F, Grahek S, Campbell D, Jariwala A, Periago MV, Enk M, Gazzinelli MF, Bottazzi ME, Hamilton R, Brelsford J, Yakovleva A, Li G, Peng J, Correa-Oliveira R, Hotez P, Bethony J. 2017. Safety and immunogenicity of the Na-GST-1 hookworm vaccine in Brazilian and American adults. *PLoS Negl Trop Dis* **11**:e0005574.

Einarsdottir T, Huygen K. 2011. Buruli ulcer. *Hum Vaccin* **7**:1198–1203.

El-Sayed NM, Myler PJ, Blandin G, Berriman M, Crabtree J, Aggarwal G, Caler E, Renauld H, Worthey EA, Hertz-Fowler C, Ghedin E, Peacock C, Bartholomeu DC, Haas BJ, Tran AN, Wortman JR, Alsmark UC, Angiuoli S, Anupama A, Badger J, Bringaud F, Cadag E, Carlton JM, Cerqueira GC, Creasy T, Delcher AL, Djikeng A, Embley TM, Hauser C, Ivens AC, Kummerfeld SK, Pereira-Leal JB, Nilsson D, Peterson J, Salzberg SL, Shallom J, Silva JC, Sundaram J, Westenberger S, White O, Melville SE, Donelson JE, Andersson B, Stuart KD, Hall N. 2005. Comparative genomics of trypanosomatid parasitic protozoa. *Science* **309**:404–409.

Gallup JL, Sachs JD. 2001. The economic burden of malaria. *Am J Trop Med Hyg* **64**(1–2 Suppl):85–96.

Guy B, Barrere B, Malinowski C, Saville M, Teyssou R, Lang J. 2011. From research to phase III: preclinical, industrial and clinical development of the Sanofi Pasteur tetravalent dengue vaccine. *Vaccine* **29**:7229–7241.

Harrison GA. 1978. *Mosquitoes, Malaria, and Man: a History of the Hostilities since 1880*. E P Dutton, New York, NY.

Hotez PJ. 2004. The National Institutes of Health roadmap and the developing world. *J Investig Med* **52**:246–247.

Hotez PJ. 2007. Control of onchocerciasis—the next generation. *Lancet* **369**:1979–1980.

Hotez P. 2011a. A handful of 'antipoverty' vaccines exist for neglected diseases, but the world's poorest billion people need more. *Health Aff (Millwood)* **30**:1080–1087.

Hotez PJ. 2011b. New antipoverty drugs, vaccines, and diagnostics: a research agenda for the US President's Global Health Initiative (GHI). *PLoS Negl Trop Dis* **5**:e1133.

Hotez P. 2011c. Enlarging the "Audacious Goal": elimination of the world's high prevalence neglected tropical diseases. *Vaccine* **29**(Suppl 4):D104–D110.

Hotez PJ, Bottazzi ME, Bethony J, Diemert DD. 2019. Advancing the development of a human schistosomiasis vaccine. *Trends Parasitol* **35**:104–108.

Hotez PJ, Ferris MT. 2006. The antipoverty vaccines. *Vaccine* **24**:5787–5799.

Hotez PJ, Strych U, Lustigman S, Bottazzi ME. 2016. Human anthelminthic vaccines: rationale and challenges. *Vaccine* **34**:3549–3555.

Hu Y, Nguyen TT, Lee ACY, Urban JF Jr, Miller MM, Zhan B, Koch DJ, Noon JB, Abraham A, Fujiwara RT, Bowman DD, Ostroff GR, Aroian RV. 2018. *Bacillus thuringiensis* Cry5B protein as a new pan-hookworm cure. *Int J Parasitol Drugs Drug Resist* **8**:287–294.

Jones K, Versteeg L, Damania A, Keegan B, Kendricks A, Pollet J, Cruz-Chan JV, Gusovsky F, Hotez PJ, Bottazzi ME. 2018. Vaccine-linked chemotherapy improves benznidazole efficacy for acute Chagas disease. *Infect Immun* **86**:e00876–e17.

Keiser J, Utzinger J. 2008. Efficacy of current drugs against soil-transmitted helminth infections: systematic review and meta-analysis. *JAMA* **299**:1937–1948.

Keitel WA, Potter GE, Diemert D, Bethony J, El Sahly HM, Kennedy JK, Patel SM, Plieskatt JL, Jones W, Deye G, Bottazzi ME, Hotez PJ, Atmar RL. 2019. A phase 1 study of the safety, reactogenicity, and immunogenicity of a *Schistosoma mansoni* vaccine with or without glucopyranosyl lipid A aqueous formulation (GLA-AF) in healthy adults from a non-endemic area. *Vaccine* **37**:6500–6509.

Kremer M, Glennerster R. 2004. *Strong Medicine: Creating Incentives for Pharmaceutical Research on Neglected Diseases*. Princeton University Press, Princeton, NJ.

Lustigman S, Makepeace BL, Klei TR, Babayan SA, Hotez P, Abraham D, Bottazzi ME. 2018. *Onchocerca volvulus*: the road from basic biology to a vaccine. *Trends Parasitol* **34**:64–79.

Mackenzie CD, Geary TG. 2011. Flubendazole: a candidate macrofilaricide for lymphatic filariasis and onchocerciasis field programs. *Expert Rev Anti Infect Ther* **9**:497–501.

Mahoney RT, Krattiger A, Clemens JD, Curtiss R III. 2007. The introduction of new vaccines into developing countries. IV: Global Access Strategies. *Vaccine* **25**:4003–4011.

Mahoney RT, Maynard JE. 1999. The introduction of new vaccines into developing countries. *Vaccine* **17**:646–652.

Moran M, Guzman J, Abela-Oversteegen L, Liyanage R, Omune B, Wu L, Chapman N, Gouglas D. 2011. *Neglected Disease Research and Development: Is Innovation under Threat?* G-FINDER Policy Cures, Sydney, Australia.

Morel CM, Acharya T, Broun D, Dangi A, Elias C, Ganguly NK, Gardner CA, Gupta RK, Haycock J, Heher AD, Hotez PJ, Kettler HE, Keusch GT, Krattiger AF, Kreutz FT, Lall S, Lee K, Mahoney R, Martinez-Palomo A, Mashelkar RA, Matlin SA, Mzimba M, Oehler J, Ridley RG, Senanayake P, Singer P, Yun M. 2005. Health innovation networks to help developing countries address neglected diseases. *Science* **309**:401–404.

Moser W, Schindler C, Keiser J. 2017. Efficacy of recommended drugs against soil transmitted helminths: systematic review and network meta-analysis. *BMJ* **358**:j4307.

Moser W, Schindler C, Keiser J. 2019. Drug combinations against soil-transmitted helminth infections. *Adv Parasitol* **103**:91–115.

Neau P, Hänel H, Lameyre V, Strub-Wourgaft N, Kuykens L. 2020. Innovative partnerships for the elimination of human African trypanosomiasis and the development of fexinidazole. *Trop Med Infect Dis* **5**:17.

Osei-Atweneboana MY, Eng JK, Boakye DA, Gyapong JO, Prichard RK. 2007. Prevalence and intensity of *Onchocerca volvulus* infection and efficacy of ivermectin in endemic communities in Ghana: a two-phase epidemiological study. *Lancet* **369**:2021–2029.

Prichard RK. 2007. Markers for benzimidazole resistance in human parasitic nematodes? *Parasitology* **134**:1087–1092.

Schwab AE, Boakye DA, Kyelem D, Prichard RK. 2005. Detection of benzimidazole resistance-associated mutations in the filarial nematode *Wuchereria bancrofti* and evidence for selection by albendazole and ivermectin combination treatment. *Am J Trop Med Hyg* **73**:234–238.

Schwab AE, Churcher TS, Schwab AJ, Basáñez MG, Prichard RK. 2007. An analysis of the population genetics of potential multi-drug resistance in *Wuchereria bancrofti* due to combination chemotherapy. *Parasitology* **134**:1025–1040.

Taylor MJ, von Geldern TW, Ford L, Hübner MP, Marsh K, Johnston KL, Sjoberg HT, Specht S, Pionnier N, Tyrer HE, Clare RH, Cook DAN, Murphy E, Steven A, Archer J, Bloemker D, Lenz F, Koschel M, Ehrens A, Metuge HM, Chunda VC,

Ndongmo Chounna PW, Njouendou AJ, Fombad FF, Carr R, Morton HE, Aljayyoussi G, Hoerauf A, Wanji S, Kempf DJ, Turner JD, Ward SA. 2019. Preclinical development of an oral anti-*Wolbachia* macrolide drug for the treatment of lymphatic filariasis and onchocerciasis. *Sci Transl Med* **11:**eaau2086.

Vlaminck J, Cools P, Albonico M, Ame S, Chanthapaseuth T, Viengxay V, Do Trung D, Osei-Atweneboana MY, Asuming-Brempong E, Jahirul Karim M, Al Kawsar A, Keiser J, Khieu V, Faye B, Turate I, Mbonigaba JB, Ruijeni N, Shema E, Luciañez A, Santiago Nicholls R, Jamsheed M, Mikhailova A, Montresor A, Mupfasoni D, Yajima A, Ngina Mwinzi P, Gilleard J, Prichard RK, Verweij JJ, Vercruysse J, Levecke B. 2020. Piloting a surveillance system to monitor the global patterns of drug efficacy and the emergence of anthelmintic resistance in soil-transmitted helminth control programs: a Starworms study protocol. *Gates Open Res* **4:**28.

Wolf J, Bruno S, Eichberg M, Jannat R, Rudo S, VanRheenen S, Coller BA. 2020. Applying lessons from the Ebola vaccine experience for SARS-CoV-2 and other epidemic pathogens. *NPJ Vaccines* **5:**51.

Young R, Bekele T, Gunn A, Chapman N, Chowdhary V, Corrigan K, Dahora L, Martinez S, Permar S, Persson J, Rodriguez B, Schäferhoff M, Schulman K, Singh T, Terry RF, Yamey G. 2018. Developing new health technologies for neglected diseases: a pipeline portfolio review and cost model. *Gates Open Res* **2:**23.

12

The Newest NTDs and a Plea to "Repair the World"

There are no better grounds on which we can meet other nations and demonstrate our own concern for peace and the betterment of mankind than in a common battle against disease.
JOHN GARDNER, FORMER SECRETARY OF HEALTH, EDUCATION, AND WELFARE

[T]ikkun olam says that, having accepted the notion that we should treat one another with respect and dignity, we come together as human beings in comity and cooperation to repair and improve the world around us.
MARIO CUOMO, FORMER GOVERNOR OF NEW YORK

The Newest NTDs

In recent years, the World Health Organization (WHO) has added several key diseases and conditions to its list of neglected tropical diseases (NTDs). Scabies is one of the most important and was added to the list in 2017.[1] It is a highly disfiguring condition of the skin affecting almost 200 million people living in extreme poverty, and is caused by an ectoparasitic "itch mite" known as *Sarcoptes scabiei*.[1] These mites burrow into the skin, depositing their eggs in the burrows or tracts they produce. The resulting hypersensitivity and allergic response cause intense itching and skin disfigurement. It also can result in dangerous secondary bacterial infections, including a condition known as impetigo, which in turn can lead to autoimmune illness of the kidney (glomerulonephritis). The highest disease rates occur in Indonesia and Papua New Guinea, although scabies is also found among aboriginal populations in Australia. Other important areas include Southeast Asia, western China, and tropical regions of the Americas, especially Brazil and French Guiana, to name a few. Scabies is amenable to preventive chemotherapy approaches that include

Forgotten People, Forgotten Diseases: The Neglected Tropical Diseases and Their Impact on Global Health and Development, Third Edition. Peter J. Hotez.
© 2022 American Society for Microbiology. DOI: 10.1128/9781683673903.ch12

ivermectin to target the itch mite and azithromycin for the treatment of impetigo.[1] The International Alliance for the Control of Scabies (https://www.controlscabies.org/) is promoting such initiatives.

Yaws is another disfiguring tropical infection of the skin, as well as a destructive illness of the bones and joints.[1] It is caused by a bacterial spirochete, *Treponema pallidum* subsp. *pertenue*. Unlike *T. pallidum* subsp. *pallidum*, the cause of syphilis, *T. pallidum* subsp. *pertenue* is not usually transmitted by sexual contact. Instead, yaws is often transmitted between children who play and interact with each other. It is therefore often referred to as a nonvenereal treponematosis. Most of the yaws cases occur in just three countries: Papua New Guinea, Solomon Islands, and Ghana. The illness is amenable to elimination with azithromycin and represents a collateral benefit to preventive chemotherapy for trachoma.

Similar to yaws, mycetoma is a destructive illness of the tissues, typically the foot, and affecting mostly young adults, especially males involved in agricultural pursuits.[1] It has both fungal and bacterial causes that produce swellings comprising small granules (referred to as "grains") representing collections of the microorganisms. The leading agent is *Madurella mycetomatis*, although there are other causes. The regions most affected are areas of extreme poverty in East and West Africa, India, Mexico, and Brazil. A related chronic destructive illness of the mouth is known as noma, although it is not currently listed by the WHO as an NTD. Finally, snakebite envenoming is the first and only noninfectious condition on the WHO list of NTDs, although I feel that podoconiosis should also be added.[1] The numbers are quite impressive, with 4 million to 6 million snakebites annually, according to the WHO and other sources, leading to roughly 2 million cases of clinical illness and 80,000 to 120,000 deaths. In this respect, snakebite envenomation may represent the leading cause of mortality among the NTDs, although it is not currently counted under the Global Burden of Disease (GBD) 2019 category of "Malaria and NTDs." Like mycetoma and yaws, snakebite envenomation mostly occurs in rural areas, with the highest rates in sub-Saharan Africa and India. There is an urgent need to expand prevention measures, including health education, as well as to improve wound care and increase access to lifesaving antitoxins.

Tikkun Olam

While in office, and for years afterward, former New York governor Mario Cuomo helped to popularize the Hebrew phrase *tikkun olam*, which has come to mean our moral obligation to finish the job that God began by repairing the world through social action.[2] Jewish scholars point to two 2nd-century sources for *tikkun olam*—the Aleinu prayer, now chanted at the conclusion of every Jewish prayer service; and the Mishnah, the first compendium of Jewish oral law. In the 16th century, the renowned Kabbalist rabbi Isaac Luria expanded the use of the term. The concept of *tikkun olam* was taught to me by a mentor and cousin, Rabbi Philip Lazowski, of Beth Hillel Synagogue in Bloomfield, CT. Phil is a Holocaust survivor and the author of *Understanding Your Neighbor's Faith: What Christians and Jews Should Know about Each Other*,[2] as well as someone who has given much thought to repair. I believe that almost no other action embodies *tikkun olam* as much as the relief of human suffering through the control of the NTDs. In aggregate, the NTDs are the most common infections of the world's poorest people, in whom they cause chronic disability and disfigurement on a massive and, at times, almost unimaginable scale. Through their poverty-promoting impact on child development, pregnancy outcome, and worker productivity, the NTDs represent a major reason why poor people cannot lift themselves out of poverty and why the low-income countries where they live cannot economically advance. Therefore, a global assault on the NTDs through both widespread deployment of rapid-impact packages and the simultaneous development and implementation of new control tools could one day become a highly productive application of medical science and public health for repairing the world. I have used the term *science tikkun* to report on this important humanitarian goal for medical science.[2]

The control and elimination of the NTDs may also provide added global benefit in light of the underappreciated link between NTDs and human conflict. I use a definition of "conflict" proposed by Lea Berrang-Ford as "the occurrence of civil war, rebel insurgency, violent governance, political or military oppression of populations, and military combat."[3] Earlier (in chapter 7), we saw how the conflicts sweeping across the African continent, including Angola, the Democratic Republic of the Congo, the Central African Republic, and Sudan, led to a reemergence of highly lethal outbreaks of human African trypanosomiasis (HAT). In some areas of these countries, the mortality rates of sleeping sickness previously exceeded those of better-known conditions, including HIV/AIDS.[3] Surveillance mapping for onchocerciasis is also incomplete in these same regions.[4] We also saw (in both chapters 5 and 7) how civil war in Sudan and later South Sudan led to serious outbreaks of kala-azar and trachoma. The basis by which conflict has promoted the reemergence of HAT and other vector-borne NTDs has been reviewed, and at least four key determinants have been identified.[3] They include (i) economic and global effects, such as abandonment or appropriation of land, collapse of local economies, and the purging of educated and business elite, all of which lead to an interruption of public health services; (ii) a decline in specific health services, especially collapse in vector control programs; (iii) forced migration and internal displacement of populations, with resulting decreased access to health facilities, as well as land abandonment, regrowth of vegetation, and increased vector habitat; and (iv) regional insecurity and restricted access for external humanitarian support.[3]

Chris Beyrer and his colleagues at Johns Hopkins University have also examined the relationships between specific NTDs and their unique epidemiology in conflict zones, using lymphatic filariasis (LF) in eastern Burma and Chagas disease and leishmaniasis in Colombia as case studies.[4] In eastern Myanmar (Burma), ethnic minority groups were engaged in a civil war with the military regime for decades. As a result of a government-led counterinsurgency campaign, up to 1 million Burmese people were displaced internally and another million fled across the border into Thailand. The government of Myanmar reduced or halted diethylcarbamazine (DEC) mass drug administration efforts in the ethnic minority areas of these regions. Based on cross-sectional surveys of migrant populations coming across the Thai border, the prevalence rates of LF became extraordinarily high (reaching 10%), while another 40% of the migrants showed evidence of previous exposure to LF. Today, Myanmar still has some of the highest prevalence rates of LF, according to the GBD 2019. As Burmese immigrants find jobs in Thailand's major cities, since the appropriate mosquito vectors are present, it is anticipated that LF could reemerge in urban areas of Thailand.[4]

In Colombia, the Beyrer group has also determined that the establishment of guerilla groups together with increased cocaine drug trafficking has created an environment of extreme violence in rural areas.[4] To counter this deterioration, the Colombian government, like the Burmese government discussed above, previously escalated its military budget, often at the expense of its public health infrastructure. Colombia's vector control programs in particular have suffered, so that at one time less than 17% of the population in areas at high risk of Chagas disease benefited from government-led vector control programs.[4] As a result, in the northeastern part of the country near the Venezuelan border, the seroprevalence of Chagas disease rose to 6% or higher.[4] Similarly, those actively involved in the Colombian conflict, including military personnel and guerillas, as well individuals who were either kidnapped or incarcerated, became infected with cutaneous leishmaniasis.[4] At one time, military stocks of antiparasitic antimonials were targets for guerilla fighters.[4] However, we recently found that as the drug trade began shifting away from Colombia to the Northern Triangle region of Honduras, El Salvador, and Guatemala in the 2010s, so did the burden of NTDs.[4]

In summary, the examples of HAT, onchocerciasis, LF, Chagas disease, and leishmaniasis illustrate an intimate link between the NTDs and long-standing conflict. The common features of this relationship include reductions in public health control programs because of violence and a shifting of limited resources in favor of military spending and buildup. Such situations are particularly common in remote and rural areas where guerilla and rebel movements geographically overlap with NTDs and their vectors.[4] Also contributing to the NTD-conflict link are human migrations into areas where disease is endemic, which increase human vulnerability and often decrease access to essential medicines.[4]

Is it also possible that the NTDs not only emerge and reemerge in settings of conflict and post-conflict turmoil, but that in addition, because the NTDs themselves are destabilizing, they could ignite a series of events that culminate in conflict? Thus, just as the NTDs occur in the setting of poverty and promote poverty, could they also promote conflict? We have previously shown high levels of the geographic overlap between areas of civil and international conflict (during the last decade of the 20th century) and areas with either high infant mortality or high under-age-5 child mortality. In the developing regions of the world, most of the infant mortality results from either prematurity (often from malaria and the NTDs hookworm infection and schistosomiasis) or sepsis (typically from bacterial infections), while under-age-5 child mortality results mostly from infectious diseases, including many tropical infections, such as malaria, and childhood viral and bacterial infections, such as pneumococcal pneumonia, respiratory syncytial virus infection, rotavirus gastroenteritis, and measles. Therefore, both maps indirectly demonstrate a tight geographic overlap between infection and conflict.

Figure 12.1 shows a semiquantitative relationship between countries grouped by their under-age-5 child mortality rates and the percentage of those countries that were engaged in armed conflict during the 1990s.[5] My interpretation of the graph is that nations with serious infectious disease problems were more likely to have been involved in armed struggles during the recent past, particularly when the under-age-5 child mortality exceeds 100 per 1,000, meaning that more than 10% of the pediatric population fails to reach its 5th birthday. Indeed, since 1990, approximately 80% of all wars have been fought in sub-Saharan Africa, Asia, and tropical regions of the Americas, especially in areas where multiple NTDs are coendemic.[5] Of interest is the observation that many of the nations with the worst health indicators are Islamic countries, as determined by their membership in the Organization of the Islamic Conference.[6]

These relationships do not distinguish between cause and effect, but it is highly plausible that a nation threatened by endemic infection and child morbidity and mortality would be destabilized by the large-scale effects of infection on families and civil society.[7] Throughout this book we have emphasized how infectious diseases and the NTDs in particular are destabilizing through (i) their impact on the agricultural workforce and their downstream effects on fields going untended, increased risks of famine, and overall lack of economic productivity; (ii) their impact on families, i.e., injuries to family breadwinners and the creation of a generation of orphans; and (iii) their impact on community governance through their disabling effects on community leaders.[7] We have seen how entire communities can be destabilized by highly endemic

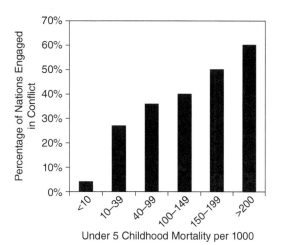

Figure 12.1 Relationship between under-age-5 child mortality (per 1,000) and areas of conflict during the 1990s. (Reproduced from Hotez PJ. 2001. Vaccines as instruments of foreign policy. *EMBO Reports* 2:862–868.)

blindness from onchocerciasis and trachoma, widespread disability from LF and dracunculiasis, impaired child development and future productivity from hookworm infection and schistosomiasis, and widespread mortality from HAT and leishmaniasis. The demonstration of a specific link between the destabilizing effects of endemic NTDs and increased risk of conflict remains elusive, but there are sufficient connecting threads to warrant a consideration of NTD control as an element of international diplomacy.

Could we incorporate NTD control as a new element for U.S. foreign policy? Beginning in the post-Sputnik era, there has been a modest though interesting American history of linking health and diplomacy. One of the best examples occurred with the development of the live attenuated oral polio vaccine (OPV) developed by Albert Sabin during the middle and late 1950s.[5] In response to polio epidemics raging in the urban centers of both the United States and the Soviet Union, both countries put aside their ideological differences and worked together to take Sabin's live poliovirus strains originating from his research laboratory at Cincinnati Children's Hospital and test them clinically as a vaccine in Communist Russia. Many Americans are unaware that the live OPV they received as children was licensed in the United States only after it was refined during large-scale clinical testing on tens of millions of schoolchildren in the USSR (Fig. 12.2). Less than a decade later, American and Soviet microbiologists collaborated in scaling up production of a vaccine for smallpox, which led to that disease's eradication during the late 1970s.[5] Today,

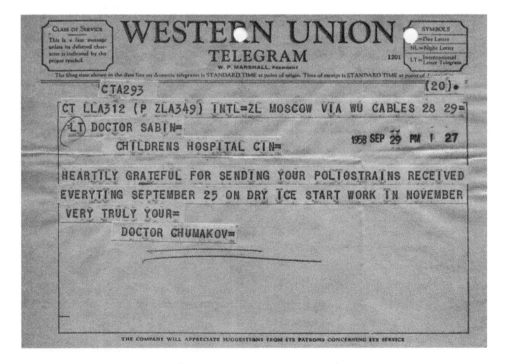

Figure 12.2 Telegram sent from Mikhail Chumakov to Albert Sabin regarding test strains for a polio vaccine. Courtesy of the Hauck Center for the Albert B. Sabin Archives, Henry R. Winkler Center for the History of the Health Professions, University of Cincinnati Libraries.

smallpox remains the only disease ever eradicated by humankind, a feat that occurred as a component of Cold War vaccine diplomacy!

During the last half of the 20th century, subsequent efforts to fully embrace vaccine diplomacy as a permanent component of U.S. foreign policy have met with mixed results. Following efforts to send medical supplies to Cuba in exchange for prisoner release after the failed Bay of Pigs invasion, the Kennedy administration took the bold step of creating through the Department of State the U.S. Agency for International Development (USAID) and the Peace Corps.[5] Subsequently, through the advocacy of then-Secretary of Health, Education, and Welfare John Gardner, the Johnson administration helped to draft legislation for the International Health Education Act.[1,5] This was an innovative proposal that would have brought medical and public health attachés to U.S. embassies worldwide. Unfortunately, the Act failed to advance through a key congressional committee and never materialized. During the 1980s and 1990s, national immunization days were a key element of cease-fires in war-torn regions of Afghanistan, Sierra Leone, and Sudan, and we saw earlier (in chapter 4) how former President Jimmy Carter was successful at brokering the guinea worm cease-fire in Sudan. In this new century, the Bush administration launched disease control efforts for HIV/AIDS (the President's Emergency Plan for AIDS Relief [PEPFAR]), malaria (the President's Malaria Initiative [PMI]), and even an NTD program for distribution of rapid-impact packages in more than 30 African, Asian, and Latin American countries (https://www.neglecteddiseases.gov/where-we-work/) (chapter 10). Each of these initiatives is administered through USAID, and they represent important first steps in infusing global disease control into U.S. foreign policy. However, for the most part, the USAID NTD Program and the other USAID disease control initiatives have until recently functioned as stand-alone programs that are not truly integrated into U.S. diplomatic efforts.

Over the last two decades, a handful of influential scholars of U.S. foreign policy have lamented a loss of international respect for U.S. leadership, while simultaneously pointing out the potential power of humanitarian assistance as a mechanism to restore American influence.[8] For instance, former U.S. National Security Advisor Zbigniew Brzezinski (who passed away in 2017) said America must "place a higher premium on a truly shared global cause" if it is to "derive any political benefit from the cultural revolution it is unleashing worldwide" and that "the United States should treat globalization less as a gospel and more as an opportunity for the betterment of the human condition." He warned, "If American policymakers do not deliberately infuse it with politically evident moral content, focused on the alleviation of the human condition, their uncritical embrace of it could backfire."[8] Similarly, former Secretary of State Henry Kissinger stated, "At the apogee of its power, the United States finds itself in an ironic position. In the face of perhaps the most profound and widespread upheavals the world has ever seen, it has failed to develop concepts relevant to the emerging realities."[8] Both Brzezinski and Kissinger, as well as Joseph Nye, Jr., the former dean of Harvard's Kennedy School of Government, have been strong advocates for American humanitarian assistance, but only if it is simultaneously sustained by American domestic opinion and if it reso-nates with the international community and has a high probability of success.[8] More recently, in an influential article published in *Foreign Affairs*, then-U.S. Secretary of State Hillary Clinton put forward a bold initiative to effectively incorporate what she terms "civilian power" (which could include disease con-trol initiatives) into U.S. foreign policy.[8]

In papers published in *PLoS Neglected Tropical Diseases* and the *Brown Journal of World Affairs*, I argued that control of the NTDs meets all of the criteria for successful humanitarian interventions.[5,8] Specifically, the biblical legacy of the NTDs, together with their enormous global health impact and the relatively low cost of poverty reduction efforts, makes NTD reduction a "best buy" not only in public health but also for U.S. international relations.[5]

I believe that there are at least two mechanisms for pursuing NTD control in the context of our national foreign policy.

The first is in the area of research and development. The United States is known and esteemed as a world-class center of international biomed-ical research. Young scientists come from all over the world to study in our laboratories, and through programs sponsored by our National Institutes of Health—including the Fogarty International Center and the Tropical Medicine Research Centers and International Centers for Infectious Diseases Research programs of the National Institute of Allergy and Infectious Diseases—as well as the overseas laboratories of the Department of Defense and Centers for Disease Control and Prevention in Indonesia, Peru, Kenya, and elsewhere, we have enormous capacity for training foreign scientists and fostering collab-oration in overseas laboratories. These initiatives, however, are underfunded and not well linked to the exciting product development partnerships for NTD drugs and vaccines, some of which are sponsored by the Bill & Melinda Gates Foundation. We are in an excellent position to build public-private partner-ships for NTDs that span government, academia, and private industry—part-nerships not too different from the military-industrial complex envisioned by Vannevar Bush in his role during the 1940s and 1950s as director of the Office of Scientific Research, the umbrella organization of the Manhattan Project, as well as the Carnegie Institution of Washington. These activities helped to transform the California universities into scientific juggernauts and overall brought scientific preeminence to the United States. Today, we could build a similar model to establish our leadership on NTDs, working with government research institutes and manufacturers in the major innovative developing countries such as Brazil, China, India, and Indonesia. I have also tried to make the case for us to consider vaccine development with nations with which we have sharp international disagreements.[9] I once estimated that American vac-cines have saved approximately 160 million lives, a number equivalent to all of the lives lost in global conflicts during the 20th century.[5] Accordingly, the U.S. President's Global Health Initiative administered by the Department of State should consider setting aside 1 to 2% of its annual budget for global health research and development. In so doing, an additional US$100 million to $200 million would be allocated to support academic and industrial scientists, possibly in association with matching support from private foundations for a pipeline of products for HIV/AIDS, malaria, tuberculosis, and NTDs.[10] The Global Health Technologies Coalition (GHTC) is exploring such possibilities (https://www.ghtcoalition.org).

At the same time, our Department of State, possibly through USAID or the new State Department Office of Global Health Diplomacy, should work closely with Uniting to Combat Neglected Tropical Diseases and other organizations to establish or expand public-private partnerships for the widespread deployment of rapid-impact packages in the countries at risk for polyparasitism and coendemic NTDs. Such an enterprise could include an appointment of a special ambassador for NTD control, just as the PEPFAR and PMI directors have ambassadorial status. The difference, however, is that an NTD equivalent of Vannevar Bush would work across the public and private sectors in order to pursue any avenue possible for NTD control. This was a component of my activities when I served as U.S. Science Envoy for the State Department and White House in 2015 to 2016 in the Obama administration.

I believe that a new government-academic-industrial enterprise devoted to NTDs offers an unprecedented new opportunity for reducing global poverty and repairing the world. We have in hand today an extraordinary toolbox of highly effective preventive chemotherapeutic drugs and the technological capacity to build a new generation of drugs and vaccines. We must heed the warning of Elie Wiesel that "Man's weakness lies not in his inability to obtain victories, but in his inability to make use of them." We have a win-win opportunity to control and eliminate humankind's major NTDs and need to act in a timely manner and with great efficiency.

Notes

1. The full list of NTDs is found at https://www.who.int/teams/control-of-neglected-tropical-diseases. The global burden of scabies is outlined at http://www.healthdata.org/results/gbd_summaries/2019/scabies-level-3-cause (Global Burden of Disease Collaborative Network, 2020), with further details in Engelman and Steer, 2018; Hotez et al., 2019; and Marks et al., 2019. Information about yaws is from Mitja et al., 2015; and Solomon et al., 2015. Information about mycetoma and related deep fungal infections can be found at https://www.who.int/health-topics/mycetoma-chromoblastomycosis-and-other-deep-mycoses#tab=tab_1 and in van de Sande, 2013. Information about snakebite envenomation is at https://www.who.int/health-topics/snakebite#tab=tab_1, while the global burden of "venomous animal contact" is at http://www.healthdata.org/results/gbd_summaries/2019/venomous-animal-contact-level-4-cause. Options for snakebite envenomation interventions are in Bhaumik et al., 2020.

2. The phrase *tikkun olam* appears in many of Mario Cuomo's writings and speeches. The quotation at the beginning of the chapter is from a Pew Forum on Religion and Public Life presentation and discussion in 2002 and is found at www.pewforum.org/Politics-and-Elections/Religion-on-the-Stump-Politics-and-Faith-in-America.aspx. The Gardner quotation is found in Carey, 1970. Lazowski's books include Lazowski, 2004. The concept of science tikkun is discussed in Hotez, 2019.

3. The role of conflict and its impact on HAT is articulated in Berrang-Ford, 2007. I expanded on some of these aspects in my presidential address to the American Society of Tropical Medicine and Hygiene; see Hotez, 2012.

4. The role of conflict and its impact on LF and Chagas disease transmission is described in Beyer et al., 2007. Information about the shift to the Northern Triangle is in Hotez et al., 2020.

5. I have explored the relationship between conflict and infectious diseases in several articles, including Hotez, 2001a; Hotez, 2001b; Hotez, 2006; Broder et al., 2002; Hotez, 2014; and Du et al., 2018.

6. The relationship between conflict and infection with regard to the Organization of the Islamic Conference is explored in Hotez and Herricks, 2015.

7. The destabilizing effects of ill health and security are nicely articulated in Schneider and Moodie, 2002; and Guha-Sapir and van Panhuis, 2002.

8. Quotations and citations are found in Kissinger, 2001, p. 18–27; Brzezinski, 2004, p. 160–217; Nye, 1999; and Clinton, 2010. These quotations are also found in Hotez, 2001b; Hotez, 2006; Hotez, 2002; and Hotez, 2011a.

9. Hotez, 2008; Hotez, 2011b; and Hotez, 2013.

10. Hotez, 2011c.

References

Berrang Ford L. 2007. Civil conflict and sleeping sickness in Africa in general and Uganda in particular. *Confl Health* **1**:6.

Beyrer C, Villar JC, Suwanvanichkij V, Singh S, Baral SD, Mills EJ. 2007. Neglected diseases, civil conflicts, and the right to health. *Lancet* **370**:619–627.

Bhaumik S, Beri D, Lassi ZS, Jagnoor J. 2020. Interventions for the management of snakebite envenoming: an overview of systematic reviews. *PLoS Negl Trop Dis* **14**:e0008727.

Broder S, Hoffman SL, Hotez PJ. 2002. Cures for the Third World's problems: the application of genomics to the diseases plaguing the developing world may have huge medical and economic benefits for those countries and might even prevent armed conflict. *EMBO Rep* **3**:806–812.

Brzezinski Z. 2004. *The Choice: Global Domination or Global Leadership.* Basic Books, New York, NY.

Carey HL. 1970. A war we can win. Health as a vector of foreign policy. *Bull N Y Acad Med* **46**:334–350.

Clinton HR. 2010. Leading through civilian power: redefining American diplomacy and development. *Foreign Aff* **89**:13–24.

Du RY, Stanaway JD, Hotez PJ. 2018. Could violent conflict derail the London Declaration on NTDs? *PLoS Negl Trop Dis* **12**:e0006136.

Engelman D, Steer AC. 2018. Control strategies for scabies. *Trop Med Infect Dis* **3**:98.

Global Burden of Disease Collaborative Network. 2020. Global Burden of Disease Study 2019 (GBD 2019). GBD Results Tool. Institute for Health Metrics and Evaluation, Seattle, WA.

Guha-Sapir D, van Panhuis WG. 2002. *Armed Conflict and Public Health: a Report on Knowledge and Knowledge Gaps.* WHO Collaborating Centre for Research on the Epidemiology of Disasters, Brussels, Belgium.

Hotez PJ. 2001a. Vaccine diplomacy. *Foreign Policy* **124**:68–69.

Hotez PJ. 2001b. Vaccines as instruments of foreign policy. The new vaccines for tropical infectious diseases may have unanticipated uses beyond fighting diseases. *EMBO Rep* **2**:862–868.

Hotez PJ. 2002. *Appeasing Wilson's ghost: the expanded role of the new vaccines in international foreign policy. Health and Security Series occasional paper no. 3.* Chemical and Biological Arms Control Institute, Washington, DC.

Hotez PJ. 2006. The "biblical diseases" and U.S. vaccine diplomacy. *Brown J World Aff* **12**:247–258.

Hotez PJ. 2008. Reinventing Guantanamo: from detainee facility to Center for Research on Neglected Diseases of Poverty in the Americas. *PLoS Negl Trop Dis* **2**:e201.

Hotez PJ. 2011a. Unleashing "civilian power": a new American diplomacy through neglected tropical disease control, elimination, research, and development. *PLoS Negl Trop Dis* **5**:e1134.

Hotez PJ. 2011b. Engaging Iran through vaccine diplomacy. *Pacific Standard* November 30, 2011. www.psmag.com/science/engaging-iran-through-vaccine-diplomacy-38029/.

Hotez PJ. 2011c. New antipoverty drugs, vaccines, and diagnostics: a research agenda for the US President's Global Health Initiative (GHI). *PLoS Negl Trop Dis* **5**:e1133.

Hotez PJ. 2012. The Four Horsemen of the Apocalypse: tropical medicine in the fight against plague, death, famine and war. *Am J Trop Med Hyg* **87**:3–10.

Hotez PJ. 2013. A reunification Rx for Korea. *Los Angeles Times* **2013**(January 24).

Hotez PJ. 2014. "Vaccine diplomacy": historical perspectives and future directions. *PLoS Negl Trop Dis* **8**:e2808.

Hotez PJ. 2019. Science tikkun: a framework embracing the right of access to innovation and translational medicine on a global scale. *PLoS Negl Trop Dis* **13**:e0007117.

Hotez PJ, Damania A, Bottazzi ME. 2020. Central Latin America: two decades of challenges in neglected tropical disease control. *PLoS Negl Trop Dis* **14**:e0007962.

Hotez PJ, Fenwick A, Molyneux DH. 2019. Collateral benefits of preventive chemotherapy—expanding the war on neglected tropical diseases. *N Engl J Med* **380**:2389–2391.

Hotez PJ, Herricks JR. 2015. Impact of the neglected tropical diseases on human development in the organisation of Islamic cooperation nations. *PLoS Negl Trop Dis* **9**:e0003782.

Kissinger H. 2001. *Does America Need a Foreign Policy? Toward a Diplomacy for the 21st Century.* Simon & Schuster, New York, NY.

Lazowski P. 2004. *Understanding Your Neighbor's Faith: What Christians and Jews Should Know about Each Other.* KTAV Publishing House, Jersey City, NJ.

Marks M, Toloka H, Baker C, Kositz C, Asugeni J, Puiahi E, Asugeni R, Azzopardi K, Diau J, Kaldor JM, Romani L, Redman-MacLaren M, MacLaren D, Solomon AW, Mabey DCW, Steer AC. 2019. Randomized trial of community treatment with azithromycin and ivermectin mass drug administration for control of scabies and impetigo. *Clin Infect Dis* **68**:927–933.

Mitjà O, Marks M, Konan DJ, Ayelo G, Gonzalez-Beiras C, Boua B, Houinei W, Kobara Y, Tabah EN, Nsiire A, Obvala D, Taleo F, Djupuri R, Zaixing Z, Utzinger J, Vestergaard LS, Bassat Q, Asiedu K. 2015. Global epidemiology of yaws: a systematic review. *Lancet Glob Health* **3**:e324–e331.

Nye JS Jr. 1999. Redefining the national interest. *Foreign Aff* **78**:22–35.

Schneider M, Moodie M. 2002. *The Destabilizing Impacts of HIV/AIDS.* Center for Strategic and International Studies, Washington, DC.

Solomon AW, Marks M, Martin DL, Mikhailov A, Flueckiger RM, Mitjà O, Asiedu K, Jannin J, Engels D, Mabey DC. 2015. Trachoma and yaws: common ground? *PLoS Negl Trop Dis* **9**:e0004071.

van de Sande WW. 2013. Global burden of human mycetoma: a systematic review and meta-analysis. *PLoS Negl Trop Dis* **7**:e2550.

Appendix: What Are the Neglected Tropical Diseases?

As outlined in chapter 1, the neglected tropical diseases (NTDs) are a group of chronic and disabling tropical infections. They also promote poverty because of their impact on child health and development, pregnancy outcome, and worker productivity. The NTDs occur primarily in areas of extreme poverty in the world's low- and middle-income countries. This book focuses primarily on a core group of 20 NTDs adopted by the World Health Organization (WHO), including the most prevalent conditions—ascariasis, trichuriasis, hookworm, schistosomiasis, food-borne trematodiases, lymphatic filariasis, trachoma, taeniasis, onchocerciasis, leishmaniasis, Chagas disease, and human African trypanosomiasis, among others. In addition, several important urban NTDs, namely leptospirosis, dengue, Zika, and rabies, are addressed, as well as some important NTDs in the United States and Canada. I also discuss the newest NTDs adopted by the WHO, including scabies, yaws, mycetoma, and snakebite envenomation. Table 1 presents a more complete list of the NTDs, most of which are topics considered for the open access journal *PLoS Neglected Tropical Diseases*.

Table 1 The major NTDs[a]

Pathogen type	Infection
Helminth infections	Ascariasis
	Trichuriasis
	Hookworm infection
	Strongyloidiasis
	Toxocariasis and larva migrans
	Lymphatic filariasis
	Onchocerciasis
	Loiasis

Table continues on next page

Table 1 *(continued)*

Pathogen type	Infection
	Dracunculiasis
	Schistosomiasis
	Food-borne trematodiases
	Taeniasis-cysticercosis
	Echinococcosis
Protozoan infections	Leishmaniasis
	Chagas disease
	Human African trypanosomiasis
	Amebiasis
	Giardiasis
	Balantidiasis
	Toxoplasmosis
	Trichomoniasis
Bacterial infections	Bartonellosis
	Bovine tuberculosis
	Buruli ulcer
	Cholera
	Leprosy
	Leptospirosis
	Melioidosis
	Noma
	Relapsing fever
	Rheumatic fever and poststreptococcal glomerulonephritis
	Trachoma
	Typhus
	Yaws and other treponematoses
Viral infections	Dengue fever and other flavivirus infections, e.g., West Nile virus and Zika virus infections, chikungunya, and yellow fever
	Rabies
	HTLV-1
Fungal infections	Mycetoma and chloroblastomycosis
	Paracoccidioidomycosis
Ectoparasitic infections	Scabies
	Myiasis
	Tungiasis
Non-infectious conditions	Podoconiosis
	Snakebite envenomation

[a] See also http://www.plosntds.org.

Index

About the Author

Peter J. Hotez, M.D., Ph.D., is the founding dean of the National School of Tropical Medicine; professor of Pediatrics and Molecular Virology and Microbiology at Baylor College of Medicine in Houston, Texas; and university professor in Biology at Baylor University. He is also the Texas Children's Hospital Endowed Chair in Tropical Pediatrics and co-director of the Texas Children's Center for Vaccine Development, and holds appointments as the Baker Institute Fellow in Disease and Poverty at Rice University, and Fellow of the Hagler Institute for Advanced Study at Texas A&M University. Professor Hotez is the author of more than 600 scientific articles, founding Editor-in-Chief of *PLoS Neglected Tropical Diseases*, and past president of the American Society of Tropical Medicine and Hygiene. He is an elected member of the National Academy of Medicine and the American Academy of Arts and Sciences, and has received recognition and awards from the Pan American Health Organization of the World Health Organization, Research!America, B'nai B'rith, the American Medical Association, the Association of American Medical Colleges, and Ronald McDonald House Charities, among others. Professor Hotez is a frequently invited guest on major cable news networks and podcasts, and is a recognized leader in both communicating science and combating anti-science movements.

Other Titles from Peter J. Hotez

Preventing the Next Pandemic
Vaccine Diplomacy in a Time of Anti-science
Johns Hopkins University Press
2021

Vaccines Did Not Cause Rachel's Autism
My Journey as a Vaccine Scientist, Pediatrician, and Autism Dad
Johns Hopkins University Press
2018

Blue Marble Health
An Innovative Plan to Fight Diseases of the Poor amid Wealth
Johns Hopkins University Press
2016